Potato virus Y: biodiversity, pathogenicity, epidemiology and management

Christophe Lacomme • Laurent Glais
Dirk U. Bellstedt • Brice Dupuis
Alexander V. Karasev • Emmanuel Jacquot
Editors

Potato virus Y: biodiversity, pathogenicity, epidemiology and management

Springer

Editors
Christophe Lacomme
Virology & Zoology
Science and Advice for Scottish Agriculture
Edinburgh, UK

Dirk U. Bellstedt
Department of Biochemistry
Stellenbosch University
Matieland, South Africa

Alexander V. Karasev
Plant, Soil and Entomological Sciences
University of Idaho
Moscow, Idaho, USA

Laurent Glais
FN3PT/RD3PT
Paris, France

IGEPP, Agrocampus Ouest, INRA
Université de Rennes 1
Le Rheu, France

Brice Dupuis
Field crops
Agroscope Field crops
Nyon 1, Switzerland

Emmanuel Jacquot
UMR-BGPI TA A-54/K
CIRAD
Montpellier Cedex 5, France

ISBN 978-3-319-58858-2 ISBN 978-3-319-58860-5 (eBook)
DOI 10.1007/978-3-319-58860-5

Library of Congress Control Number: 2017947049

Printed on acid-free paper

This Springer imprint is published by Springer Nature
The registered company is Springer International Publishing AG
The registered company address is: Gewerbestrasse 11, 6330 Cham, Switzerland

Preface

Plant viruses are very important pathogens causing significant direct and indirect losses to crop production and threaten global food sustainability. Increase in human population and balancing the demand for more sustainable ways of crop and food production, while maintaining crop productivity and quality, pose continuous challenges to scientists, agronomists and farmers worldwide. Over the past decades, considerable progress has been made in the understanding of the molecular basis of plant pathogen interactions, epidemiology of diseases and their causal agents and the deployment of this knowledge to design suitable control and management methods. Plants have the ability to defend themselves against most types of pathogens including viruses. Breeding programs have successfully introgressed resistance genes in numerous plant species such as potato to provide means to minimise the impact of viruses. However, as for any biological entities, the continuous evolution of pathogens, in particular viruses such as *Potato virus Y* (PVY), to escape host defence mechanisms and to adapt to different environments represents a constant threat. Potato was recently ranked as the fourth most important crop in the world and the most important non-grain crop, while PVY was identified as one of the top ten most important pathogens due to its economic impact in all potato-growing areas worldwide. In 2009, the international "PVY-Wide" network was created with the aim to share and disseminate knowledge on different aspects of PVY research focussing on the PVY-potato pathosystem and on the interactions of PVY with other solanaceous and non-solanaceous plant species. This informal network initially comprised 26 laboratories from 20 countries and has expanded over the years to include up to 40 laboratories. The participants are from different types of organisations including academia, agricultural research organisations, plant health organisations, laboratories involved in certification schemes mainly on seed potato production and private companies involved in pathogen diagnostics, breeding, and so on. The objectives of this book is to review and disseminate information communicated by colleagues of the PVY-Wide network (including yet unpublished and many other published data) on PVY research worldwide spanning the past few decades, to report the most up-to-date research outputs of basic and of more applied nature and to identify knowledge gaps with the view to stimulate future research.

This book should appeal to plant virologists, plant pathologists and the broad diagnostic, breeding and agronomical industries. The nine chapters of the book cover the essential aspects of PVY research including structure-function and diversity of the PVY genome, plant responses, evolution, diagnostic, epidemiology and transmission, control and management, resistance and the interactions of PVY with other plant species. The editors and authors of this book are indebted to all our colleagues of the PVY-Wide network as well as colleagues from the European Association of Potato Research Virology section for their input. Finally, last but by no means least, we would like to thank our colleague Stuart Carnegie for his contribution, valuable comments and suggestions.

Edinburgh, UK Christophe Lacomme
Le Rheu, France Laurent Glais
March 2017

Contents

Contributors

Dirk U. Bellstedt Department of Biochemistry, University of Stellenbosch, Stellenbosch, South Africa

Claude Bragard Earth and Life Institute – Applied Microbiology – Phytopathology, Université catholique de Louvain UCL, Louvain-la-Neuve, Belgium

Thierry Candresse INRA, UMR1332 Biologie du Fruit et Pathologie, Villenave d'Ornon, France

University Bordeaux, UMR1332 Biologie du Fruit et Pathologie, Villenave d'Ornon, France

Stuart Carnegie Science and Advice for Scottish Agriculture (SASA), Edinburgh, UK

Mohamad Chikh Ali Department of Plant Soil and Entomological Sciences, University of Idaho, Moscow, ID, USA

Kim Davie Science and Advice for Scottish Agriculture (SASA), Edinburgh, UK

Kürt Demeulemeester INAGRO, Rumbeke-Beitem, Belgium

Brice Dupuis Agroscope, Institute for Plant Production Sciences IPS, Nyon 1, Switzerland

Earth and Life Institute – Applied Microbiology – Phytopathology, Université catholique de Louvain UCL, Louvain-la-Neuve, Belgium

Chantal Faure INRA, UMR1332 Biologie du Fruit et Pathologie, Villenave d'Ornon, France

University Bordeaux, UMR1332 Biologie du Fruit et Pathologie, Villenave d'Ornon, France

Adrian Fox Fera Science Ltd, Sand Hutton, York, UK

Christiane Gebhardt Max-Planck-Institute for Plant Breeding Research, Cologne, Germany

Laurent Glais FN3PT/RD3PT, Paris, France

IGEPP, Agrocampus Ouest, INRA, Université de Rennes 1, Le Rheu, France

Maurice Hullé IGEPP, Agrocampus Ouest, INRA, Université de Rennes 1, Le Rheu, France

Emmanuel Jacquot UMR Biologie et Génétique des Interactions Plante-Parasite, Campus International de Baillarguet, Montpellier Cedex 5, France

Alexander V. Karasev Department of Plant Soil and Entomological Sciences, University of Idaho, Moscow, ID, USA

John Kerr Science and Advice for Scottish Agriculture (SASA), Edinburgh, UK

Kerstin Kruger Department of Zoology & Entomology, University of Pretoria, Pretoria, South Africa

Denis Kutnjak Department of Biotechnology and Systems Biology, National Institute of Biology, Ljubljana, Slovenia

Christophe Lacomme Science and Advice for Scottish Agriculture (SASA), Edinburgh, UK

Armelle Marais INRA, UMR1332 Biologie du Fruit et Pathologie, Villenave d'Ornon, France

University Bordeaux, UMR1332 Biologie du Fruit et Pathologie, Villenave d'Ornon, France

Natasa Mehle National Institute of Biology, Ljubljana, Slovenia

Spela Modic Agricultural Institute of Slovenia, Ljubljana, Slovenia

Benoît Moury Pathologie Végétale, INRA, Montfavet, France

Xianzhou Nie Agriculture and Agri-Food Canada, Fredericton, NB, Canada

Phillip Nolte University of Idaho, Idaho Falls, USA

Jon Pickup Science and Advice for Scottish Agriculture (SASA), Edinburgh, UK

Maruša Pompe-Novak Department of Biotechnology and Systems Biology, National Institute of Biology (NIB), Ljubljana, Slovenia

Maja Ravnikar National Institute of Biology, Ljubljana, Slovenia

Jean-Louis Rolot Walloon Agricultural Research Centre, Libramont, Belgium

Vincent Simon Pathologie Végétale, INRA, Montfavet, France

Mathuresh Singh Agricultural Certification Services, Fredericton, NB, Canada

Thomas Steinger Agroscope ACW, Nyon, Switzerland

Laurence Svanella-Dumas INRA, UMR1332 Biologie du Fruit et Pathologie, Villenave d'Ornon, France

University Bordeaux, UMR1332 Biologie du Fruit et Pathologie, Villenave d'Ornon, France

Jari P.T. Valkonen Department of Agricultural Sciences, University of Helsinki, Helsinki, Finland

Kazuo N. Watanabe Gene Research Center, University of Tsukuba, Tsukuba, Ibaraki, Japan

Ewa Zimnoch-Guzowska Plant Breeding and Acclimatization Institute (IHAR), Młochów Research Center, Młochów, Poland

Chapter 1
General Characteristics of *Potato virus Y* (PVY) and Its Impact on Potato Production: An Overview

Christophe Lacomme and Emmanuel Jacquot

Abstract Diseases caused by plant viruses can have significant and devastating impacts on many cultivated crops worldwide. The impact of disease caused by a virus depends on the virus species, strains, type of inoculum, host plant characteristics, vector pressure, climatic conditions, trade, changes in agricultural landscape and intensive production practices. Viruses affect plants by causing a large variety of symptoms such as alteration of shape, pigmentation, necrosis on different parts of the plant, thus affecting plant development. In most of the cases, these lead to a decrease in crop yield and quality. There are numerous viruses that affect potato; among them, *Potato virus Y* is considered to be one of the ten most important plant viruses of crops, because of its worldwide distribution and economic impact. Some PVY isolates are able to cause potato ringspot necrotic disease in infected tubers rendering them unmarketable. Understanding the genetic diversity and molecular biology of PVY is essential to understand its infectious cycle, epidemiology and developing efficient methods of control and management for the virus itself and its vector. In spite of an ever-increasing wealth of data in these topics, several major scientific challenges remain in understanding the molecular nature of the interaction between PVY, its hosts, aphid vector in different environments and the epidemiology of PVY. This and following chapters will present the context and current state of our knowledge for these different topics and attempt to provide some answers to these important questions.

C. Lacomme (✉)
Science and Advice for Scottish Agriculture (SASA),
1 Roddinglaw Road, Edinburgh EH12 9FJ, UK
e-mail: Christophe.Lacomme@sasa.gsi.gov.uk

E. Jacquot
UMR Biologie et Génétique des Interactions Plante-Parasite, Campus International de Baillarguet, 34398 Montpellier Cedex 5, France

C. Lacomme et al. (eds.), *Potato virus Y: biodiversity, pathogenicity, epidemiology and management*, DOI 10.1007/978-3-319-58860-5_1

1 Introduction

Changes in the agricultural landscape, crop management, crop intensification, introduction of foreign plant material via increased trade and climate change favor the emergence of infectious diseases of plants (Fargette et al. 2006). Plant viruses, as causal agents of diseases, can have significant and devastating impacts on many cultivated crops worldwide. These impacts depend, among other parameters, on virus inoculum, host plant characteristics (genotype and development stage), vector pressure and climatic conditions (Anderson et al. 2004). Most viruses affect plants by altering their development, causing in most of the cases a range of symptoms such as alteration of shape, modification of pigmentation, elicitation of necrosis on leaves, fruit or tubers, and reduction in plant growth. These different symptoms lead to a decreased crop yield and/or crop quality. These effects can, however, vary greatly for each virus/host combination, and it is not uncommon for crop losses to be either moderate or dramatically high, as exemplified by tomato spotted wilt virus disease on lettuce in USA generating losses of 30–90% (Sherwood et al. 2003) and tomato yellow leaf curl disease on common bean and tomato reducing yield up to 80% (Anderson et al. 2004). Occurrences of virus diseases can sometimes spread over large areas within a relatively short timeframe, as it was the case of bunchy top virus disease that was introduced in Australia in 1913, wiping out banana production in New South Wales by 1927 (Magee 1927, reviewed in Smith et al. 1998). Epidemics of virus disease in a new ecological niche, especially in a suitable environment, can often be very difficult to control and regular outbreaks are likely to occur. The *Groundnut rosette virus* is a good example of pathogen associated to regular outbreaks as more than 15 epidemics of this plant virus were reported on groundnut since the beginning of the twentieth century with losses up to £200 million in sub-Saharan Africa (Sastry and Zitter 2014).

In order to develop effective virus management strategies, it is necessary to diagnose accurately the virus(es) associated with the disease and to understand the disease life cycle of etiological agents (Sastry and Zitter 2014). An accurate assessment of agronomical impacts of a virus disease will require further knowledge on its epidemiology by studying the dynamics and distribution of the disease in hosts and alternative hosts (including wild plants) that act as reservoir of inoculum (Sastry and Zitter 2014). The agronomical impact of a virus depends on the intended use and economic importance of its host plants (grown either as ornamental, staple crop, or cash crop). In 2013, potato was ranked as the fourth most important crop in the world behind corn, wheat, and rice and was ranked the most important non-grain crop with an annual production of over 364 million tons. The importance of potatoes as a staple food worldwide has increased in the past few decades (World Potato Statistics 2015). There has been a dramatic increase in production and demand in Asia, Africa, and Latin America. For the first time in 2005, the developing world's production exceeded that of the developed world. This trend is continuing and reached 52% of global potato output in 2013. China and India are now the greatest producers of potato with about 96 and 45 million tons in 2013, respectively (Table 1.1).

Table 1.1 Top 25 potato producing countries in 2013

Rank	Country	Potato production [tons]
1	China	95,987,500
2	India	45,343,600
3	Russian Federation	30,199,100
4	Ukraine	22,258,600
5	The United States of America	19,843,900
6	Germany	9,669,700
7	Bangladesh	8,603,000
8	France	6,975,000
9	The Netherlands	6,801,000
10	Poland	6,334,200
11	Belarus	5,913,710
12	The United Kingdom	5,580,000
13	Iran (Islamic Republic of)	5,560,000
14	Algeria	4,928,030
15	Egypt	4,800,000
16	Canada	4,620,000
17	Peru	4,570,670
18	Malawi	4,535,960
19	Turkey	3,948,000
20	Pakistan	3,802,200
21	Brazil	3,553,770
22	Belgium	3,479,600
23	Kazakhstan	3,343,600
24	Romania	3,289,720
25	Nepal	2,690,420

Source: World Potato Statistics, FAOSTAT, 2014

2 Viruses Infecting Potato

The importance of viruses as agents of infectious disease of plants was emphasized by Anderson et al. (2004). Viruses represent almost half (47%) of emerging infectious diseases surveyed between 1996 and 2002. A virus can infect many different plant species and a single plant can be infected by many different virus species, strains or isolates. Viruses are submicroscopic obligate intracellular parasites living and replicating in host cells. With some rare exceptions, viruses are assembled into particles made of a nucleic acid core that can be of different nature (see Table 1.2) and encapsidated into a matrix essentially composed of coat protein (CP) and in some cases, additional viral-encoded "accessory" proteins facilitating virus movement and/or transmission. Many diseases of potato are caused by viruses and can be transmitted to succeeding crops through infected seed tubers. Virus disease leads to an ongoing decline in health of a propagated crop, which in early descriptions was generically reported as "degeneration". These pathological phenotypes were further distinguished by the names of leaf roll, mosaic and streak (reviewed by Salaman 1949).

Table 1.2 List of viruses infecting potato and their prevalence in cultivated potato. (*tentative species)

Acronym	Species	Family/(Subfamily)/Genus	Type of genome	Geographical distribution	Transmission/Vector
AMV	*Alfalfa mosaic virus*	*Bromoviridae/Alfamovirus*	ssRNA+	Worldwide	Aphids
APLV	*Andean potato latent virus*	*Tymoviridae/Tymovirus*	ssRNA+	South America	TPS, Beetles.
APMoV	*Andean potato mottle virus*	*Secoviridae/Comovirinae/Comovirus*	ssRNA+	Latin America	Contact, Beetles
AVB	*Arracha virus B*	*Secoviridae/Comovirinae/Cheravirus*	ssRNA+	South America	TPS
BCTV	*Beet curly top virus*	*Geminiviridae/Curtovirus*	ssDNA	Worldwide (arid regions)	Leafhoppers
CMV	*Cucumber mosaic virus*	*Bromoviridae/Cucumovirus*	ssRNA+	Worldwide (rare)	Aphids
EMDV	*Eggplant mottle dwarf nucleorhabdovirus*	*Rhabdoviridae/Nucleorhabdovirus*	ssRNA−	Iran	Aphids
GBNV	*Groundnut bud necrosis virus*	*Bunyaviridae/Tospovirus*	ssRNA−	India	Thrips
GRSV	*Groundnut ringspot virus*	*Bunyaviridae/Tospovirus*	ssRNA−	Argentina	Thrips
INSV	*Impatiens necrotic spot virus*	*Bunyaviridae/Tospovirus*	ssRNA−	Iran	Thrips
PAMV	*Potato aucuba mosaic virus*	*Alphaflexiviridae/Potexvirus*	ssRNA+	Worldwide (uncommon)	Contact, Aphids
PBRSV	*Potato black ringspot virus*	*Secoviridae/Comovirin ae/Nepovirus*	ssRNA+	Peru, others Andean countries?	TPS, Contact, Nematodes?
PLRV	*Potato leaf roll virus*	*Luteoviridae/Polerovirus*	ssRNA+	Worldwide	Aphids
PMTV	*Potato mop top virus*	*Virgaviridae/Pomovirus*	ssRNA+	Europe, America, Asia	Fungi (*Spongospora subterranea*)
PotLV	*Potato latent virus*	*Betaflexiviridae/Quinvirina e/Carlavirus*	ssRNA+	North America	Aphids
PVA	*Potato virus A*	*Potyviridae/Potyvirus*	ssRNA+	Worldwide	Aphids

PVH	*Potato virus H*	*Betaflexiviridae/Quinvirinae/Carlavirus*	ssRNA+	China	Aphids?
PVM	*Potato virus M*	*Betaflexiviridae/Quinvirinae/Carlavirus*	ssRNA+	Worldwide	Aphids
PVP	*Potato virus P*	*Betaflexiviridae/Quinvirinae/Carlavirus*	ssRNA+	South America	Aphids
PVS	*Potato virus S*	*Betaflexiviridae/Quinvirinae/Carlavirus*	ssRNA+	Worldwide	Aphids
PVT	*Potato virus T*	*Betaflexiviridae/Trivirinae/Tepovirus*	ssRNA+	South America	Contact, TPS, Pollen
PVU	*Potato virus U*	*Secoviridae/Comovirinae/Nepovirus*	ssRNA+	Peru	TPS, Contact, Nematodes?
PVV	*Potato virus V*	*Potyviridae/Potyvirus*	ssRNA+	Worldwide	Aphids
PVX	*Potato virus X*	*Alphaflexiviridae/Potexvirus*	ssRNA+	Worldwide	Contact
PVY	*Potato virus Y*	*Potyviridae/Potyvirus*	ssRNA+	Worldwide	Aphids
PYDV	*Potato yellow dwarf nucleorhabdovirus*	*Rhabdoviridae/Nucleorhabdovirus*	ssRNA−	North America	Leafhoppers
PYMV	*Potato yellow mosaic virus*	*Geminiviridae/Begomovirus*	ssDNA	Carribean, Latin America	Whiteflies
PYV	*Potato yellowing virus	*Bromoviridae/Ilarvirus*	ssRNA+	South America	TPS, Aphids
PYVV	*Potato yellow vein virus*	*Closteroviridae/Crinivirus*	ssRNA+	South America	Whiteflies
SALCV	*Solanum apical leaf curl virus	*Geminiviridae/Begomovirus*	ssDNA	South America (Peru)	?
SoMV	*Sowbane mosaic virus*	Unassigned/*Sobemovirus*	ssRNA+	Worldwide (rare)	?
TBRV	*Tomato black ring virus*	*Secoviridae/Comovirinae/Nepovirus*	ssRNA+	Europe, Asia	TPS, Nematodes
TCSV	*Tobacco chlorotic spot virus*	*Bunyaviridae/Tospovirus*	ssRNA−	Argentina/Brazil?	Thrips
ToMoTV	*Tomato mottle Taino virus*	*Geminiviridae/Begomovirus*	ssDNA	Cuba	Whiteflies
ToMV	*Tomato mosaic virus*	*Virgaviridae/Tobamovirus*	ssRNA+	Hungary	Contact
TMV	*Tobacco mosaic virus*	*Virgaviridae/Tobamovirus*	ssRNA+	Worldwide (rare)	Contact
TNV	*Tobacco necrosis virus A*	*Tombusviridae/Alphanecrovirus*	ssRNA+	Europe, South America, Tunisia	Fungi (*Olpidium brassicacae*)

(continued)

Table 1.2 (continued)

Acronym	Species	Family/(Subfamily)/Genus	Type of genome	Geographical distribution	Transmission/Vector
ToLCNDV	*Tomato leaf curl New Delhi virus*	*Geminiviridae/Begomovirus*	ssDNA	India	Whiteflies
ToYVSV	*Tomato yellow vein streak virus*	*Geminiviridae/Begomovirus*	ssDNA	South America (Brazil, Argentina?)	Whiteflies
TRSV	*Tobacco ringspot virus*	*Secoviridae/Comovirinae/Nepovirus*	ssRNA+	Worldwide (except Europe)	TPS, Nematodes
TRV	*Tobacco rattle virus*	*Virgaviridae/Tobravirus*	ssRNA+	Worldwide	Free living nematodes
TSV	*Tobacco streak virus*	*Bromoviridae/Ilarv rus*	ssRNA+	Worldwide	?
TSWV	*Tomato spotted wilt tospovirus*	*Bunyaviridae/Tospovirus*	ssRNA−	Worldwide (hot climates)	Thrips
WPMV	*Wild potato mosaic virus*	*Potyviridae/Potyvirus*	ssRNA+	South America (Peru)	Aphids

Adapted from Jeffries (1998), Valkonen (2007) and updated according to ICTV (2015). TPS: true potato seed. (+)ssRNA, single stranded RNA positive sense; (−)ssRNA, single stranded RNA negative sense; ssDNA: single stranded DNA

2.1 General Properties and Disease Symptoms Caused by Viruses Infecting Potato

Cultivated potatoes can be infected naturally by at least 39 viruses that are classified into 13 families (Table 1.2). The incidence, impact and geographical distribution of these virus species are extremely variable and largely depend on the occurrence of vectors, climatic conditions and management of host crops/plant species.

The foliar symptoms caused by potato viruses include leaf rolling, mosaic (severe or mild), stunting, rugosity, chlorosis, mottling and necrosis (Fig. 1.1). Many potato diseases are often complexes of related or unrelated viruses causing a specific type of foliar symptom (Fig. 1.1). However, symptom severity caused by a single virus isolate can significantly vary between cultivars (Fig. 1.1d, e). In some cases of mixed infections (*e.g.* potex- and luteovirus, potex- and potyviruses, Barker 1987; Vance 1991; Pruss et al. 1997), symptoms can be even more severe than those associated with single infections (Fig. 1.1c).

In addition to foliar symptoms, some viruses can also cause symptoms in tubers, appearing usually as internal and/or external (superficial) necrosis, ringspots and growth cracks (Fig. 1.2). In some cases, tuber necrosis is only observed in specific interactions between one virus and a potato cultivar. Indeed, PLRV infection can elicit tuber net necrosis in the cultivar Russet Burbank (Douglas and Pavek 1972). On the other hand, potato tuber necrotic ringspot disease (PTNRD) (Fig. 1.2f, g, h and i) caused by some *Potato virus Y* (PVY) isolates can be observed in a relatively wide range of potato cultivars. PTNRD caused serious losses in potato crops in several central European countries (Slovenia, Hungary and Germany) and the Middle East (Lebanon) in the 1980s–1990s (Le Romancer et al. 1994). The impact of the disease was dramatic (*i.e.* 18,000 ha or 60% of potato area with more than 50% of frequency of necrotic tubers were reported) and was largely associated with the emergence of PVY isolates with tuber necrotic properties and by a large proportion of acreage in these countries being occupied by a small number of potato cultivars (*e.g.* cvs. Igor, Lola, Monalisa, Rosalie and Hela) which were susceptible to PTNRD (Le Romancer and Nedellec 1997) (see Chaps. 3 and 5). Several studies have shown that PTNRD development depends on the potato genotype, virus genotype and particularly environmental conditions (optimal conditions for PTNRD expression are 20°C during both crop growth and storage). Consequently, PTNRD may develop only in a small proportion of infected tubers if the environmental conditions are less than optimal. In addition, some cultivars such as Spunta, Thalassa and Maris Piper either do not develop PTNRD or only develop relatively mild PTNRD symptoms, while susceptible cultivars such as Hermes, Igor, Lola, Nadine, Nicola, Pentland Crown, and Romano are prone to severe PTNRD development (Le Romancer and Nedellec 1997). Information relating to genetic resources against PVY present in potato germplasm and to potato-virus interactions are presented in Chaps. 2, 3 and 8.

Fig. 1.1 Examples of foliar symptoms caused by some viruses infecting potato. (**a**): Leaf roll caused by *Potato leaf roll virus* (PLRV). (**b**): Mottling caused by *Potato mop top virus* (PMTV). (**c**): Severe mosaic on cv. Red Pontiac caused by *Potato virus A* (PVA) and *Potato virus X* (PVX) mixed infection (stunting and rugosity). D and E: Range of symptoms caused by *Potato virus Y* (PVY) on different potato cultivars. Severe mosaic (leaf distortion and well-defined chlorotic patches) on cv King Edward (**d**) and mild mosaic (mild mottle, not well-defined chlorotic patches) on cv Pentland Crown (**e**) elicited by the same PVY isolate. Photos are courtesy of SASA (Crown copyright©, UK)

2.2 *Potato virus Y*

2.2.1 Economic Impact

Potato virus Y (PVY) was considered to be one of the ten most important plant viruses of crops, because of its worldwide distribution and economic impacts (Scholthof et al. 2011). PVY is the most important virus infecting potatoes due to its worldwide prevalence, and being the main cause of crop degeneration (both yield and quality) (De Bokx and Huttinga 1981). Some PVY isolates are able to cause PTNRD in infected tubers (see Chap. 3) rendering them unmarketable and thus reducing the marketable yield of tubers. PVY is also a major threat for tobacco and pepper crops and, to a lesser extent, for tomato and eggplant productions (Bhat et al. 1999; Aramburu et al. 2006; Mascia et al. 2010) (see Chap. 9).

Because of its important economic impact, extensive programs have been developed to control PVY epidemics by applying prophylactic measures, controlling aphid vectors and breeding for resistance in potato cultivars. Losses due to virus

Fig. 1.2 Examples of tuber symptoms caused by some viruses infecting potato. (**a**): Net necrosis caused by PLRV on cv. Russet Burbank (Photo courtesy of SA Slack©). (**b**): Spraing on cv. Bute caused by PMTV. (**c**): Spraing and internal necrosis on cv. Habibi caused by PMTV. (**d**): Spraing on cv. Valor caused by *Tobacco rattle virus* (TRV). (**e**): Growth cracks caused by PVA on cv. Estima. (**f–g**): Potato tuber necrotic ringspot disease (PTNRD) symptoms caused by PVY (circular ringspot with sunken necrotic skin) on cv. Nadine. (**h**): Severe PTNRD symptoms on cv. Nadine caused by PVY. (**i**): Isolated PTNRD blisters on cv. Maris Piper caused by PVY. Unless specified photos are courtesy of SASA ©Crown copyright (UK)

diseases are not only restricted to direct losses of plant products but are also associated with indirect financial losses such as increased production costs (*e.g.* breeding, training and machinery), cost of control and management of disease (virus control, certification, inspection, virus testing and management tools) and sometimes social and environmental costs (loss of resources, cultural change and contamination of the environment). It has been reported that both direct and indirect estimated losses incurred to PVY to be about $34 million per year for the Idaho state (USA) economy (McIntosh and O'Connell 2014). It was estimated that for each 1% increase of PVY incidence in seed crops (cvs. Russet Burbank and Russet Norkotah), this could result in a reduction of yield of about 180 kg per hectare representing a gross revenue loss of about $18 per hectare (Nolte et al. 2004).

The greatest losses associated to PVY are experienced when the seed tubers being planted are already infected (secondary infection) (De Bokx and van der Want 1987; Whitworth et al. 2006). When plants become infected from virus in seed tubers, reductions in their tuber yield can range from 10 up to 80% in very extreme cases (De Bokx and van der Want 1987). A study of more than 30 cultivars grown in pots demonstrated yield reductions between 50% and 85% compared with plants derived from uninfected seed tubers (reviewed in Valkonen 2007). However, yield losses in field grown potato crops with low incidences of PVY are often less marked because neighboring healthy plants have a competitive advantage for space and nutrients and, therefore, compensate for affected plants with an increase in tuber yield. A yield of reduction of 10–15% would be expected for an incidence of PVY-infected seed tubers of 30% in Spain (Valkonen 2007); while in a separate study, yield losses from a crop with 10–20% PVY-infected seed tubers in Finland were found to be negligible (Kurppa and Hassi 1989). However, although crop yield may be only minimally affected, tuber growth can be altered resulting in a wide range of tuber size, grade and shape, thus affecting marketability.

2.2.2 Demarcation Between Virus Genera and Virus Species

Potato virus Y is a member of the *Potyvirus* genus and one of eight genera (*Brambyvirus*, *Bymovirus*, *Ipomovirus*, *Macluravirus*, *Poacevirus*, *Potyvirus*, *Rymovirus* and *Tritimovirus*) belonging to the family *Potyviridae*. The *Potyviridae* family is the second largest plant virus family after *Geminiviridae*, encompassing about 30% of all described plant viruses (ICTV 2015, Berger et al. 2005). *Potyvirus* is one of the largest genus of plant viruses with 162 virus species currently identified (ICTV 2015). Potyviruses share similar properties in relation to their mode of transmission (aphid transmitted in a non-persistent manner) and genome relatedness. Adams et al. (2005a, b) defined criteria for the demarcation of species and set a threshold of 75–76% of nucleotide identity and 81–82% of amino acid identity, for which higher values represent comparisons between full genome sequences of different isolates of the same species. Alternatively, sequence comparison of the CI gene (RNA helicase, see below) most accurately reflects analysis of the full potyviral genome, suggesting that the RNA helicase is best suited for taxonomic studies when it comes to discrimination between virus genera and species (Adams et al. 2005a).

2.2.3 Genome Structure

Potato virus Y, as for other members of the *Potyvirus* genus, have rod shaped, flexuous particles (about 700 nm in length, 11–13 nm in diameter, helix pitch 3.4–3.5 nm) (Fig. 1.4a) encapsidating the viral RNA with multiple copies (2000 units) of a single coat protein (CP) of 30 kDa. The genome of PVY is a positive (+)-sense single stranded RNA of approximately 9700 nucleotides in length, linked at its 5′ end to a viral protein genome-linked (VPg) and ending with a poly-A tail at its 3′ end (Fig. 1.3). The PVY genome contains one open reading frame (ORF) which is translated as a large polyprotein (about 340–370 kDa), that is then cleaved into 10 (multi) functional proteins (Fig. 1.3): P1, HC-Pro, P3, 6K1, CI, 6K2, VPg, NIa, NIb and CP (reviewed in Danci et al. 2009). More recently, an additional protein P3N-PIPO (Pretty Interesting Potyvirus ORF) has been identified in potyviruses. P3N-PIPO is generated by either a ribosomal slippage creating a +2 frameshift within the P3 ORF

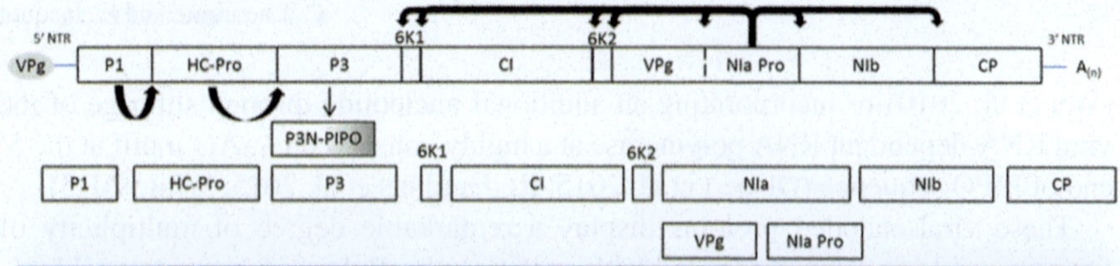

Protein	Size	Function	Cellular Location / Interaction / Comments	References
P1	35 kDa	Serine proteinase. Cleave its own terminus. Stimulate genome amplification. Stimulate HC-Pro silencing suppression activity. Stabilise CP.	Interacts with nucleic acids and binds RNA. Interacts with chloroplastic Rieske Fe/S proteins..	Verchot et al (1991), Verchot & Carrington (1995) Anandalakshmi et al (1998), Shi et al (2007)
HC-Pro	52 kDa	Helper-component proteinase. Cysteine protease. Cleave its own terminus. Required for aphid transmission. Involved in viral replication and pathogenicity. Silencing suppressor activity.	Localise at the ends of virions. Interacts with virus-encoded proteins CP, CI, P1, VPg, NIa. Binds RNA. Interacts with a large number of host proteins (calmodulin-related protein, small RNA methyltransferase, proteasome components, microtubule associated protein, *etc*...). PTK motif interacts with DAG motif of CP to promote binding to aphids mouthparts.	Carrington et al (1989), Dunoyer et al (2014), Kasshau & Carrington (2001), Revers & Garcia (2015), Blanc et al (1997), Torrance et al (2006), Valli et al (2014)
P3, 6K1, P3N-PIPO	50 kDa 6 kDa 25 kDa	P3: required for viral replication, pathogenicity and symptomatology. 6K1: exact function unknown. Role in potyviral infection. P3N-PIPO: required for viral cell-to-cell movement.	P3 associate with cylindrical inclusion formed by CI and with nuclear inclusion formed by NIa and NIb. Localize to endoplasmic reticulum membranes, Golgi apparatus and replication vesicles. Interacts with host Rubisco subunits. PIPO is believed to promote movement of CI and virions/RNA-CP complexes through plasmodesmata.	Guo et al (2001), Rodriguez Cerezo et al (1993), Langenberg & Zhang (1997), Lin et al (2011), Hong et al (2007), Wei et al (2010), Chung et al (2008), White (2015), Ollspert et al (2015), Rodamilans et al (2015),
CI	71 kDa	Forms the Cylindrical cytoplasmic Inclusion protein and pinwheels. RNA helicase. Cell-to-cell movement.	NTP binding, NTPase and ATPase activities. Acts in conjunction with P3N-PIPO. Associated with end of virions, may provide motor function to help virus to traffic through plasmodesmata. Interacts with eIF4E, chloroplast PSI, dsRNA-dependant Protein Kinase inhibitor.	Sorel et al (2014), Fernandez et al (1997), Wei et al (2010), Rodriguez Cerezo et al (1997), Gabrenaite et al (2008), Tavert et al (2012), Jimenez et al (2006), Bilgin et al (2003).
6K2	6 kDa	Membrane anchoring to ER-types membranes.	Crystalline inclusions in cytoplasm and nuclei. Associates with NIa. Strong association with ER vesicles, anchoring of the viral replication complex to ER membranes.	Leonard et al (2004), Beauchemin et al (2007) Schaad et al (1997)
VPg	21 kDa	Virus genome-linked protein. Initiation of replication complex. Nuclear Inclusion protein.	Interacts with host proteins (eukaryotic translation initiation factor eIF4E and eIF(iso)4E, ...) to initiate replication. Interacts with most of potyviral proteins including 6K2 to associate with vesicular ER-membranes	Wittman et al (1997), Leonard et al (2000), Beauchemin et al (2007), Jiang & Laliberte (2011),
NIa-Pro	27 kDa	Small Nuclear Inclusion protein A. Serine-like protease activity. Main potyviral protease : cleave P3-6K1-CI-6K2-VPg-NIa-NIb-CP . Prime RNA synthesis. Required for systemic infection.	Aggregates in the nucleus and cytoplasm to form nuclear inclusion bodies. Binds RNA on its own or as a NIa-6K2 complex. DNAse activity putatively involved in host gene expression regulation.	Adams et al (2005b), Anindya & Savithri (2004).
NIb	58 kDa	Large Nuclear Inclusion protein B. RNA-dependant RNA polymerase (RNA replicase). Viral replication.	Aggregates in the nucleus to form nuclear inclusion bodies. Interacts with NIa. Interacts with host proteins eEF1A, PABP, Hsc70-3, SCE1 to form functional replication complexes. and NIb nucleocytoplasmic transport	Kassanis (1939), Hong & Hunt (1996), Anindya et al (2005), Dufresne et al (2008), Xiong & Wang (2013)
CP	30 kDa	Coat Protein. Role in viral RNA encapsidation, virus movement, aphid transmission and regulation of viral RNA amplification.	In virions CP subunits arranged as helix wrapping genomic RNA. DAG motif at the CP N-terminus interacts with the PTK motif of the HC-pro to promote binding to aphids mouthparts. Interacts with NIb to promote genome amplification. Interacts with hosts Rubisco large subunit.	Lopez et al (2009), Carrington et al (1998), Rojas et al (1997), Blanc et al (1997), Feki et al (2005),

Fig. 1.3 Genomic map of a representative member of Potyvirus genus. Cleavage sites of the potyviral polyprotein by viral-encoded proteases are indicated (bold arrows) and the generated mature proteins are presented below the genomic map. The main characteristics of potyviral proteins are summarized in the table (reviewed in Revers and Garcia 2015)

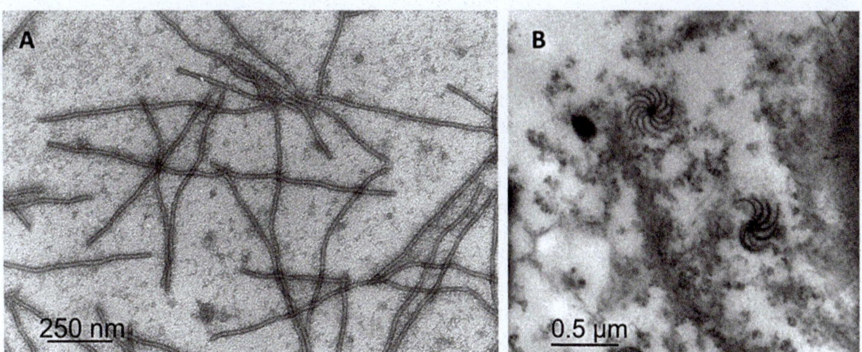

Fig. 1.4 Electron microscopy micrographs of PVY particles (virions) (**a**) and "pinwheels" structures (**b**) in infected plant cells. Courtesy of C. Kerlan (INRA, France) and M.T. Znidaric (NIB, Slovenia)

(Wei et al. 2010) or incorporating an additional nucleotide through slippage of the viral RNA-dependent RNA polymerase at a highly conserved $G_{1\text{-}2}A_{6\text{-}7}$ motif at the 5′ end of PIPO sequence (Olspert et al. 2015, Rodamilans et al. 2015, White 2015).

These viral-encoded proteins display a remarkable degree of multiplicity of functions and are often associated with specific subcellular compartments (chloroplasts, Endoplasmic Reticulum and Golgi apparatus). Viral proteins interact with numerous other viral-encoded proteins and, in some cases, with host proteins which will allow potyviruses to perform all basic viral functions and complete their life cycle (Fig. 1.5). The function of these proteins is summarized in Fig. 1.3 and their roles will be discussed in the following chapters.

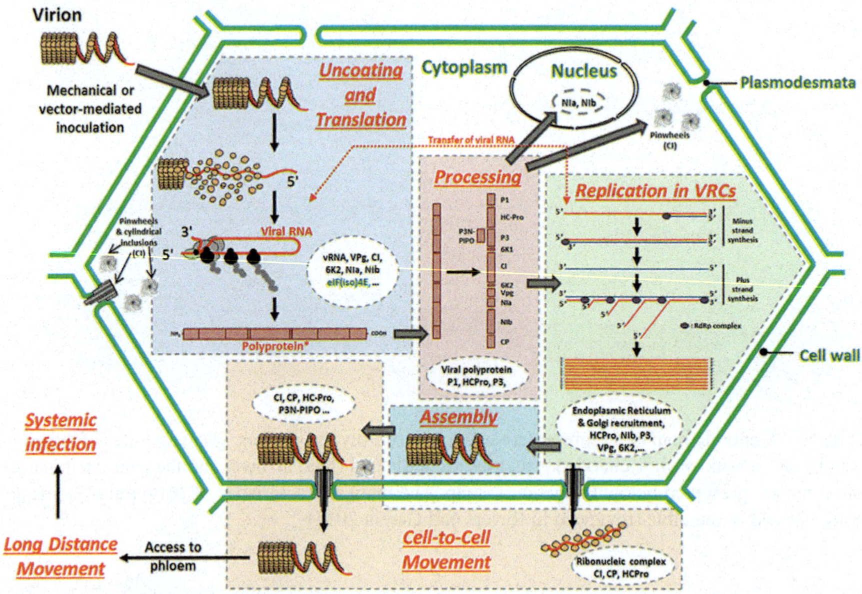

Fig. 1.5 Schematic representation of the major steps in potyvirus infection (adapted from Ivanov et al. 2014). Once virions have infected a plant cell (*i.e.* through aphid feeding or mechanical inoculation), uncoating occurs and expose the viral genomic RNA (5′-3′ molecule illustrated in red) which is then recruited by the host translation machinery (eIF(iso)4E, ribosomes, *etc*...) to synthesize the viral proteins. Processing and maturation of proteins occur and Viral Replication Complex (VRCs) are generated. VRCs are associated to the membrane of host organelles (endoplasmic reticulum, Golgi and vesicles; not illustrated). Replication is initiated by synthesizing a (−) ssRNA (5′-3′ molecule illustrated in blue) by the viral replicase using the (+) ssRNA as a template. The newly synthesized (−) ssRNA is in turn used as a template by the viral replicase to produce numerous (+) ssRNA, which are either encapsidated to produce new virions or recruited by the viral and host proteins to form ribonucleic complexes. Virions and/or ribonucleic complexes recruit the host cytoskeleton (not illustrated) and are transported into the neighboring cells through plasmodesmata via the coordinated action of viral and host proteins. (*): illustration of translation products are simplified as the polyprotein leading to the production of P3N-PIPO is not illustrated

3 Major Properties of PVY

During PVY infection, host cells undergo cytopathological changes that can be observed by electron microscopy. PVY virions (Fig. 1.4a) have been observed to be associated with plasmodesmata (see Sect. 3.2), endoplasmic reticulum (ER) and Golgi apparatus. PVY induce typical cellular inclusion know as cylindrical inclusion (CI) bodies and "pinwheels" structures (Fig. 1.4b). As obligate cellular parasites, viruses must highjack host cellular components in order to perform basic viral functions such as replication, local-systemic movement, transmission and inhibition of host defense mechanisms. These events involve complex molecular mechanisms regulated by host and viral proteins, leading to extensive host gene expression reprogramming events (see Chap. 2) that are spatially and temporally closely associated (for recent reviews on the molecular biology of potyviruses see Ivanov et al. 2014 and Revers and Garcia 2015).

3.1 Replication

After a viral particle has entered a cell through either probing by an aphid vector or by wounding of an epidermal cell, uncoating of the viral particle occurs exposing the viral RNA that will recruit ribosomes and host factors to initiate translation of the viral genes and replication (Fig. 1.5). The (+) sense single stranded (ss)RNA is copied into a complementary (−) sense single RNA strand which in turn is used as a template to synthesize new (+) strands by the action of the RNA-dependent RNA polymerase and RNA helicase. The (+) ssRNA molecules produced during the replication process will then be encapsidated to form new virions (Figs. 1.4a and 1.5).

3.2 Local and Systemic Movement

Once encapsidated, PVY virions move within the initial cell to reach plasmodesmata (PD, symplastic pores between cells), which they cross to enter the neighboring cell and initiate another cycle of replication. A wealth of data supports a direct role for CI (RNA helicase) and CP in the cell-to-cell movement of the viral RNA through plasmodesmata (PD) where CI could form conical structures that facilitate the movement of virions (or CP-coated ribonucleic particles) across the cell wall (Roberts et al. 1998, reviewed in Sorel et al. 2014). The coordinated action of P3N-PIPO (in anchoring the ribonucleic complex to PD), CP and HC-Pro (in increasing the size exclusion limit of PDs) and host proteins will promote virus movement (Fig. 1.5 and Wei et al. 2010).

Successive intra and intercellular movements will occur until the virions reach phloem vessels to be transported throughout the plant in sap from source tissues (*e.g.* tubers and leaves) to sink tissues (*e.g.* newly formed leaves, stems, roots and tubers). This long distance phloem-mediated movement of virus is tightly regulated, because rate of virus translocation can be reduced or blocked in specific tissues/organs depending on the developmental stage of host plant (Revers and Garcia 2015).

3.3 Transmission

To survive in their environment, viruses can be transmitted either from the infected plant to the progeny (i) through propagation (*i.e.* seeds or storage organs such as tubers), and hence could be present in the next crop generation (vertical transmission) or (ii) mechanically or by means of a vertebrate or invertebrate vector such as animals, insect, fungi or bacteria (horizontal transmission). Aphids are the most common vector of plant viruses accounting for more than 60% of viruses transmitted by invertebrates (reviewed in Wilson 2014 and in Katis et al. 2007). Aphid-mediated transmissions mainly occur in persistent or non-persistent modes.

In the persistent mode of transmission, virus is acquired by an aphid feeding on phloem sap, ingested and internalized by the vector. The virus then either replicates (propagative) or do not (circulative) inside the aphid (Katis et al. 2007). The acquisition of virus can last several hours and is often specific (*e.g.* PLRV transmission is essentially performed by the peach-potato colonizing aphid *Myzus persicae* Sulzer). In persistent transmission, a latency period occurs during which time the virus invades the salivary glands before it can be transmitted to a new host.

Non-persistent transmission (non-circulative) is the most common mode of transmission of plant viruses. PVY is transmitted by aphids in a non-persistent manner (see Chap. 5). The helper component protein (HC-Pro) facilitates the binding of virus particles to the aphid's stylet during brief periods of probing of an infected leaf. Subsequently, this virus is transferred mechanically to a new host during further feeding. Unlike persistent transmission, non-persistently transmitted viruses are acquired in less than a minute, do not need latency period and can be transmitted almost immediately after acquisition. However, aphids lose their infectivity rapidly following subsequent feeding. The association between aphid vector and virus is believed to be relatively non-specific, with a wide range of virus species-strains being able to be transmitted with different efficiency by a wide range of aphid species (see Chap. 6).

In addition to persistent and non-persistent transmission mode, semi-persistent transmission has been described for some viruses (*e.g. Citrus tristeza virus* – CTV). Semi-persistent transmission requires periods of acquisition and inoculation longer than non-persistent mode, but does not include a latency period as described for persistent transmission. Usually semi-persistent transmission is efficient about 15 min after acquisition period (Katis et al. 2007).

All types of interactions between aphid, plant and virus involve very complex molecular mechanisms which regulate a wide array of events (*e.g.* virus retention-infection efficiency, suppression of host defense mechanisms, virus-vector host range [for a review, see Giordanengo et al. 2010] and virus-induced chemical changes in infected plants) that impact on aphid behavior and performance (for recent reviews, see Eigenbrode and Bosque-Perez 2016; Fereres 2016). These characteristics of transmission have major implications for the epidemiology and management of viruses and their vectors worldwide. These aspects will be addressed in Chaps. 6 and 7.

4 Current and Future Challenges in PVY Research and Management

For the past 20 years, the vast majority of viral infections in cultivated potato are mainly caused by PVY. In spite of strict management and prophylactic measures, recombinant PVY variants have become prevalent in most of the potato-growing areas worldwide (see Chaps. 3 and 4). To provide assurance of the quality of seed potatoes being planted, seed potato certification schemes have been established to produce seed potatoes containing as limited as practicable amounts of viruses, including PVY (see Chap. 7). However, controlling and managing non-persistent viruses remain an ongoing challenge (Gray et al. 2010, Karasev and Gray 2013). Chemical control of the aphid vectors by insecticides is not effective for non-persistent viruses (Kirchner et al. 2014). Consequently, efficient control of PVY requires the development of different researches programs addressing the following topics: (i) the identification of host resistance genes and the consequences of their deployment (Chap. 8), (ii) the characterization of PVY diversity and pathogenicity (Chaps. 2, 3 and 4), (iii) the epidemiology of PVY in different environments (Chap. 6), (iv) the innovation in diagnosis methods and their deployment (Chap. 5) and (v) the development of suitable control measures and crop management (Chap. 7).

In spite of an increasing wealth of data, several major scientific challenges remain in understanding the complexity of the interaction between PVY with its host(s), aphids vector in their environment, and more broadly, of other vector-borne virus diseases in plants. The current challenges and questions for the scientific community include the following: (i) Why some PVY variants/biotypes are becoming more prevalent in some geographical area? (ii) What is the biological significance of the genomic variability of PVY? (iii) What are the genetic and molecular bases of PVY/host/vector interactions? (iv) How diverse is the epidemiology of PVY in various ecological niches? and (v) Can we integrate environmental and epidemiological data to develop accurate predictive model(s) of PVY incidence? The following chapters will present the current state of our knowledge in these different topics and attempt to provide some answers to these important questions.

References

Adams MJ, Antoniw JF, Fauquet CM (2005a) Molecular criteria for genus and species discrimination within the family Potyviridae. Arch Virol 150:459–479

Adams MJ, Antoniw JF, Beaudoin F (2005b) Overview and analysis of the polyprotein cleavage sites in the family Potyviridae. Mol Plant Pathol 6:471–487

Anandalakshmi R, Pruss GJ, Ge X, Marathe R, Mallory AC, Smith TH et al (1998) A viral suppressor of gene silencing in plants. Proc Natl Acad Sci U S A 95:13079–13084

Anandalakshmi R, Marathe R, Ge X, Herr JM Jr, Mau C, Mallory A et al (2000) A calmodulin-related protein that suppresses posttranscriptional gene silencing in plants. Science 290:142–144

Anderson PK, Cunningham AA, Patel NG, Morales FJ, Epstein PR, Daszak P (2004) Emerging infectious diseases of plants: pathogen pollution, climate change and agrotechnology drivers. Trends Ecol Evol 19:535–544

Anindya R, Savithri HS (2004) Potyviral NIa proteinase, a proteinase with novel deoxyribonuclease activity. J Biol Chem 279:32159–32169

Aramburu J, Galipienso L, Matas M (2006) Characterization of potato virus Y isolates from tomato crops in northeast Spain. Eur J Plant Pathol 115:247–258

Barker H (1987) Invasion of non-phloem tissue in Nicotiana clevelandii by Potato leafroll luteovirus is enhanced in plants also infected with Potato virus Y. J Gen Virol 68:1223–1227

Beauchemin C, Boutet N, Laliberte J-F (2007) Visualization of the interaction between the precursors of VPg, the viral protein linked to the genome of Turnip mosaic virus, and the translation eukaryotic initiation factor iso 4E in planta. J Virol 81:775–782

Berger PH et al (2005) In: Fauquet CM, Mayo MA, Maniloff J, Desselberger U, Ball LA (eds) Virus taxonomy: eighth report of the International Committee on the Taxonomy of Viruses. Elsevier Academic, San Diego, pp 819–841

Bhat AI, Varma A, Pappu HR, Rajamannar M, Jain RK, Praveen S (1999) Characterization of a potyvirus from eggplant (Solanum melongena) as a strain of Potato virus Y by N-terminal serology and sequence relationships. Plant Pathol 48:648–654

Bilgin DD, Liu Y, Schiff M, Dinesh-Kumar SP (2003) P58IPK, a plant ortholog of double-stranded RNA-dependent protein kinase PKR inhibitor, functions in viral pathogenesis. Dev Cell 4:651–661

Blanc S, Lopez-Moya JJ, Wang RY, Garcıa-Lampasona S, Thornbury DW, Pirone TP (1997) A specific interaction between coat protein and helper component correlates with aphid transmission of a potyvirus. Virology 231:141–147

Business Insider The 10 most important crops in the world. New York (USA). http://www.businessinsider.com/10-crops-that-feed-the-world-2011-9?op=1/%20-%20-plantains-1&IR=T#4-potatoes-7

Carrington JC, Freed DD, Sanders TC (1989) Autocatalytic processing of the potyvirus helper component proteinase in Escherichia coli and in vitro. J Virol 63:4459–4463

Carrington JC, Jensen PE, Schaad MC (1998) Genetic evidence for an essential role for potyvirus CI protein in cell-to-cell movement. Plant J 14:393–400

Chung BYW, Miller WA, Atkins JF, Firth AE (2008) An overlapping essential gene in the Potyviridae. Proc Natl Acad Sci U S A 105:5897–5902

Danci O, Ziegler A, Torrance L, Gasemi S, Danci M (2009) Potyviridae family-short review. J Horticult For Biotechnol 13:410–420

Douglas DR, Pavek JJ (1972) Net necrosis of potato tubers associated with primary, secondary and tertiary infection of leafroll. Am Potato J 49:330–333

De Bokx JA, Huttinga H (1981) Potato virus Y. CMI/AAB Descr Plant Viruses 242:6

De Bokx JA, van der Want JPH (1987) Viruses of potatoes and seed potato production. Centre for Agricultural Publishing and Documentation (Pudoc), Wageningen, pp 1–259

Dufresne PJ, Ubalijoro E, Fortin MG, Laliberte JF (2008) Arabidopsis thaliana class II poly(A)-binding proteins are required for efficient multiplication of turnip mosaic virus. J Gen Virol 89:2339–2348

Dunoyer P, Thomas C, Harrison S, Revers F, Maule A (2004) A cysteine-rich plant protein potentiates Potyvirus movement through an interaction with the virus genome linked protein VPg. J Virol 78:2301–2309
Eigenbrode SD, Bosque-Perez NA (2016) Chemical ecology of aphid-transmitted plant viruses. In: Brown JK (ed) Vector-mediated transmission of plant pathogens. APS Press, St. Paul, pp 3–19
Fargette D, Konaté G, Fauquet C, Muller E, Peterschmitt M, Thresh JM (2006) Molecular ecology and emergence of tropical plant viruses. Annu Rev Phytopathol 44:235–260
Feki S, Loukili MJ, Triki-Marrakchi R, Karimova G, Old I, Ounouna H et al (2005) Interaction between tobacco ribulose-l,5-biphosphate carboxylase/oxygenase large subunit (Rubisco-LSU) and the PVY coat protein (PVY-CP). Eur J Plant Pathol 112:221–234
Fereres A (2016) Aphid behavior and the transmission of noncirculative viruses. In: Brown JK (ed) Vector-mediated transmission of plant pathogens. APS Press, St. Paul, pp 31–45
Fernandez A, Guo HS, Saenz P, Simon-Buela L, Gomez de Cedron M, Garcıa JA (1997) The motif V of plum pox potyvirus CI RNA helicase is involved in NTP hydrolysis and is essential for virus RNA replication. Nucleic Acids Res 25:4474–4480
Gabrenaite-Verkhovskaya R, Andreev IA, Kalinina NO, Torrance L, Taliansky ME, Ma¨kinen K (2008) Cylindrical inclusion protein of potato virus A is associated with a subpopulation of particles isolated from infected plants. J Gen Virol 89:829–838
Giordanengo P, Brunissen L, Rusterucci C, Vincent C, van Bel A, Dinant S, Girousse C, Faucher M, Bonnemain JL (2010) Compatible plant-aphid interactions: how aphids manipulate plant responses. C R Biol 333(6-7):516–523. doi:10.1016/j.crvi.2010.03.007
Gray S, De Boer S, Lorenzen J et al (2010) Potato virus Y: an evolving concern for potato crops in the United States and Canada. Plant Dis 94:1384–1397
Guo DY, Rajamaki ML, Saarma M, Valkonen JPT (2001) Towards a protein interaction map of potyviruses: protein interaction matrixes of two potyviruses based on the yeast two-hybrid system. J Gen Virol 82:935–939
Hong Y, Hunt AG (1996) RNA polymerase activity catalyzed by a potyvirus-encoded RNA-dependent RNA polymerase. Virology 226:146–151
Hong XY, Chen J, Shi YH, Chen JP (2007) The '6K1' protein of a strain of Soybean mosaic virus localizes to the cell periphery. Arch Virol 152:1547–1551
International Committee on Taxonomy of Viruses (ICTV) (2015) Virus taxonomy. EC47, London
Ivanov KI, Eskelin K, Lohmus A, Makinen K (2014) Molecular and cellular mechanisms underlying potyvirus infection. J Gen Virol 95:1415–1429
Jeffries C (1998) Potato. In FAO/IPGRI technical guidelines for the safe movement of germplasm. No. 19 Food and Agriculture Organization of the United Nations/International Plant Genetic Resources Institute, Rome. ISBN 92-9043-390-6
Jiang J, Laliberte J-F (2011) The genome-linked protein VPg of plant viruses—A protein with many partners. Curr Opin Virol 1:347–354
Jimenez I, Lopez L, Alamillo JM, Valli A, Garcia JA (2006) Identification of a Plum pox virus CI-interacting protein from chloroplast that has a negative effect in virus infection. Mol Plant-Microbe Interact 19:350–358
Karasev AV, Gray SM (2013) Continuous and emerging challenges of Potato virus Y in potato. Annu Rev Phytopathol 51:571–586
Kassanis B (1939) Intracellular inclusions in virus infected plants. Ann Appl Biol 26:705–709
Kasschau KD, Carrington JC (2001) Long-distance movement and replication maintenance functions correlate with silencing suppression activity of potyviral HC-Pro. Virology 285:71–81
Katis N, Tsitsipis JA, Stevens M, Powell G (2007) Transmission of plant viruses. In: van Emden H, Harrington R (eds) Aphids as crop pests. CABI, Wallingford, pp 353–390
Kirchner SM, Hiltunen LH, Santala J et al (2014) Comparison of straw mulch, insecticides, mineral oil, and birch extract for control of transmission of potato virus Y in seed potato crops. Potato Res 57:59–75
Kurppa A, Hassi A (1989) Reaction of four table potato cultivars to primary and secondary infection by potato viruses Yo and Yn. Ann Agric Fenn 28:297–307

Langenberg WG, Zhang LY (1997) Immunocytology shows the presence of tobacco etch virus P3 protein in nuclear inclusions. J Struct Biol 118:243–247

Leonard S, Plante D, Wittmann S, Daigneault N, Fortin MG, Laliberte JF (2000) Complex formation between potyvirus VPg and translation eukaryotic initiation factor 4E correlates with virus infectivity. J Virol 74:7730–7737

Leonard S, Viel C, Beauchemin C, Daigneault N, Fortin MG, Laliberte JF (2004) Interaction of VPg–Pro of Turnip mosaic virus with the translation initiation factor 4E and the poly(A)-binding protein in planta. J Gen Virol 85:1055–1063

Le Romancer M, Kerlan C, Nedellec M (1994) Biological characterization of various geographical isolates of Potato virus Y inducing superficial necrosis on potato tubers. Plant Pathol 43:138–144

Le Romancer M, Nedellec M (1997) Effect of plant genotype, virus isolate and temperature on the expression of the potato tuber necrotic ringspot disease (PTNRD). Plant Pathol 46:104–111

Lin L, Luo Z, Yan F, Lu Y, Zheng H, Chen J (2011) Interaction between potyvirus P3 and ribulose-1,5-bisphosphate carboxylase/oxygenase (Rubisco) of host plants. Virus Genes 43:90–92

Lopez-Moya JJ, Valli A, Garcıa JA (2009) Potyviridae. Encyclopedia of Life Sciences (ELS). Wiley, Chichester, http://www.els.net/

Magee CJM (1927) Investigations of the bunchy top disease of bananas. Council for Scientific and Industrial Research Bulletin no. 30. Melbourne

Mascia T, Finetti-Sialer MM, Cillo F, Gallitelli D (2010) Biological and molecular characterization of a recombinant isolate of Potato virus Y associated with a tomato necrotic disease occurring in Italy. J Pl Pathol 92:131–138

McIntosh C, O'Connell J (2014) Extract from Potato Grower (February 2014 edition), Capital Press. http://www.potatogrower.com/2014/02/study-shows-pvy-costs-idaho

Nolte P, Whitworth J, Thornton MK, McIntosh CS (2004) Effect of seedborne *Potato virus Y* on performance of Russet Burbank, Russet Norkotah, and Shepody potato. Plant Dis 88:248–252

Olspert A, Chung BYW, Atkins JF, Carr JP, Firth AE (2015) Transcriptional slippage in the positive-sense RNA virus family Potyviridae. EMBO Rep 16:995–1004

Pruss G, Ge X, Shi XM, Carrington JC, Bowman Vance V (1997) Plant viral synergism: the potyviral genome encodes a broad-range pathogenicity enhancer that transactivates replication of heterologous viruses. Plant Cell 9(6):859–868

Revers F, García JA (2015) Molecular biology of potyviruses. Adv Virus Res 92:101–199. doi:10.1016/bs.aivir.2014.11.006. Epub 2015 Jan 21

Roberts IM, Wang D, Findlay K, Maule AJ (1998) Ultrastructural and temporal observations of the potyvirus cylindrical inclusions (Cls) show that the Cl protein acts transiently in aiding virus movement. Virology 245:173–181

Rodamilans B, Valli A, Mingot A, San León D, Baulcombe D, López-Moya JJ, García JA (2015) RNA polymerase slippage as a mechanism for the production of frameshift gene products in plant viruses of the Potyviridae family. J Virol 89:6965–6967. doi:10.1128/JVI.00337-15

Rodriguez-Cerezo E, Ammar E, Pirone ED, Shaw JG (1993) Association of the non-structural P3 viral protein with cylindrical inclusions in potyvirus-infected cells. J Gen Virol 74:1945–1949

Rodrıguez-Cerezo E, Findlay K, Shaw JG, Lomonossoff GP, Qiu SG, Linstead P et al (1997) The coat and cylindrical inclusion proteins of a potyvirus are associated with connections between plant cells. Virology 236:296–306

Rojas MR, Zerbini FM, Allison RF, Gilbertson RL, Lucas WJ (1997) Capsid protein and helper component-proteinase function as potyvirus cell-to-cell movement proteins. Virology 237:283–295

Salaman RN (1949) The history and social influence of the potato. Cambridge University press, Cambridge, MA

Sastry KS, Zitter TA (2014) Plant virus and viroid diseases in the tropics. Springer Science/Business Media B.V, Dordrecht, doi: 10.1007/978-94-007-7820-7_2

Schaad MC, Jensen PE, Carrington JC (1997) Formation of plant RNA virus replication complexes on membranes: role of an endoplasmic reticulum-targeted viral protein. EMBO J 16:4049–4059

Scholthof KB, Adkins S, Czosnek H, Palukaitis P, Jacquot E, Hohn T, Hohn B, Saunders K, Candresse T, Ahlquist P, Hemenway C, Foster GD (2011) Top 10 plant viruses in molecular plant pathology. Mol Plant Pathol 12:938–954

Sherwood JL, German TL, Moyer JW, Ullman DE (2003) Tomato spotted wilt. Plant Health Instructor. doi:10.1094/PHI-I-2003-0613-02

Shi Y, Chen J, Hong X, Chen J, Adams MJ (2007) A potyvirus P1 protein interacts with the Rieske Fe/S protein of its host. Mol Plant Pathol 8:785–790

Smith MC, Holt J, Kenyon L, Foot C (1998) Quantitative epidemiology of Banana Bunchy Top Virus Disease and its control. Plant Pathol 47:177–187

Sorel M, Garcia JA, German-Retana S (2014) The Potyviridae cylindrical inclusion helicase: a key multipartner and multifunctional protein. Mol Plant-Microbe Interact 27:215–226

Tavert-Roudet G, Abdul-Razzak A, Doublet B, Walter J, Delaunay T, German-Retana S et al (2012) The C terminus of lettuce mosaic potyvirus cylindrical inclusion helicase interacts with the viral VPg and with lettuce translation eukaryotic initiation factor 4E. J Gen Virol 93:184–193

Torrance L, Andreev IA, Gabrenaite-Verhovskaya R, Cowan G, Makinen K, Taliansky ME (2006) An unusual structure at one end of potato potyvirus particles. J Mol Biol 357:1–8

Valkonen JPT (2007) Viruses: economical losses and biotechnological potential. In: Vreugdenhil D (ed) Potato biology and biotechnology, advances and perspectives. Elsevier, Oxford, pp 619–633

Valli A, Gallo A, Calvo M, Perez, J. d. J., & Garcia, J. A. (2014) A novel role of the potyviral helper component proteinase contributes to enhance the yield of viral particles. J Virol 88:9808–9818

Vance VB (1991) Replication of potato virus X RNA is altered in coinfections with potato virus Y. Virology 182(2):486–494

Verchot J, Koonin EV, Carrington JC (1991) The 35-kDa protein from the N-terminus of a potyviral polyprotein functions as a third virus-encoded proteinase. Virology 185:527–535

Verchot J, Carrington JC (1995) Evidence that the potyvirus P1 proteinase functions in trans as an accessory factor for genome amplification. J Virol 69:3668–3674

Wei T, Zhang C, Hong J, Xiong R, Kasschau KD, Zhou X et al (2010) Formation of complexes at plasmodesmata for potyvirus intercellular movement is mediated by the viral protein P3N-PIPO. PLoS Pathog 6:e1000962

White KW (2015) The polymerase slips and PIPO exists. EMBO Rep 16:885–886

Whitworth JL, Nolte P, McIntosh C, Davidson R (2006) Effect of Potato virus Y on yield of three potato cultivars grown under different nitrogen levels. Plant Dis 90(1):73–76

Wilson CR (2014) Applied plant virology. CABI, Wallingford, ISBN 978-1-78064-425-7

Wittmann S, Chatel H, Fortin MG, Laliberte JF (1997) Interaction of the viral protein genome linked of turnip mosaic potyvirus with the translational eukaryotic initiation factor (iso) 4E of Arabidopsis thaliana using the yeast two-hybrid system. Virology 234:84–92

World Potato Statistics (2015) The potato sector (FAOSTAT). http://www.potatopro.com/world/potato-statistics

Xiong R, Wang A (2013) SCE1, the SUMO-conjugating enzyme in plants that interacts with NIb, the RNA-dependent RNA polymerase of Turnip mosaic virus is required for viral infection. J Virol 7:4704–4715

Chapter 2
Molecular and Cellular Events During Infection of Potato by PVY

Maruša Pompe-Novak and Christophe Lacomme

Abstract Host plants that can be infected by pathogens, including viruses, exhibit either a compatible or an incompatible interaction. In an incompatible interaction, plants resist a virus by restricting cell invasion and virus replication. Plants exhibiting a compatible interaction are said to be susceptible and the virus can replicate and invade the plant. These interactions depend on the host genotype, the nature of the virus isolate and the degree of host sensitivity to infection. The response of a plant towards virus infection can as well vary throughout the various developmental and physiological growth stages of a plant, which are influenced by environmental conditions. To successfully invade a plant, a virus will hijack cellular functions of the host to its own benefit to promote local and systemic movement. In all types of interactions, infection by PVY causes a vast cellular re-programming of the host transcriptome resulting in numerous cytological, biochemical and physiological changes. Despite significant advances in the understanding of the cellular and molecular bases of their pathogenicity, many questions remain about the mechanisms by which viruses manipulate host defences and create an optimal intracellular environment to complete their infectious cycle. This chapter will present some case studies of cellular and molecular re-programming in various types of interactions between potato cultivars and PVY strains, that are also likely to have an influence on other multitrophic interactions.

M. Pompe-Novak (✉)
Department of Biotechnology and Systems Biology, National Institute of Biology (NIB), Vecna pot 111, SI-1000 Ljubljana, Slovenia
e-mail: marusa.pompe.novak@nib.si

C. Lacomme
Science and Advice for Scottish Agriculture (SASA), 1 Roddinglaw Road, Edinburgh EH12 9FJ, UK

C. Lacomme et al. (eds.), *Potato virus Y: biodiversity, pathogenicity, epidemiology and management*, DOI 10.1007/978-3-319-58860-5_2

1 Introduction

The development and severity of the disease symptoms on potato plants infected by PVY depend on the host genotype, the nature of the virus isolate and the degree of host sensitivity to infection. The latter can vary throughout the various developmental and physiological growth stages of a plant which are influenced by environmental conditions. Environment will also play an important role in the phenology of vectors especially with aphid-transmitted viruses such as PVY. This multitrophic interaction of host, virus, vector and environment is complex and will impact on the incidence and severity of plant disease from year to year (Draper et al. 2002; Scholthof 2007). This chapter aims to provide a summary of the current knowledge of the molecular, biochemical and cellular events associated with the various types of PVY-potato interactions.

2 Types of Potato-PVY Interactions

In nature, the interaction between a plant virus and its host fall broadly into two categories depending on the capacity of the virus to colonise a plant (Cooper and Jones 1983) (Fig. 2.1). In the majority of cases, a virus will lack the pathogenicity to infect a plant unless it is a true host plant. Plants that are immune to virus infection cannot be infected and do not support virus multiplication, even in directly inoculated cells, and virus titre does not increase in these cells (Valkonen 1994). Immunity per se, defined as an absolute state of exclusion of a virus at the cellular level, has not been found in potato-PVY interactions (Valkonen et al. 1996).

Host plants that can be infected by viruses (Cooper and Jones 1983) exhibit either a compatible or an incompatible interaction (Hinrich-Berger et al., 1999). Plants exhibiting a compatible interaction are said to be susceptible and the virus can replicate and invade the plants (Cooper and Jones 1983). Susceptible potato genotypes can be either sensitive or tolerant to PVY infection. Sensitive potato plants develop disease symptoms, while tolerant plants develop no or very mild symptoms, although they can accumulate high amounts of virus (Ravnikar 2005) (Fig. 2.2b).

In an incompatible interaction, plants resist a virus by restricting cell invasion and virus replication. During an incompatible interaction, plants can respond to virus infection with extreme resistance (ER) or a hypersensitive response (HR) (see Chap. 8). In the case of ER, potato plants show no symptoms as with susceptible tolerant cultivars or very limited necrosis (in the form of pinpoint lesions) in some genotypes (Valkonen et al. 1996). Virus titres remain extremely low, below the limit of detection, due to either inhibition of virus multiplication in the infected cells or restricted cell-to-cell movement of the virus (Valkonen 1994; Valkonen et al. 1996; Solomon-Blackburn and Barker 2001; Valkonen 2015). Mechanical inoculation of cvs Pentland Squire (tolerant) and Santé (ER, Ry_{sto}) by PVY^{NTN} results in no or

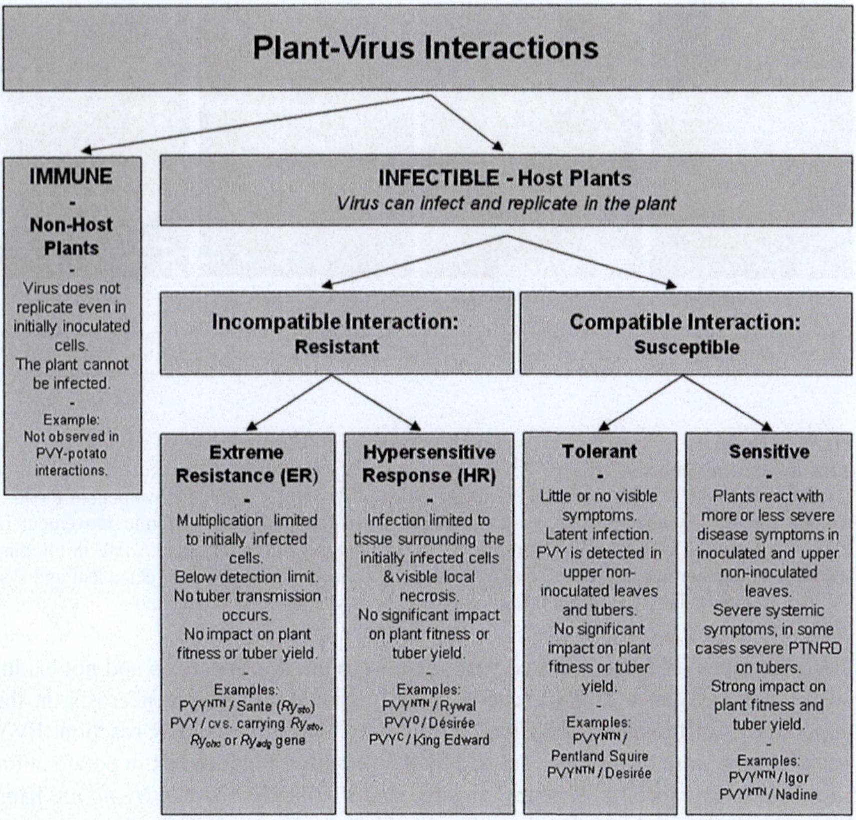

Fig. 2.1 Types of plant-virus interactions (adapted from Hull 2001). Examples of interactions between PVY strains and potato cultivars are presented

discrete symptoms on the inoculated leaves; however, while in cv. Pentland Squire, PVY^{NTN} spreads and accumulates to high level through the whole plant (tolerance), in cv. Santé, multiplication and spread of PVY^{NTN} is severely restricted (ER).

In the case of hypersensitive resistance response (HR), virus loading into the phloem in the initially infected leaf and translocation to other parts of the plant is prevented by an as yet unknown mechanism; however, virus replication and initial cell-to-cell movement are not affected. Later, most of the infected cells die, which result in a localised necrotic lesion at the site of infection (Valkonen 2015). For example, infection of cv. Rywal by PVY^{NTN} results in necrotic lesions on the inoculated leaves (Fig. 2.2a) and inhibition of systemic virus movement, with no significant impact on yield.

Although HR is generally defined as a mechanism to prevent pathogen invasion by the rapid death of cells at the infection site preventing the spread of pathogens to other parts of the plant (Pallas and Garcia 2011); in the case of potato-PVY interactions, the term HR is also used even when necrosis is observed in upper leaves. In

Fig. 2.2 Diversity of symptom expression observed on different potato cvs and genetic background following infection by PVY^{NTN}. (**a**) cv. Rywal (inoculated leaf, local necrosis resulting in restriction of PVY^{NTN} movement). (**b**) cv. Desiree (inoculated leaf, systemic movement of PVY^{NTN} without obvious symptoms). (**c**) cv. Desiree::*NahG* (inoculated leaf, systemic movement of PVY^{NTN}). Note the effect of *NahG* transgene overexpression on cv. Desiree::*NahG* in eliciting severe symptoms on leaves. (**d**) cv. Igor (inoculated leaf, severe necrosis on infected leaf and systemic movement). The type of interaction and severity of symptoms are indicated

these instances, HR is associated with the development of necrosis and not exclusively with resistance as PVY can eventually spread and trigger necrosis on the upper leaves of the plant (see Chap. 5, Fig. 5.1). Based on this HR reaction, PVY strain groups were historically defined by their ability to induce HR in potato cultivars harbouring specific *N* genes (Desiree and Pentland Crown – *Ny:nc:nz*, King Edward – *ny:Nc:nz*, Maris Bard and Pentland Ivory – *Ny:Nc:Nz*) where the term HR was associated with either local or systemic necrosis (Singh et al. 2008, Karasev and Gray 2013) (see Chap. 3).

However, virus-strain specificity of HR mediated by *Ny*, *Nc* and *Nz* genes makes this type of resistance inefficient towards PVY variants that are no longer "recognized" by their cognate *N* gene(s) and therefore are able to overcome this type of resistance (Valkonen 2015). Another limitation is the temperature sensitivity of some HR genes. There are many instances in which higher temperatures (usually above 20° C) break the resistance, which results in systemic movement of PVY. In cvs harbouring *Ny* resistance genes, resistance breakdown can result in some cases in systemic leaf necrosis (so-called leaf-drop) and mosaic symptoms in upper parts of the plant. This is exemplified by PVY^{O} on cv. Pito (Valkonen et al. 1998). Nonetheless, there is evidence that the formation of necrotic lesions and resistance, while being related, are yet independent phenomena. In the case of PVY^{NTN} on cv. Igor, while HR-like necrotic lesions develop on the inoculated leaves together with local and systemic chlorosis (Fig. 2.2d), PVY^{NTN} spreads systemically and infects the whole plant, including tubers, which develop severe PTNRD symptoms, resulting in significant yield reduction. This interaction is defined as sensitive (Pompe-Novak et al. 2006).

In general, the HR has been associated with an array of defence responses mediated through the initial recognition event between an avirulence gene from the pathogen (PVY in this case) and a resistance gene from the plant (reviewed in Pallas and Garcia 2011). These defence responses can be activated with different efficacy and timing and involves a range of molecular signalling pathways. Salicylic acid (SA) mediates the expression of a vast array of local and systemic defence responses (Huang et al. 2005). Disrupting the level of SA by overexpressing *NahG* gene which metabolises endogenous SA (Fig. 2.2c) will affect the expression and efficacy of these defence responses, which as a direct consequence will modify the outcome of the PVY-potato interaction as exemplified in Fig. 2.2 (panels b and c). These different types of host reactions to infection by PVY highlight the diversity and complexity of the molecular mechanisms associated with these events (see Sect. 4).

3 PVY Distribution in Cells, Tissues and Organs

In a compatible interaction, systemic infection of a plant by PVY will occur in different stages. After the virus has infected a single cell of a susceptible plant, it then moves into neighbouring cells, leaf veins and vascular system; once into the vasculature, the virus infects distant parts of the plant (Vuorinen et al. 2010). In an incompatible interaction, this process is stopped in the primary infected (inoculated) leaf. For efficient transmission of virus to new hosts by a vector, the presence of the virus throughout a plant increases the likelihood that an aphid will acquire virus during probing and feeding. The distribution of PVY within an infected plant was, therefore, investigated in the various parts and cells.

3.1 Kinetic of PVY Movement and Distribution in Plants

3.1.1 Movement Kinetic of PVY Strains in Susceptible and Resistant Potato Cultivars

Following mechanical inoculation of the surface of leaves, the movement of PVY^{O}, PVY^{NTN} and $PVY^{N\text{-}Wi}$ was studied in potato cultivars displaying either resistance or susceptibility using immunostaining or ELISA methods (Hinrichs et al. 1998; Rusetsky and Blotskaya 2001; Mehle et al. 2004). The results are summarised in Table 2.1. One day after inoculation, PVY was detected in single cells or groups of two neighbouring epidermal cells in both susceptible and resistant potato cultivars, spreading within a few days to form clusters of four to ten epidermal cells. However, 10 days after inoculation, PVY^{O} was detected in larger clusters of epidermal and mesophyll cells in susceptible cultivars, while in resistant cultivars, movement was severely restricted (Hinrichs et al. 1998) (Table 2.1). Similarly, PVY^{NTN} could be

Table 2.1 Movement of PVY^{O}, PVY^{NTN} and $PVY^{N\text{-}Wi}$ in potato cultivars at various intervals after mechanical inoculation

Cultivars	1 day	2 days	4 days	7–9 days	10 days	1–3 weeks
Susceptible cvs: Quarta, Igor, Yavar, Axamit, Desiree, Pentland squire	Single cells & neighboring cells	Clusters of 2–5 epidermal cells	Clusters of 6–10 epidermal cells. PVY^{NTN} detected by ELISA	PVY^{NTN} detected by ELISA on systemic leaves, apices and roots.	Larger clusters of cells infected, 10–15% mesophyll cells.	PVY^{NTN} Systemic infection in all aerial parts & roots
Resistant cvs: Pirola, Bettina, Rywal	Single cells and neighbouring cells	Clusters of 2–5 epidermal cells	Clusters of 6–10 epidermal cells	Same as 4 days, no systemic infection by PVY^{O} and PVY^{NTN} detected by ELISA	No systemic infection	No systemic infection

detected by ELISA only in inoculated leaves of susceptible cultivars 4–5 days following inoculation. Subsequently, the virus moved upwards and downwards in the stem and could be detected in upper uninoculated leaves, apices and roots 7–9 days after inoculation (Mehle et al. 2004). Systemic movement of PVY^{NTN} and $PVY^{N\text{-}Wi}$ was not observed in cultivars exhibiting extreme resistance (cv. Santé) or hypersensitivity (cv. Rywal) (Mehle et al. 2004).

The amount of PVY accumulated in plants of susceptible cultivars 1, 2 and 3 weeks after mechanical inoculation differed amongst lower, middle and upper leaves (Rusetsky and Blotskaya 2001). Three weeks after inoculation, the amount of PVY^{NTN} was lower in the roots than in the foliage of the sensitive cvs Igor and Désirée, and the tolerant cv. Pentland Squire. Between these three susceptible cultivars, there was no statistically significant difference in the amount of PVY^{NTN} in stems and roots, but a statistically significant difference was found in apical shoots, where the amount of PVY^{NTN} was greater in the tolerant cv. Pentland Squire and significantly lower in the most sensitive cv. Igor (Mehle et al. 2004). The amount of virus which accumulated in the three potato cultivars did not appear to correlate with their relative sensitivity (Draper et al. 2002; Mehle et al. 2004).

3.1.2 Live Imaging of PVY::GFP Movement in Plants

The cell-to-cell movement of PVY was investigated using fluorescently labelled PVY^{N}605::GFP which allows live cells to be examined non-destructively by confocal laser scanning microscopy. PVY^{N}605::GFP multiplied in patches in mechanically inoculated leaves of tobacco (*Nicotiana tabacum*) cv. Xanthi. These appeared as small foci 3–4 days after inoculation which expanded the next few days. The mean rate of cell-to-cell movement was approximately 7 cells per hour (Rupar et al.

2015). PVY^N605::GFP was detected in uninoculated leaves 4–5 days after inoculation and became unevenly distributed in these systemically infected leaves 1–5 days later when PVY^N605::GFP was unloaded from plant leaf veins of class I, II, III and VI (Rupar et al. 2015). In inoculated leaves of the susceptible potato cv. Bintje, PVY^N605::GFP appeared as small fluorescent foci 6–8 days after inoculation and was detected in systemically infected leaves 18–20 days after inoculation suggesting that PVY^N605::GFP was relatively stable under these experimental conditions and PVY can potentially be accessible to probing aphids within 20 days of a plant being infected (Rupar et al. 2015).

3.1.3 Kinetics of PVY Movement Following Mechanical or Aphid-Borne Infection

The process of infection by PVY following transmission by aphids was monitored in order to determine whether mechanical and aphid-mediated PVY infection differed. Three and 4 weeks after inoculation by viruliferous aphids, the distribution pattern of PVY^{NTN}, PVY^O and $PVY^{N:O}$ in hairy nightshade (*Solanum sarrachoides*) and the susceptible potato cv. Russet Burbank was similar to that with mechanical inoculation. The virus titre for all three strains was smaller in the upper foliar parts than in the lower parts and inoculated leaves. The amount of PVY^O and $PVY^{N:O}$ virus which accumulated within the various parts of a plant was similar for the strains tested. The amount of PVY in plants inoculated 15 days after planting increased between 3 and 4 weeks after inoculation, but the amount of PVY did not increase in older plants inoculated 30 days after planting (Cervantes and Alvarez 2011), suggesting that infection, and hence transmission, of PVY may be influenced by the stage of growth of plants at the time of infection.

3.2 *Distribution of PVY in Organs and Tissues in Susceptible Cultivars*

PVY^{NTN} was shown to have accumulated in leaves, stems, shoot tips, roots and tubers by qPCR 3 weeks after mechanical inoculation; however, the amount of viral RNA varied greatly amongst the various parts of plants (Kogovšek et al. 2011). High amounts of viral RNA were detected in all tissues of symptomatic leaves, stems, shoot tips, roots and petioles of senescent leaves. However, the amount of viral particles differed within infected leaves being greatest in chlorotic ringspots compared with necrotic ringspots, and being undetectable in surrounding yellowing tissue (Kogovšek et al. 2011).

The localisation of PVY^{NTN} viral RNA was also analysed at the cellular level by in situ hybridisation. PVY RNA was detected in leaf trichomes, petioles and stem parenchyma cells (Kogovšek et al. 2011). In stem tissues of potato plants and tobacco cv. Xanthi, PVY^O coat protein (CP) was found by tissue print immunoblot technique to be localised on the whole surface of the cross-sections with the highest concentrations in the epidermal and phloem tissues. In petioles of tobacco plants, PVY was confined to the peripheral layers of parenchyma, epidermis and phloem tissue (Krzymowska and Hennig 1997). Localisation of CP of PVY in vascular tissues of leaves was observed four to six times more frequently in parenchyma cells than in companion cells. This suggests that PVY first enters the vascular parenchyma cells and then the companion cells during the invasion process (Ding et al. 1998).

3.3 Distribution of PVY and PVY-Encoded Proteins in Infected Cells

Cytoplasmic inclusion bodies, seen as pinwheels, were identified in necrotic and neighbouring cells in PVY^O-inoculated leaves of the susceptible cv. Quarta (Hinrichs-Berger et al. 1999). In the sensitive cv. Igor, cytoplasmic inclusion bodies were located in the cytoplasm of epidermal cells, parenchymal stem cells and leaf trichomes 3 weeks after mechanical inoculation with PVY^{NTN}. The cytoplasmic inclusion bodies were localised together with the viral RNA and viral particles within the same type of cells or in close vicinity (Pompe-Novak et al. 2001; Kogovšek et al. 2011).

PVY^N was immunolocalised in chloroplasts of infected potato plants (Poljšak-Prijatelj and Ravnikar 1992). In tobacco leaves infected by PVY^O, PVY-encoded P1 protein was immunolocalised in the cytoplasm and associated with pinwheel protein inclusions (Arbatova et al. 1998).

In PVY-infected transgenic tobacco plants overexpressing the PVY helper component proteinase (HC-Pro), HC-Pro was detected only in chloroplasts of inoculated leaves (Tu et al. 2015). Contrastingly, when HC-Pro was transiently expressed in non-infected *Nicotiana benthamiana* epidermal cells, HC-Pro was distributed throughout the cytoplasm and localised in a small number of large, amorphous cytoplasmic inclusions that contained α-tubulin. In non-infected cells, HC-Pro appears in numerous dot-like inclusions distributed regularly throughout the cytoplasm and associated with the cortical endoplasmic reticulum and microtubule cytoskeleton (Del Toro et al. 2014). Viral infection, therefore, appears to trigger re-localisation of HC-Pro within chloroplasts of PVY-infected cells.

4 Cellular Re-programming During Potato-PVY Interactions

Infection by PVY causes a vast cellular re-programming of the host transcriptome, resulting in numerous cytological, biochemical and physiological changes.

4.1 Cytological Changes Caused by PVY Infection in Potato

PVY^{NTN} secondary infected potato plants are altered in a numbers of ways: they are smaller and the number of cells in the apical meristem is reduced, especially in the peripheral zone involved in leaf primordial formation. In the apical meristem of infected plants, the mitotic activity of the peripheral zone was decreased, leading to a smaller number of leaves and a reduction in total leaf surface area (Dolenc et al. 2000). In both resistant and susceptible cvs Quarta and Pirola, similar cytological and histological responses occurred in necrotic streaks along the veins of the lower surface of PVY^{O}-inoculated leaves (Hinrichs-Berger et al. 1999). Necrosis was observed in collenchyma, epidermal and mesophyll cells, while it was not seen in the vascular system (Hinrichs et al. 1998). The necrotic cells contained very dense cytoplasm with almost no vacuole, enlarged amoeboid nuclei, numerous vesicles, ruptured tonoplast membrane, and phenolic and cell wall appositions (Hinrichs-Berger et al. 1999).

Cytological changes were monitored in potato plants of the resistant cv. Rywal and the susceptible cv. Igor (Table 2.2). In cv. Rywal, phenolic compounds (*e.g.* lignins) were deposited at the edge of lesions 6 days after inoculation with $PVY^{N\text{-}Wi}$. Tissue structures at the site of lesions were observed under light and electron microscopy to be disorganised because of a profound loss of cellular integrity (Table 2.2) and the production of reactive oxygen species (ROS) associated with the

Table 2.2 Cytological changes occurring in potato cultivars infected by PVY^{NTN} and $PVY^{N\text{-}Wi}$ at various times after inoculation

Cultivars	3 days	6–8 days	3 weeks
Susceptible cv. Igor	NT	Swelling of chloroplasts, loosening of thylakoid structure in the necrotic area of the inoculated leaves. Cell wall wrinkling. Increase in number of chloroplasts in a cell in systemically infected leaves. Thickening of outer cell wall.	Accumulation of peroxisomes, invagination of nuclei and increased number of plasmodesmata between sieve elements and companion.
Resistant cv. Rywal	Nucleoli significantly smaller. Chromatin partially condensed. ROS associated with HR development. Callose deposition.	Nuclei and nucleoli partially degraded. Chromatin partially condensed, invagination of nuclear envelope. No significant changes in the sizes and shapes of the chloroplasts or mitochondria. Apposition of phenolic compounds (lignins) at the lesion edge. ROS associated with HR development. Callose deposition.	NT

NT not tested

hypersensitive response. In addition, callose apposition was found in lesions (Baebler et al. 2014). In susceptible cultivar, swelling of chloroplasts, loosening of thylakoid structure and changes in optical density of chloroplasts were observed (Table 2.2) (Pompe-Novak et al. 2001; Kogovšek et al. 2011). Chloroplast enlargement could also be reproduced in HC-Pro transgenic tobacco plants. A model of the inhibition of chloroplast division by HC-Pro was proposed in which HC-Pro binds to MinD (a calcium-dependent ATPase), resulting in the failure of Z-ring assembly and the consequent arrest of chloroplast division and enlargement of chloroplast (Tu et al. 2015). The conformational differences of HC-Pro between PVY^{O} and PVY^{N} strains (Tian and Valkonen 2013) were shown to be crucial in the recognition of the strains by potato resistance genes and consequently for the development of necrotic local lesions and the systemic virus spread (Tian and Valkonen 2015).

The β-1,3-glucanase class III (Glu-III) and the mitogen-activated protein kinase StMKK6 were shown to be involved in potato response to PVY infection. Glu-III was found to promote faster spread of PVY^{NTN} in potato and shown to localise in patches in the cell wall in a similar fashion to plasmodesmata in susceptible potato plants and in *Nicotiana benthamiana* plants (Dobnik et al. 2013). In contrast, the mitogen-activated protein kinase StMKK6 was localised in the nucleus of *Nicotiana benthamiana* cells infected by PVY^{NTN}, while no fluorescence was observed in uninoculated plants (Lazar et al. 2014).

These data illustrate the fact that a broad range of complex histological and cytological changes are caused by PVY infection in potato. One could assume that the nature and structure of HC-Pro resulting from infection by various PVY strains, its interactions with specific cellular host factors and other pleiotropic effects might play an important role in PVY infection.

4.2 *Biochemical Changes in Potato Caused by PVY*

The effect of infection by PVY on primary metabolites and other biochemical pathways was investigated (Table 2.3). In potato plants of cv. Igor infected by PVY^{NTN}, the total amount of photosynthetic pigment was not statistically different from that in uninoculated plants up to 5 days post-inoculation (dpi) (Milavec et al. 1999). However, by 5 dpi the amount of chlorophyll was lower in leaves with local lesions than in symptomless leaves of infected plants and healthy control plants (Milavec et al. 2001a). In leaves of cv. Désirée inoculated with PVY^{NTN}, photosynthetic activity as well as integrated net photosynthetic rate had decreased by 5 dpi (Stare et al. 2015).

The concentration of primary metabolites relating to sugar and aminoacid metabolism, the TCA cycle, the GABA shunt, ROS scavengers, and phenylpropanoids 1 day after inoculation was less than that in uninoculated plants, but this was followed by a pronounced increase after 6 days, coinciding with the onset of viral multiplication in infected leaves (Table 2.3). There was also a more rapid onset of ROS scavengers accumulating in leaves inoculated with PVY^{N} than in those inoculated with PVY^{NTN} (Kogovšek et al. 2016).

Table 2.3 Biochemical changes occurring at various intervals after inoculation in potato cultivars infected by PVY^{NTN} at various time after infection

Cultivars	1–4 dpi	5–6 dpi	14 dpi	20, 40 and 60 dpi	secondary infection
Sensitive cv. Igor, Chunzao	Lower concentration of primary metabolites of the sugar and amino-acid metabolism, TCA cycle, GABA shunt, ROS scavengers, and phenylpropanoids.	Lower amount of chlorophylls in leaves with local lesions.	Lower ratio between active and inactive cytokinins.	Reduced net photosynthetic rate, stomatal conductance and chlorophyll content.	Higher total amount of cytokinins.
	Inhibition of rutin accumulation.	Higher concentration of primary metabolites connected to sugar and amino-acid metabolism, TCA cycle, GABA shunt, ROS scavengers, and phenylpropanoids.	Higher amount of free salicylic acid, salicylic acid conjugates and gentisic acid.	Decreased maximum carboxylation velocity of ribulose-1,5-bisphosphate carboxylase/oxygenase, decreased maximum electron transport rate contributing to ribulose-1,5-bisphosphate regeneration, decreased relative quantum efficiency of photosystem 2 electron transport, decreased efficiency of excitation energy capture by open photosystem 2 reaction centres and photochemical quenching.	Lower ratio between active and inactive cytokinins.
	Higher activity of ionically-bound and covalently-bound peroxidases.	Higher activity of soluble and ionically-bound peroxidases.	Inhibited accumulation of rutin.		
			Three-times higher total protein concentration.		

(continued)

Table 2.3 (continued)

Cultivars	1–4 dpi	5–6 dpi	14 dpi	20, 40 and 60 dpi	secondary infection
Tolerant cvs. Désirée, Pentland squire	No influence on the accumulation of rutin.	Lower photosynthetic efficiency and the integrated net photosynthetic rate.	No influence on the accumulation of rutin.	NT	Lower amount of photosynthetic pigments.
			One third higher total protein concentration.		Altered ratio between cysteine proteinases and their endogenous inhibitors.
Resistant cv. Sante	Higher content of jasmonic acid and its precursor 12-oxo phytodienoic acid.		No effect on the ratio between active and inactive cytokinins.	NT	No effect on the ratio between active and inactive cytokinins.
	No effect on rutin accumulation.		No influence on the accumulation of rutin.		
			One third higher total protein concentration.		

NT not tested, *GABA* gamma-aminobutyric acid, *TCA* tricarboxylic acid, *ROS* reactive oxygen species

In addition, the total amount of phytohormones (jasmonic acid, salicylic acid, cytokinins), their precursor and derivatives were found to increase during PVY infection (Table 2.3) (Dermastia et al. 1995; Dermastia and Ravnikar 1996; Anžlovar et al. 1996; Milavec et al. 2001a, b, 2008; Pompe-Novak et al. 2002; Krečič-Stres et al. 2005; Kovač et al. 2009).

These observations highlight the crucial role of plant hormones in local and systemic responses of potato to PVY infection and their role in producing a range of disease symptoms in potato cultivars differing in their relative resistance.

The importance of the salicylic acid (SA) pathway in the response of potato to PVY infection was also demonstrated in transgenic potatoes of cvs Desiree and Rywal (Désirée::*NahG* and Rywal::*NahG*) expressing the *NahG* transgene responsible for endogenous SA depletion. PVY multiplied at a faster rate in plants of Désirée::*NahG* than in non-transgenic plants (Stare et al. 2015). Severe foliar symptoms of PVY infection developed in plants of Désirée::*NahG*, suggesting that the mechanisms for PVY tolerance had been disrupted (Baebler et al. 2011). The failure to accumulate SA in transgenic Rywal::*NahG* plants resulted in the expression of disease symptoms because of the unrestricted spread of PVY and the inhibition of SA-mediated defense responses (Baebler et al. 2014). Grafting experiments showed that SA has a critical role in inhibiting the spread of PVY in parenchymal tissue of cv. Rywal. The central role of SA in orchestrating *Ny*-1-mediated and other associated defence responses was confirmed by transcriptome analysis and showed that the absence of SA leads to considerable changes in the expression of genes involved in hormonal signalling in particular (Baebler et al. 2014).

When a plant responds to a challenge by a pathogen, phenolic compounds can act as a chemical barrier, an anti-oxidant or a signalling molecule. The accumulation of the flavonoid glycoside rutin between 4 h and 14 days after inoculation with PVY^{NTN} was inhibited in plants of susceptible cultivars, but was not inhibited in the moderately sensitive cv. Désirée or the extremely resistant cv. Santé (Kreft et al. 1999). Infection by PVY^{O} also reduced the amounts of flavonoids found in plants of nightshade, *Physalis angulata* (Nagai et al. 2015), highlighting the importance of flavonoids in a plant's response to PVY infection.

In addition, higher amounts of Na+/K+ ATPase was observed in extremely resistant cultivar Santé, resistant cultivar Carlingford and moderately sensitive cultivar Désirée as compared to highly sensitive cultivar Igor, but the amount of Na+/K+ ATPase did not differ between PVY^{NTN}-infected and healthy plants of any of the investigated cultivars (Gruden et al. 2000).

4.3 Transcriptome Changes During PVY-Potato Interactions

Transcriptome profiling of potato has been conducted following infection by PVY^{NTN}, PVY^{N}, $PVY^{N\text{-}Wi}$ and PVY^{O} of various cultivars, representing very different types of interactions ranging from compatible to extreme resistance.

4.3.1 Compatible Interactions

Transcriptome profiling of the sensitive cv. Igor identified several groups of genes that were regulated following PVY^{NTN} infection 0.5 h post-inoculation (hpi), 12 hpi and 7 dpi in inoculated leaves, 14 dpi in systemic non-inoculated leaves, and in leaves of secondary infected plants (Pompe-Novak et al. 2006; Baebler et al. 2009, 2011). A summary is presented in Table 2.4. Changes in gene expression 0.5 h after inoculation were less prominent than 12 h after inoculation (Baebler et al. 2009). Hierarchical clustering showed that gene expression pattern was the most similar in inoculated leaves 7 dpi and non-inoculated leaves 7 dpi. The highest number of common regulated genes was observed in non-inoculated leaves 14 dpi and leaves of secondary infected plants (Pompe-Novak et al. 2006).

The changes in transcriptome were compared in the sensitive cvs Igor and Nadine infected with two differently aggressive strains of PVY: PVY^{NTN} and PVY^{N}. Both cultivars responded similarly to the infection. The number of differently expressed genes in inoculated leaves increased between 0.5 h after inoculation and 48 h. In

Table 2.4 Relative expression of genes in PVY^{NTN}-infected plants of cv

	Primary infected plants				
	Inoculated leaves			Non-inoculated leaves	Secondary infected plants
Genes	0.5 hpi	12 hpi	7 dpi	14 dpi	
Involved in photosynthesis	↑	↓	↔	↓	↓
Heat shock proteins	↑	↑			
Heat shock protein 70			↓	↓	↓
Heat shock protein 80			↑		↓
β-1,3-glucanase		↑			↑
WRKY domain transcription factor families	↓	↑			
Several homologues of putative disease resistance genes		↑			
Involved in calcium signalling (calnexin, calcineurin, Calmodulin binding)		↑			
Involved in light signalling		↑			
Pathogenesis-related (PR) proteins (*e.g.* chitinase)		↑			
Involved In the metabolism and action of plant hormones		Mostly ↓			
Involved in cell wall degradation		Mostly ↓			
For proteinase inhibitors		Mostly ↓			

Igor compared with those in healthy uninoculated plants for (i) inoculated leaves 0.5 and 12 h after inoculation and after 7 days, (ii) uninoculated leaves of infected plants 14 days after inoculation and (iii) secondary infected plants

↑ – up-regulated; ↓ – down-regulated; ↔ – no significant change

dpi days post-inoculation, *hpi* hours post-inoculation

both cultivars, the expression of genes associated with photosynthesis was greater for PVY^{NTN} than for PVY^{N} and this difference was greatest 0.5 h after inoculation but less after 12 and 48 h. Pectin methylesterase inhibitor (PMEI) genes in PVY^{NTN}-inoculated plants were expressed much less than in PVY^{N}-inoculated plants. Differences between PVY^{NTN} and PVY^{N} were also observed in the expression of genes connected to the accumulation of sugars, genes for Fe-superoxide dismutase and genes encoding β-1,3-glucanase classes I and II (Kogovšek et al. 2010). While the expression of cell-wall invertase had decreased in the PVY^{N}-inoculated leaves, the expression of fructokinase had increased in the PVY^{NTN}-inoculated leaves 1 day after inoculation compared with healthy controls. By 6 days the PVY^{NTN}-inoculated leaves showed an increase in the expression of cell-wall invertase, hexokinase and β-amylase relative to the healthy controls. It can be concluded that, when comparing plant responses to PVY^{N} and PVY^{NTN}, the latter strain appears to be more effective in overcoming the first signal induction, thus delaying a plant's defence responses. In addition, PVY^{NTN} has the ability to alter the defence response, leading to successful infection (Kogovšek et al. 2016). In a different pathosystem, the transcriptome response of cv. Premier Russet, resistant to PVY^{O} and susceptible to PVY^{N}, and cv. Russet Burbank, susceptible to all PVY strains, to infection by PVY^{O} and PVY^{NTN} was analysed 4 and 10 h after inoculation. In this study, more similarities were found between incompatible and compatible reactions within a cultivar than between compatible reactions involving two different cultivars (Goyer et al. 2015).

There is much less data available about gene expression in potato plants of the tolerant cv. Désirée inoculated mechanically with PVY^{NTN}. In contrast to the highly sensitive cv. Igor, genes for β-1,3-glucanase classes I, II and III are uniformly expressed in inoculated leaves between 3 and 9 days after inoculation, and more expressed after 10 days in uninoculated leaves than earlier in the process of infection. Although the idea of glucanases facilitating spread of viruses by regulating the size exclusion limit and the permeability of plasmodesmata was proposed 40 years ago, so far little is known about the effect of its overexpression on viral infection (Zavaliev et al. 2011). In transgenic lines of cv. Désirée overexpressing Glu-III, PVY^{NTN} was found to spread faster than in non-transgenic control lines, while some multiplication of PVY^{NTN} was observed in transgenic lines of cv. Santé overexpressing Glu-III (Dobnik et al. 2013).

In cv. Désirée inoculated mechanically with PVY^{NTN}, genes involved in photosynthesis were highly expressed in inoculated leaves and systemically infected leaves 9 days after inoculation and this expression was greater than at earlier sampling times. The intensity of defence-related gene expression was much weaker in plants of cv. Desiree than in transgenic Desiree::*NahG* plants, suggesting that the strength of the gene expression is associated with the severity of the symptoms. In addition, it suggests that resistance can be at least partly uncoupled from SA-related defence pathways and in this case that while SA affect viral replication in the inoculated leaf, it does not prevent further spread of PVY (Baebler et al. 2011). This data highlights the complexity of these regulatory pathways in mediating resistance to PVY infection, expression of symptoms and downstream expression of defence genes.

4.3.2 Incompatible Interactions – Hypersensitive Response

In cv. Rywal infected by PVY^{N-Wi}, the more obvious changes were transiently observed 1 and 3 days after inoculation but were much less apparent after 6 days when induction prevailed over repression, confirming the transient nature of early changes. The functional classes of the up-regulated genes encompass cell-wall-, stress-, and secondary-metabolism-associated pathways. Down-regulation of gene expression was associated with carbohydrate metabolism, signalling, and RNA silencing. In contrast, in PVY^{N-Wi}-infected transgenic Rywal::*NahG* plants (which fail to accumulate salicylic acid and restrict PVY in the inoculated leaves), the changes were more pronounced 6 days after inoculation compared with those at 1 and 3 days. Genes associated with stress, carbohydrate metabolism and signalling were weakly down-regulated 1 day after inoculation. By 3 days, the number of up-regulated genes increased (*e.g.* secondary metabolism, stress, signalling), while photosynthesis-related genes were down-regulated. By 6 days, more than 25% of up-regulated genes were related to disease development and virus multiplication (Baebler et al. 2014).

4.3.3 Incompatible Interactions – Extreme Resistance

Gene expression changes were more pronounced in cv. Santé than in the sensitive cv. Igor 0.5 h after inoculation. This effect was transient because changes in cv. Santé were much less pronounced after 12 h. This suggests that the faster response to PVY infection resulting from an extreme resistance reaction might contribute a more efficient defence reaction overall. However, 12 h after inoculation, there was no notable trend of up-regulation of the genes associated with plant defence or resistance, virus recognition or signalling. Genes encoding heat shock proteins were down-regulated, while genes involved in cell wall degradation, secondary metabolism, lignin biosynthesis and PR proteins were mostly up-regulated. Genes involved in brassinosteroid (BR) biosynthesis and 2-ODD gene involved in secondary metabolism were strongly up-regulated and were shown to be amongst the features determining the extreme resistance of cv. Santé to infection by PVY^{NTN}. Amongst the functional classes identified, the highest proportion were associated with transport (14% of all hits), followed by hormone metabolism, DNA-RNA-nucleotide metabolism and protein biosynthesis (Baebler et al. 2009; van Dijk et al. 2009).

5 Multitrophic PVY Interactions

In the field, plants are often challenged by several biotic stressors. PVY-infected plants are frequently attacked by aphids and other herbivores. As described earlier, the vast array of biochemical changes in plants caused by PVY infection is also likely to have an influence on herbivores attacking an infected plant. Transcriptome studies have shown that PVY^{NTN}-infected and healthy potato plants respond differently to larval infestations of Colorado potato beetle (*Leptinotarsa decemlineata* Say) (Petek et al. 2014).

Feeding on PVY-infected plants can have a negative effect on the growth of the aphid vector *Myzus persicae*; the body and cornicle length, body and head width, and distance between compound eyes were significantly smaller in adult *M. persicae* reared on tobacco plants infected by PVY compared with healthy plants (Ren et al. 2015). In contrast, the numbers of the aphid *Macrosiphum euphorbiae* on tomato plants (*Solanum lycopersicum*) of cv. Castlemart were greater on those infected by PVY^{NTN}, $PVY^{N:O}$ and PVY^{O} than on healthy plants (Kersch-Becker and Thaler 2013). The growth of larvae of Colorado potato beetle (*Leptinotarsa decemlineata* Say) and cabbage looper (*Trichoplusia ni*) was enhanced when reared on PVY^{NTN}-infected plants, but not when reared on $PVY^{N:O}$- and PVY^{O}-infected plants (Kersch-Becker and Thaler 2013; Petek et al. 2014). The enhancement of larval growth in Colorado potato beetle on PVY^{NTN}-infected plants was linked to a reduction in the accumulation of transcripts associated with the anti-nutritional properties of potato (Petek et al. 2014). The impact of PVY infection on other type of interactions, including the effect on its own vector in relation to tropism and fitness, may provide new tools for future control and management of viruses in plants and crops.

6 Potato-PVY Interaction Data Integration

To understand the processes that lead to different types of interactions between potato and PVY, detailed spatiotemporal studies of the infection of various susceptible and resistant by PVY strains need to be conducted and integrated. This will lead to an understanding of the physiology of the interactions within the plant, by integrating biochemical responses and gene expression patterns for different types of interactions, host genotypes and strains of PVY. One of the main challenges for research of plant-pathogen interactions lies in the computer-assisted integrative study of large and increasingly complex combinations of data in order to understand host molecular mechanisms in response to pathogen infection. To tackle complex scientific questions, experimental data sets from different sources often need to be harmonised in regard to structure, formatting and annotation in order to open their content to integrative analysis (Sansone et al. 2012). Scientific workflows have been increasingly used in the last decade as an instrument for data-intensive science. They involve the use of multiple software tools and data resources in a staged fashion, with the output of one tool being passed as input to the next, leading to detailed structural, functional, and evolutionary results (Abouelhoda et al. 2012).

With the application of high-throughput data technologies in the studies of potato-PVY interaction and consequent rise in the amounts of experimental data available, suitable tools for data analyses and integration have been developed. For functional analysis of differentially expressed genes, MapMan ontology (Thimm et al. 2004) was adapted for potato (Rotter et al. 2007; Usadel et al. 2009). The

MapMan ontology was developed to cover plant-specific pathways and processes. It also enables the visualisation of gene expression together with metabolite accumulation organised in metabolic pathways (Kogovšek et al. 2016). MapMan ontology tree is linked to Gene Ontology MapMan (GoMapMan). GoMapMan was designed to expand Gene Ontology (GO) (The Gene Ontology Consortium 2000; The Gene Ontology Consortium 2013) to new plant species, including potato. In contrast to hierarchical tree structure of the MapMan ontology, GoMapMan is a collection of three non-overlapping ontologies, namely, molecular function, biological process and cellular component, allowing accessibility to knowledge stored in different biological databases (Ramsak et al. 2014). Plaza is a platform using comparative genomic resources to study gene evolution in plants and is a valuable source of extraction of functional gene relations between species (Proost et al. 2009). In the studies of potato-PVY interactions, it is especially useful for translation of known functions of Arabidopsis genes to potato genes. Gene set enrichment analysis (GSEA) (Subramanian et al. 2005) was used to search for groups of potato genes involved in the same processes that were significantly altered by PVY infection, using MapMan ontology as the source of the gene sets (Baebler et al. 2014; Stare et al. 2015).

Finally, all these datasets data should be integrated in a model of potato response to PVY infection that would allow further interpretation of biological data and prediction of the infection outcome. Up until now, only a draft model of protein-protein interactions between *Potyvirus* and *Arabidopsis thaliana* (Elena and Rodrigo 2012), and the network of salicylic acid, jasmonic acid and ethylene signal transduction pathways for modelling *Arabidopsis* defence response signalling have been constructed (Miljkovic et al. 2012). The importance of a model plant like *A. thaliana* is essential to creating a blueprint for modelling the responses of plants to pathogens. However, due to the complexity of the multiplicity of interactions between virus species, and even each virus strains, and host plant genotypes, an understanding of the interactions between potato and PVY should prove valuable, both for fundamental and applied sciences that could lead to novel biotechnological uses.

References

Abouelhoda M, Issa SA, Ghanem M (2012) Tavaxy: integrating Taverna and Galaxy workflows with cloud computing support. BMC Bioinformatics 13:77

Anzlovar S, Kovac M, Ravnikar M (1996) Photosynthetic pigments in healthy and virus-infected potato plantlets grown in vitro. Phyton 36:221–230

Arbatova J, Lehto K, Pehu E, Pehu T (1998) Localization of the P1 protein of potato Y potyvirus in association with cytoplasmic inclusion bodies and in the cytoplasmic bodies and in the cytoplasm of infected cells. J Gen Virol 79:2319–2323

Baebler Š, Krečič-Stres H, Rotter A, Kogovšek P, Cankar K, Kok EJ, Gruden K, Kovač M, Žel J, Pompe-Novak M, Ravnikar M (2009) PVY^{NTN} elicits a diverse gene expression response in different potato genotypes in the first 12 h after inoculation. Mol Plant Pathol 10:263–275

Baebler Š, Stare K, Kovač M, Blejec A, Prezelj N, Stare T, Kogovšek P, Pompe-Novak M, Rosahl S, Ravnikar M, Gruden K (2011) Dynamics of responses in compatible potato – Potato virus Y interaction are modulated by salicylic acid. PLoS One 6:e29009. doi:10.1371/journal.pone.0029009

Baebler Š, Witek K, Petek M, Stare K, Tušek-Žnidarič M, Pompe-Novak M, Renaut J, Syajko K, Strzelczyk-Żyta D, Marczewski W, Morgiewicz K, Gruden K, Hennig J (2014) Salicylic acid is an indispensable component of the Ny-1 resistance-gene-mediated response against Potato virus Y infection in potato. J Exp Bot 65:1095–1109

Cervantes FA, Alvarez JM (2011) Within plant distribution of Potato virus Y in hairy nightshade (*Solanum sarrachoides*): an inoculum source affecting PVY aphid transmission. Virus Res 159:194–200

Cooper JI, Jones AT (1983) Responses of plants to viruses: proposals for the use of terms. Phytopathology 73:127–128

Del Toro F, Tena FF, Tilsner J, Wright KM, Tenllado F, Chung BN, Praveen S, Canto T (2014) Potato virus Y HCPro localization at distinct, dynamically-related and environment-influenced structures in the cell cytoplasm. Mol Plant-Microbe Interact 27:1331–1343

Dermastia M, Ravnikar M (1996) Altered cytokinin pattern and enhanced tolerance to potato virus Y^{NTN} in the susceptible potato cultivar (*Solanum tuberosum* cv. Igor) grown *in vitro*. Physiol Mol Plant Pathol 48:65–71

Dermastia M, Ravnikar M, Kovač M (1995) Increased cytokinin-9-glucosylation in roots of susceptible *Solanum tuberosum* cultivar infected by Potato virus Y^{NTN}. Mol Plant-Microbe Interact 8:327–330

Ding XS, Carter SA, Deom CM, Nelson RS (1998) Tobamovirus and Potyvirus accumulation in minor veins of inoculated leaves from representatives of the *Solanaceae* and *Fabaceae*. Plant Physiol 116:125–136

Dobnik D, Baebler Š, Kogovšek P, Pompe-Novak M, Štebih D, Panter G, Janež N, Morisset D, Žel J, Gruden K (2013) β-1,3-glucanase class III promotes spread of PVYNTN and improves in planta protein production. Plant Biotechnol Rep 7:547–555

Dolenc J, Vilhar B, Dermastia M (2000) Systemic infection with potato virus Y^{NTN} alters the structure and activity of the shoot apical meristem in a susceptible potato cultivar. Physiol Mol Plant Pathol 56:33–38

Draper MD, Pasche JS, Gudmestad NC (2002) Factors influencing PVY development and disease expression in three potato cultivars. Amer J of Potato Res 79:155–165

Elena SF, Rodrigo G (2012) Towards an integrated molecular model of plant-virus interactions. Curr Opin Virol 2:719–724

Goyer A, Hamlin L, Crosslin JM, Buchanan A, Chang JH (2015) RNA-Seq analysis of resistant and susceptible potato varieties during the early stages of potato virus Y infection. BMC Genomics 16:472

Gruden K, Štrukelj B, Ravnikar M, Herzog-Velikonja B (2000) A putative viral resistance-connected protein isolated from potato cultivar Sante resistant to PVY^{NTN} infection. Phyton 40:191–200

Hinrichs J, Berger S, Shaw JG (1998) A hypersensitive response-like mechanism is involved in resistance of potato plants bearing the Ry_{sto} gene to the potyviruses potato virus Y and tobacco etch virus. J Gen Virol 79:167–176

Hinrichs-Berger J, Harford M, Berger S, Buchenauer H (1999) Cytological responses of susceptible and extremely resistant potato plants to inoculation with potato virus Y. Physiol Mol Plant Pathol 55:143–150

Huang Z, Yeakley JM, Garcia EW, Holdridge JD, Fan JB et al (2005) Salicylic acid-dependent expression of host genes in compatible Arabidopsis-virus interactions. Plant Physiol 137:1147–1159

Hull R (2001) Matthews' Plant Virology, 4th edn. Academic Press, London/San Diego

Karasev AV, Gray SM (2013) Continuous and emerging challenges of in potato. Ann Rev Phytopathol 51(1):571–586

Kersch-Becker MF, Thaler JS (2013) Virus strains differentially induce plant susceptibility to aphid vectors and chewing herbivores. Oecologia 174:883–892

Kogovšek P, Pompe-Novak M, Baebler Š, Rotter A, Gow L, Gruden K, Foster GD, Boonham N, Ravnikar M (2010) Aggressive and mild Potato virus Y isolates trigger different specific responses in susceptible potato plants. Plant Pathol 59:1121–1132

Kogovšek P, Kladnik A, Mlakar J, Tušek ŽM, Dermastia M, Ravnikar M, Pompe-Novak M (2011) Distribution of Potato virus Y in potato plant organs, tissues, and cells. Phytopathology 101:1292–1300

Kogovšek P, Pompe-Novak M, Petek M, Fragner L, Weckwerth W, Gruden K (2016) Primary metabolism, phenylpropanoids and antioxidant pathways are regulated in potato as a response to Potato virus Y infection. PLoS One 11(1):e0146135. doi:10.1371/journal.pone.0146135

Kovač M, Muller A, Milovanovič JD, Milavec M, Duchting P, Ravnikar M (2009) Multiple hormone analysis indicates involvement of jasmonate signalling in the early defence of potato to potato virus Y^{NTN}. Biol Plant 53:195–199

Krečič-Stres H, Vučak C, Ravnikar M, Kovač M (2005) Systemic Potato virus Y^{NTN} infection and levels of salicylic and gentisic acids in different potato genotypes. Plant Pathol 54:441–447

Kreft S, Pompe M, Ravnikar M, Umek A, Štrukelj B (1999) Accumulation rate of rutin is decreased after infection of susceptible potato cultivar with PVY^{NTN}. Phyton 39:259–264

Krzymowska M, Hennig J (1997) Simple and rapid technique to detect PVY presence in some Solanaceae plants. Acta Physiol Plant 19:95–99

Lazar A, Coll A, Dobnik D, Baebler Š, Bedina-Zavec A, Žel J, Gruden K (2014) Involvement of potato (*Solanum tuberosum* L.) MKK6 in Response to Potato virus Y. PLoS One 9(8):e104553. doi:10.1371/journal.pone.0104553

Mehle N, Kovač M, Petrovič N, Pompe-Novak M, Baebler Š, Krečič-Stres H, Gruden K, Ravnikar M (2004) Spread of potato virus Y^{NTN} in potato cultivars (*Solanum tuberosum* L.) with different levels of sensitivity. Physiol Mol Plant Pathol 64:293–300

Milavec M, Kovač M, Ravnikar M (1999) Photosynthetic pigments in potato plants (*Solanum tuberosum* L.) cv. Igor after primary infection with Potato virus Y^{NTN}. Plant Physiol 39:265–269

Milavec M, Ravnikar M, Kovač M (2001a) Peroxidases and photosynthetic pigments in susceptible potato infected with potato virus Y^{NTN}. Plant Physiol Biochem 39:891–898

Milavec M, Ravnikar M, Kovač M (2001b) Peroxidases in the early response of potato (*Solanum tuberosum* L. cv. Igor) susceptible to potato virus Y^{NTN}. Acta Biologica Slovenica 44:3–11

Milavec M, Gruden K, Ravnikar M, Kovač M (2008) Peroxidases in the early responses of different potato cultivars to infection by Potato virus Y^{NTN}. Plant Pathol 57:861–869

Miljkovic D, Stare T, Mozetič I, Podpečan V, Petek M, Witek K, Dermastia M, Lavrač N, Gruden K (2012) Signalling network construction for modelling plant defence response. PLoS One 7(12):e51822. doi:10.1371/journal.pone.0051822

Nagai A, Duarte LML, Chaves ALR, dos Santos DYAC (2015) *Potato virus Y* infection affects flavonoid profiles of *Physalis angulata* L. Braz J Bot 38(4):729–735

Pallas V, Garcia JA (2011) How do plant viruses induce disease/Interactions and interference with hosts components. J Gen Virol 92:2691–2705

Petek M, Rotter A, Kogovšek P, Baebler Š, Mithofer A, Gruden K (2014) Colorado potato beetle benefits from Potato virus Y-induced changes in potato defence signalling. Mol Ecol 23:5378–5391

Poljšak-Prijatelj M, Ravnikar M (1992) Detection of PVY^{N} with decoration and gold immuno labeling of thin sections in some secondary infected potato cultivars. Electron Microsc 3:465–466

Pompe-Novak M, Wrischer M, Ravnikar M (2001) Ultrastructure of chloroplasts in leaves of potato plants infected by Potato virus Y^{NTN}. Phyton 41:215–226

Pompe-Novak M, Poljšak-Prijatelj M, Popovič T, Štrukelj B, Ravnikar M (2002) The impact of potato cysteine proteinases in plant growth and development. Physiol Mol Plant Pathol 60:71–78

Pompe-Novak M, Gruden K, Baebler S, Krečič-Stres H, Kovač M, Jongsma M, Ravnikar M (2006) Potato virus Y induced changes in the gene expression of potato (*Solanum tuberosum* L.) Physiol Mol Plant Pathol 67:237–247

Proost S, Van Bel M, Sterck L, Billiau K, Van Parys T, Van de Peer Y, Vandepoele K (2009) PLAZA: a comparative genomics resource to study gene and genome evolution in plants. Plant Cell 21:3718–3731

Ramsak Ž, Baebler Š, Rotter A, Korbar M, Mozetic I, Usadel B, Gruden K (2014) GoMapMan: integration, consolidation and visualization of plant gene annotations within the MapMan ontology. Nucleic Acids Res 42(Database issue):D1167–D1175

Ravnikar M (2005) Potato virus Y and its interaction with potato. In: Freitag J (ed) ETNA Plant genomics and bioinformatics: expression microarrays and beyond – a course book, 1st edn. MPI-MPP, Postdam

Ren G, Wang X, Chen D, Wang X, Fan X, Liu X (2015) Potato virus Y-infected tobacco affects the growth, reproduction, and feeding behavior of a vector aphid, Myzus persicae (Hemiptera: Aphididae). Appl Entomol Zool 50:239–243

Rotter A, Usadel B, Baebler Š, Stitt M, Gruden K (2007) Adaptation of the MapMan ontology to biotic stress responses: application in solanaceous specie. Plant Methods 3:1–9

Rupar M, Faurez F, Tribodet M, Gutiérrez-Aguirre I, Delaunay A, Glais L, Kriznik M, Dobnik D, Gruden K, Jacquot E, Ravnikar M (2015) Fluorescently tagged Potato virus Y: a versatile tool for functional analysis of plant-virus interactions. Mol Plant-Microbe Interact 28:739–750

Rusetsky NV, Blotskaya ZV (2001) Distribution and accumulation of PVM and PVY, PVX in infected potato plants. Arch Phytopathol Plant Prot 34:275–282

Sansone SA, RoccaSerra P, Field D, Maguire E, Taylor C, Hofmann O, Fang H, Neumann S, Tong W, Amaral ZL, Begley K, Booth T, Bougueleret L, Burns G, Chapman B, Clark T, Coleman LA, Copeland J, Das S, de Daruvar A, de Matos P, Dix I, Edmunds S, Evelo CT, Forster MJ, Gaudet P, Gilbert J, Goble C, Griffin JL, Jacob D, Kleinjans J, Harland L, Haug K, Hermjakob H, Sui SJH, Laederach A, Liang S, Marshall S, McGrath A, Merrill E, Reilly D, Roux M, Shamu CE, Shang CA, Steinbeck C, Trefethen A, WilliamsJones B, Wolstencroft K, Xenarios I, Hide W (2012) Toward interoperable bioscience data. Nat Genet 44(2):121–126

Scholthof KB (2007) The disease triangle: pathogens, the environment and society. Nat Rev Microbiol 5:152–156

Singh RP, Valkonen JPT, Gray SM, Boonham N, Jones RAC, Kerlan C, Schubert J (2008) Discussion paper: the naming of Potato virus Y strains infecting potato. Arch Virol 153:1–13

Solomon-Blackburn RM, Barker H (2001) Breeding virus resistant potatoes (Solanum tuberosum): a review of traditional and molecualr approaches. Heredity 86:17–35

Stare T, Ramšak Ž, Blejec A, Stare K, Turnšek N, Weckwerth W, Wienkoop S, Vodnik D, Gruden K (2015) Bimodal dynamics of primary metabolism-related responses in tolerant potato-Potato virus Y interaction. BMC Genomics 16:716

Subramanian A, Tamayo P, Mootha VK, Mukherjee S, Ebert BL, Gillette MA, Paulovich A, Pomeroy SL, Golub TR, Lander ES (2005) Gene set enrichment analysis: a knowledge-based approach for interpreting genome-wide expression profiles. Proc Natl Acad Sci 102:15545–15550

The Gene Ontology Consortium (2000) Gene ontology: tool for the unification of biology. Nat Genet 25:25–29

The Gene Ontology Consortium (2013) Gene ontology annotations and resources. Nucleic Acids Res 41:D530–D535

Thimm O, Blasing O, Gibon Y, Nagel A, Meyer S, Kruger P, Selbig J, Muller LA, Rhee SY, Stitt M (2004) MAPMAN: a user-driven tool to display genomics data sets onto diagrams of metabolic pathways and other biological processes. Plant J 37:914–939

Tian Y-P, Valkonen PTJ (2013) Genetic Determinants of Potato virus Y required to overcome or trigger hypersensitive resistance to PVY strain group O controlled by the gene Ny in potato. MPMI 26:297–305

Tian Y-P, Valkonen PTJ (2015) Recombination of strain O segments to HCpro-encoding sequence of strain N of Potato virus Y modulates necrosis induced in tobacco and in potatoes carrying resistance genes Ny or Nc. Mol Plant Pathol 16:735–747

Tu Y, Zhang Z, Li D, Li H, Dong J, Wang T (2015) Potato virus Y HC-Pro reduces the ATPase activity of NtMinD, which results in enlarged chloroplasts in HC-Pro transgenic tobacco. PLoS One 10(8):e0136210. doi:10.1371/journal.pone.0136210

Usadel B, Poree F, Nagel A, Lohse M, Czedik-Eysenberg A, Stitt M (2009) A guide to using mapman to visualize and compare omics data in plants: a case study in the crop species, maize. Plant Cell Environ 32:1211–1229

Valkonen JPT (1994) Natural genes and mechanisms for resistance to viruses in cultivated and wild potato species (*Solanum* spp.) Plant Breed 112:1–16

Valkonen JPT (2015) Elucidation of virus-host interactions to enhance resistance breeding for control of virus diseases in potato. Breed Sci 65:69–76

Valkonen JPT, Jones RAC, Slack SA, Watanabe KN (1996) Resistance specificities to viruses in potato: standardization of nomenclature. Plant Breed 115:433–438

Valkonen JPT, Rokka VM, Watanabe KN (1998) Examination of the leaf-drop symptom of virus-infected potato using anther culture-derived haploids. Phytopathology 88:1073–1077

van Dijk JP, Cankar K, Scheffer SJ, Beenen HG, Shepherd LV, Stewart D, Davies HV, Wilkockson SJ, Leifert C, Gruden K, Kok EJ (2009) Transcriptome analysis of potato tubers-effects of different agricultural practices. J Agric Food Chem 57:1612–1623

Vuorinen AL, Gammerlgard E, Auvinen P, Somervuo P, Dere S, Valkonen JPT (2010) Factors underpinning the responsiveness and higher levels of virus resistance realised in potato genotypes carrying virus-specific R genes. Ann Appl Biol 157:229–241

Zavaliev R, Ueki S, Epel BL, Citovsky V (2011) Biology of callose (beta-1,3-glucan) turnover at plasmodesmata. Protoplasma 1:117–130

Chapter 3
Diversity, Characterisation and Classification of PVY

Laurent Glais, Dirk U. Bellstedt, and Christophe Lacomme

Abstract Since the discovery of PVY in the 1930s, many strain groups and variants have been described although consensus on an agreed international nomenclature is still a matter of debate. The challenge for plant pathologists studying PVY remains to establish the biological significance of the genomic diversity of PVY arising from the new, diverse PVY populations and how it impacts on crop production in practice. The distribution and range of PVY variants identified over the years highlight the ongoing threat that PVY poses to solanaceous crops worldwide. While genome sequencing provides accurate information on the genomic structure of an isolate and has become the method of choice as a first step to studying the phylogeny of novel isolates and their affiliation to a genotypic group, it is not a sufficient criterion to assign a PVY isolate to a specific strain group, as groups are identified by a defined set of reactions or symptoms that develop following infection of a range of potato cultivars harbouring known resistance genes and on tobacco plants. The purpose of this chapter is to give an update on the current status of knowledge of PVY diversity identified in different geographical regions and insights into identification and classification of PVY variants.

1 Introduction

One of the characteristics of viruses is their extraordinary ability to mutate and recombine at much higher rates than other organisms, approx. 10^3–10^4 times more than bacteria and eukaryotes (Gago et al. 2009), with an estimated rate of between

L. Glais (✉)
FN3PT/RD3PT, 43-45 rue de Naples, F-75008 Paris, France

IGEPP, Agrocampus Ouest, INRA, Université de Rennes 1, F-35650 Le Rheu, France
e-mail: laurent.glais@fnpppt.fr

D.U. Bellstedt
Department of Biochemistry, University of Stellenbosch, Stellenbosch 7602, South Africa

C. Lacomme
Science and Advice for Scottish Agriculture (SASA),
1 Roddinglaw Road, Edinburgh EH12 9FJ, UK

C. Lacomme et al. (eds.), *Potato virus Y: biodiversity, pathogenicity, epidemiology and management*, DOI 10.1007/978-3-319-58860-5_3

0.1 and 1 mutation per genome per generation (Tromas and Elena 2010). Since the discovery of PVY in the 1930s, many strain groups and variants have been described although consensus on an agreed international nomenclature is still a matter of debate. The challenge for plant pathologists studying PVY lies in assessing the extent to which the spread and infection by PVY isolates will impact on crop production in practice. A key part will be to establish the biological significance of the PVY genomic diversity arising from the new, diverse PVY populations. The characterisation and classification of PVY into strain groups is essentially based on their pathogenicity, that is, symptoms elicited on a range of specified host plants; however, this aspect is often less studied because the detection and identification of PVY isolates is generally based on serological (ELISA) or molecular (PCR-based and sequencing) assays. PVY classification may, therefore, be affected by the absence, in some cases, of a thorough biological characterisation (*i.e.* development of symptoms such as hypersensitive response, ringspot necrosis on tubers and accumulation of PVY in inoculated and upper uninoculated leaves). This could potentially lead to misinterpretation and mistaken assignment of an isolate to a strain group. The serological characteristics of PVY isolates have been widely used to classify PVY into serotypes because ELISA has often been the method of choice when it comes to testing plant material. However, reaching an unequivocal conclusion on the serotype of an isolate can be difficult because of the diversity of available polyclonal and monoclonal antibodies with different affinity characteristics, and sometimes lacking fully validated data. Finally, while genome sequencing provides accurate information on the genomic structure of an isolate and has become the method of choice as a first step to studying the phylogeny of novel isolates and their affiliation to a genotypic group, it is not a sufficient criterion to assign a PVY isolate to a specific strain group because the latter requires biological characterisation. The geographical spread and range of PVY variants highlight the ongoing threat that PVY poses to solanaceous crops worldwide (Chap. 4). The purpose of this chapter is to inform plant virologists and the potato industry of the current status of knowledge of PVY diversity and how PVY isolates can be identified and classified. In addition, we propose some modifications to the current nomenclature of PVY by focusing on the essential criteria that should be used to define PVY strain groups or variants.

2 The Diversity of PVY: A Constant Challenge for Its Classification and Characterisation

As reviewed by Singh et al. (2008), PVY nomenclature relies on the characterisation of strain groups, reflecting their pathogenicity and impact on a host plant. It is also essential to study the relationships between pathogenicity and genomic traits to ultimately identify the viral genetic determinants responsible for symptoms in the various hosts of PVY and to design accurate diagnostic tools (see Chap. 5).

2.1 *The Biological Characterisation of PVY Isolates: Definition of PVY Strains and Strain Groups*

PVY can infect a wide range of host plants, including solanaceous and nonsolanaceous species. Historically, the first characteristic used to classify a PVY isolate was the host plant from which an isolate was collected. This resulted in four PVY strains being defined: potato, tobacco, tomato and pepper strains. However, this strain concept has been found to be less meaningful over time because some isolates were capable of infecting more than one host species. For example, some potato and tobacco PVY isolates can infect various pepper cultivars (Horvath 1966; Marte et al. 1991; McDonald and Kristjansson 1993; Le Romancer et al. 1994). Stobbs et al. (1994) also reported that tomato plants could be infected by nontomato PVY isolates. Although it is useful to know the host from which an isolate was collected, more effective characterisation can be achieved by assessing an isolate's capacity to elicit symptoms on *Nicotiana tabacum,* to overcome resistance genes in *Solanum tuberosum* and to cause necrosis in potato tubers. These three essential criteria have allowed potato PVY isolates to be differentiated so far into seven strain groups (Table 3.1) (Singh et al. 2008; Karasev and Gray 2013).

2.1.1 PVY^{O} and PVY^{C} Strain Groups

Isolates of the PVY^{O} and PVY^{C} strain groups induce similar mosaic symptoms on infected leaves of tobacco (de Bokx and Huttinga 1981; Beemster and de Bokx 1987) (Fig. 3.1). However, symptom expression differs in potato. PVY^{C} causes mild disease symptoms of stipple streak on leaves and stems on susceptible potato cultivars, and induces a hypersensitive response (HR) in cultivars carrying the Nc_{tbr} gene (*e.g.* cvs King Edward, Duke of York/Eersteling and Maris Piper) (Cockerham 1970; de Bokx and Huttinga 1981). PVY^{O} induces more severe symptoms such as necrosis, leaf rugosity, crinkling and stunting in potato plants. PVY^{O} overcomes the Nc_{tbr} resistance gene but elicits a hypersensitive response in cultivars carrying the Ny_{tbr} gene (Jones 1990) and also *Ny-1* or *Ny-2* genes (Szajko et al. 2014). More recently within the PVY^{O} strain group, some North American isolates, referred to as PVY^{O}-O5, were found to induce in cvs Désirée and Maris Bard a HR response comparable to that produced by PVY^{O} isolates, but symptoms developed earlier and the necrotic reaction was more severe (Karasev et al. 2011). Despite these differences in symptoms, PVY^{O}-O5 isolates have been classified as isolates of the PVY^{O} strain group. Infection by isolates of PVY^{C} strain group has never been found to cause Potato Tuber Necrotic Ringspot Disease (PTNRD) in potato cultivars susceptible to PTNRD (Glais et al. 2015). However, some PVY^{O} isolates can cause PTNRD in the susceptible cvs Nadine and Yukon Gold but not in cv Béa, following foliar inoculation of plants grown in controlled conditions such as greenhouses (Gray et al. 2010; Davie 2014) (Fig. 3.2 and Table 3.1).

Table 3.1 Biological reactions or symptoms on leaves and tubers of the main indicator-host plants in response to infection by PVY strain groups and/or variants; Ny_{tbr}, Nc_{tbr}, Nz_{tbr}, *Ny-1*, *Ny-2*: potato hypersensitive resistance genes against PVY^{O}, PVY^{C}, PVY^{Z} and for the two last genes against PVY^{O} and PVY^{N} strain groups, respectively

			PVY strain group (typical isolate)								**Reported by**	
			O	**C**	**N**	**Z**	**E**	**N-Wi**	**NTN**	**Other**		
Plant species	**Cultivar**	**Genotype**	(SASA-110)	(Adgen)	(Mont)	(L26)	(MON)	(SASA-207)	(NZ)	(AST)		
Nicotiana tabacum	Xanthi, White Burley, Samsun		Mo	Mo	VN	Mo	Mo	VN	VN	VN	Chikh Ali et al. (2008), Singh et al. (2008) and Barker et al. (2009)	**On leaves**
Solanum tuberosum	Désirée, Pentland Crown, Alturas	Ny_{tbr}	HR	S	S	S	S	S	S	S	Jones (1990), Kerlan et al. (1999) and Rowley et al. (2015)	
	King Edward, Duke of York (Eersteling), Maris Piper	Nc_{tbr}	S	HR	S	S	n.d.	S	S	S	Jones (1990), Kerlan et al. (1999), Chikh Ali et al. (2014) and Glais unpublished	
	Pentland Ivory, Maris Bard, Yukon Gold	$Ny_{tbr,}$ $Nc_{tbr,}$ Nz_{tbr}	HR	HR	S	HR	S	S	S	HR	Jones (1990), Valkonen (1997), Kerlan et al. (1999, 2011), Chikh Ali et al. (2014), Tomczynska et al. (2014) and Rowley et al. (2015)	
	Rywal (Romula)	*Ny-1 (Ny-2)*	HR	n.d.	HR	n.d.	n.d.	HR	HR	HR	Szajko et al. (2008, 2014)	
	Béa		o	o	PTNRD*	PTNRD	PTNRD	o	PTNRD	PTNRD	Kerlan and Tribodet (1996), Karasev and Gray (2013) and Glais et al. (2015)	**On tubers**
	Nadine		PTNRD*	o	PTNRD*	n.d.	n.d.	PTNRD*	PTNRD	PTNRD	Browning et al. (2004) and Barker et al. (2009)	
	Yukon Gold		PTNRD*	n.d.	o	PTNRD	PTNRD	PTNRD*	PTNRD	PTNRD	Piche et al. (2004), Hu et al. (2009), Gray et al. (2010) and Galvino-Costa et al. (2012)	

Mo mosaic, *VN* veinal necrosis, *HR* hypersensitive response, *S* susceptible, *PTNRD* Potato Tuber Necrotic Ringspot Disease, *PTNRD** PTNRD observed only in greenhouse conditions and/or in specific potato cultivars, *o* no symptom, *n.d.* not documented. GenBank accession number of the listed PVY isolates can be found in Table 5.3. Notes: (1) isolate AST is proposed to define a different strain group; (2) Duke of York and Eersteling refer to the same potato cultivar

Fig. 3.1 Symptoms produced on tobacco (*Nicotiana tabacum* cv. Xanthi) and potato (*Solanum tuberosum*) leaves following PVY infection; (**a**): Mosaic induced by PVY^{O} isolates on tobacco leaves (similar symptoms are caused by PVY^{C}, PVY^{Z} and PVY^{E} isolates); (**b**): veinal necrosis and leaf deformation caused by PVY^{N} isolates on tobacco leaves (similar symptoms are caused by PVY^{NTN} and $PVY^{N\text{-}Wi}$ isolates); (**c**): necrotic, local lesions on inoculated leaves of potato cv. Eersteling (Nc_{tbr}) induced by the inoculation with a PVY^{C} isolate; (**d–e**): mosaic and veinal necrosis in apical leaves of the susceptible cv. Bintje caused by PVY^{NTN} isolate; (**f**): necrosis on potato stem, often called stipple streak, induced by PVY^{C}, and also by PVY^{O} or PVY^{N} isolates depending on the cultivar. Arrows indicate the position of the necrosis on potato leaves and stem. Photos are courtesy of © FN3PT-RD3PT / L. Glais

2.1.2 PVY^{N} Strain Group

Isolates of the PVY^{N} strain group differ from PVY^{O} and PVY^{C} strain groups because they induce veinal necrosis in tobacco and cause very mild mosaic symptoms on most potato cultivars. PVY^{N} is able to overcome the hypersensitive resistance genes Ny_{tbr}, Nc_{tbr} and Nz_{tbr} (Fig. 3.1). However, PVY^{N} isolates induce an HR in cvs Rywal and Romula that carry the *Ny-1* and *Ny-2* genes respectively (Jones 1990; Szajko et al. 2008, 2014). In field conditions, infection of potato plants by PVY^{N} isolates does not usually cause PTNRD in tubers, but some isolates may produce PTNRD in optimal conditions such as greenhouses (Fig. 3.2) (Kerlan and Tribodet 1996).

2.1.3 PVY^{NTN} Strain Group

The PVY^{NTN} strain group is closely related phenotypically to PVY^{N} (Table 3.1). Like PVY^{N}, PVY^{NTN} isolates induce veinal necrosis in tobacco, overcome Ny_{tbr} and Nc_{tbr} resistance genes in potato, and trigger HR in cvs Rywal and Romula carrying *Ny-1* and *Ny-2* resistance genes, respectively. However, isolates such as PVY^{NTN} NZ

Fig. 3.2 Symptoms of Potato Tuber Necrotic Ringspot Disease (PTNRD) caused by PVY isolates of various strain groups and on various cultivars; (**a**–**b**): PVY^{NTN} isolate NZ on cv. Rua; (**c**): PVY$^{N\text{-}Wi}$ (Polish isolate) on cv. Nadine; (**d**): PVY^{NTN} on cv. Chieftain; (**e**): PVY^{O} isolate on cv Nadine; (**f**, **g** and **h**): PVY^{NTN} isolates (European and North American) on cv. Nadine. Photos are courtesy of: © FN3PT-RD3PT / L. Glais (**a**–**b**); © SASA Crown Copyright (**c**–**h**)

can overcome these resistance genes (Kerlan et al. 2011), as do PVY^{N} isolates. The main characteristic which defines isolates of PVY^{NTN} strain group is their capacity to induce PTNRD in susceptible cultivars such as Béa, Nadine and Yukon Gold (Table 3.1; Fig. 3.2). Currently, PVY^{NTN} appears to be a very heterogeneous strain group, as unlike PVY^{N}, some PVY^{NTN} isolates such as PVY^{NTN}-H, PVY^{NTN}-NIB, 423.3, N4 (Kerlan et al. 2011) and PVY-AST (Galvino-Costa et al. 2012) elicit an HR reaction on potato cvs Maris Bard and Pentland Ivory carrying the Ny_{tbr}, Nc_{tbr} and Nz_{tbr} resistance genes (Chikh Ali et al. 2014). Consequently, we suggest that these isolates should be part of a separate strain group so far called 'Other' (see Sect. 2.1.6 and Table 3.1).

The propensity of PVY isolates to cause PTNRD can be assessed either in 'natural' (*i.e.* field grown) or in controlled conditions (*e.g.* glasshouse) that are more conducive to causing PTNRD (Le Romancer et al. 1994) (Fig. 3.1). Importantly, the development of PTNRD symptoms is strongly dependent on environmental conditions (see Chap. 5). Indeed, the inoculation of plants of three potato cultivars with PVY isolates belonging to different strain groups resulted in significant differences in severity of PTNRD, even in a controlled environment (Browning et al. 2004). This indicates that a biological assessment of an isolate's capacity to cause PTNRD might be prone to inherent variability and require appropriate controls. It is, therefore, recommended that a set of 'reference' PVY isolates belonging to a specific strain group, for example PVY^{O}, PVY^{N}, $PVY^{N\text{-}Wi}$, PVY^{NTN}, and a set of 'reference' potato cultivars susceptible to PTNRD, for example Béa, Nadine, Yukon Gold, should be included in any assay. In addition to these susceptible cultivars, we recommend that the pathogenicity of PVY isolates should be assessed on a range of cultivars known to be 'less susceptible' or resistant to PTNRD, such as Maris Piper, Spunta, Thalassa, Stella, BF15, but susceptible to PVY infection without displaying PTNRD (Le Romancer and Nedellec 1997). Such an approach would provide more extensive information on the pathogenicity of PVY isolates and would identify those posing a significantly higher risk for the potato industry (see Chap. 5).

2.1.4 $PVY^{N\text{-}Wi}$ Strain Group

Isolates of the $PVY^{N\text{-}Wi}$ strain group produce symptoms similar to PVY^{N} isolates on *N. tabacum* and *S. tuberosum.* However, isolates of this group appear to cause different reactions in another less commonly used indicator plant, *Solanum brachycarpum.* Isolates originally from Europe have been named $PVY^{N\text{-}Wi}$ and induce mosaic and necrotic spots, as with PVY^{O} isolates. In contrast, isolates originally from North America (Manitoba, Canada) were named $PVY^{N:O}$ and cause venous necrosis leading to plant death, as with PVY^{N} isolates (Singh and Singh 1994). In addition, some $PVY^{N:O}$ isolates induced roughly circular, sunken, necrotic lesions on the surface of tubers of cvs Yukon Gold, Alturas and Caribe. These symptoms were considered to be "atypical" PTNRD (Fig. 3.2) (Piche et al. 2004; Lorenzen et al. 2006). In the current classification the same acronym ($PVY^{N\text{-}Wi}$) is used to define different subgroups within this strain group, namely $PVY^{N\text{-}Wilga}$ and $PVY^{N:O}$ (Singh et al. 2008) (see Sect. 3.5; Fig. 3.5).

Differentiation of isolates $PVY^{N\text{-}Wi}$, PVY^{N} and PVY^{NTN} strain groups is essentially based on their molecular characterisation. This will be discussed in Sect. 2.3.

2.1.5 PVY^{Z} and PVY^{E} Strain Groups

Two rarer strain groups, PVY^{Z} and the PVY^{E}, have been described (Jones 1990; Kerlan et al. 2011). Isolates of the PVY^{Z} strain group exemplified by isolate L26 (Kerlan et al. 2011) produce similar symptoms on inoculated potato plants as some PVY^{NTN} isolates (Table 3.1). They also overcome Ny_{tbr} and Nc_{tbr} resistance genes, elicit an HR reaction in cv. Maris Bard carrying Nz_{tbr} gene and cause PTNRD in tubers (Kerlan et al. 2011). However, these isolates differ from PVY^{NTN} because they induce mosaic symptoms in tobacco leaves, such as with PVY^{O} isolates (Fig. 3.1). For the PVY^{E} strain group, only the Brazilian PVY-MON isolate is currently assigned into this group (Galvino-Costa et al. 2012). This isolate causes similar responses as isolates of the PVY^{Z} strain group except that it is able to overcome the Nz_{tbr} HR gene (Table 3.1). We also propose that the Syrian PVY-12 isolate be assigned to the PVY^{E} strain group and not considered to be a variant of the PVY^{NTN} strain group (Chikh Ali et al. 2008) as the symptoms elicited (*i.e.* mosaic on tobacco, no HR reaction on potato, and causing PTNRD) are similar to those of the PVY-MON isolate.

2.1.6 Exceptions: PVY Isolates That Do Not Fit Within the Current Strain Groups

So far, several reported PVY isolates display distinct biological characteristics that do not fall within the previously defined strain groups. The PVY isolates AST, H, NIB and 423.3 would seem to constitute a different strain group called ‘Other’ because of their biological characteristics. These isolates elicit veinal necrosis on tobacco and an HR reaction on plants of cv. Maris Bard (Table 3.1) (Kerlan et al. 2011; Galvino-Costa et al. 2012). As further knowledge and characterisation of new PVY isolates is gathered, it seems likely that new strain groups with novel biological characteristics will be defined.

2.2 *Serological Characterisation of PVY Isolates*

The antigenic properties of PVY strain groups have formed a very important part of their classification to complement biological characterisation. Initially, the antibodies produced against virus particles of PVY for the detection of viral infections were polyclonal, but in the early 1980s, monoclonal antibodies (MAbs) were developed as the preferred method for the identification and characterisation of PVY isolates (see Chap. 5; Table 5.2). MAbs provided the first evidence that PVY, considered to be a single virus strain, was actually composed of several strains that differed in the antigenicity of the coat or capsid protein (CP) (Fig. 3.3). CP is composed of one hydrophobic region corresponding to the central part of the protein, and two hydrophilic regions, the N- and C- terminal regions, which are exposed on the particle surface. These latter two regions correspond to the first 31 and the last 19 amino

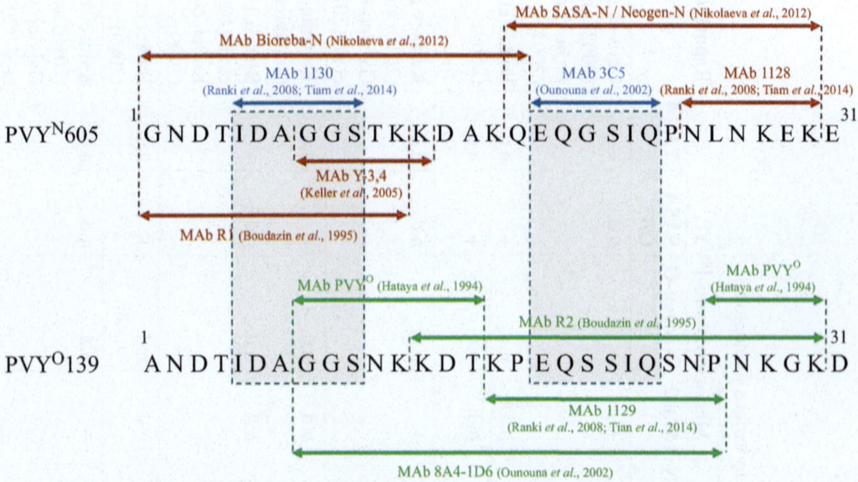

Fig. 3.3 Localisation of some epitopes for all PVY isolates (in *blue*), for isolates of PVY^N (in *red*), and for isolates of PVY^O (in *green*) within the first 31 amino acids of the N-terminal start of the PVY capsid protein; peptide sequences were based on PVY^{N}605 (Jakab et al. 1997-accession number X97895) and PVY^{O}139 (Singh and Singh 1996; accession number U09509)

acid residues of the capsid protein respectively, and carry the PVY antigenic structures (Shukla et al. 1988). However, the amino acid sequence of the CP N-terminal region for PVY^O and PVY^N isolates was found to be relatively variable (average 78% identity), compared with that for the C-terminal region (average 96% of identity) (Shukla and Ward 1988). The N-terminal region was found to contain the antigenic structures for differentiating between these strains groups (Fig. 3.3). The C-terminal carries the specific antigenic structures for PVY (Shukla et al. 1988; Vuento et al. 1993; Hataya et al. 1994; Boudazin et al. 1995; Ounouna et al. 2002; Keller et al. 2005; Ranki et al. 2008; Nikolaeva et al. 2012; Tian et al. 2014).

Two main serotypes defined using the immunological properties of PVY isolates were serotype-N which included PVY^N, PVY^{NTN}, PVY^Z and PVY^E, and serotype--O/C which included PVY^O, PVY^C and PVY$^{N\text{-}Wi}$ (Table 3.2). Over time, serotyping of a large number of PVY isolates revealed a more complex antigenicity than was originally thought. Comparative testing with various PVY^O and PVY^N specific monoclonal antibodies in 1980s and 1990s revealed that these detected most isolates as either PVY^O or PVY^N, but a few isolates were not detected (Gugerli and Fries 1983; Ellis et al. 1996). Up to nine different serological profiles were distinguished within the O/C-serotype and five within the N-serotype. However, there were exceptions, as exemplified with isolates PVY-AGA and PVY-AST, which were not detected by the PVY^N-specific monoclonal antibody SASA-N, but reacted with MAb 1F5 considered to be specific to isolates of serotype-N (Galvino-Costa et al. 2012). Suitable selection of antibodies known to detect PVY^N and PVY$^{O/C}$ serotypes is essential when assessing PVY populations and, when relevant, their characterisation (see Chap. 5; Fig. 5.3).

Table 3.2 Summary of the serological diversity of *Potato virus Y* (PVY) based on the reactions of isolates to various monoclonal

	Strain group	Isolate	Polyclonal antiserum				Monoclonal anti-PVY^N				Monoclonal anti-PVY^O				Reported by
			UID8	4C3	Y123-1	10E3	1F5	Bioreba-N	Neogen-N	SASA-N	Mab2	Neogen-O	SASA-O/C	Y38–4	
Serotype-O	**PVY^O**	139	n.d.	+	+	n.d.	–	–	n.d.	–	+	+	n.d.	+	Glais et al. (2002) and Piche et al. (2004)
		SCRI-O	n.d.	n.d.	n.d.	n.d.	n.d.	n.d.	n.d.	–	n.d.	n.d.	+	n.d.	Barker et al. (2009)
		SASA-110	n.d.	n.d.	n.d.	n.d.	n.d.	n.d.	n.d.	–	n.d.	n.d.	+	n.d.	
		O5	+	+	n.d.	n.d.	+	–	n.d.	–	+	n.d.	n.d.	n.d.	Karasev et al. (2010) and Galvino-Costa et al. (2012)
	PVY^C	Gelderse Rode	n.d.	n.d.	n.d.	n.d.	n.d.	n.d.	n.d.	n.d.	n.d.	n.d.	+	n.d.	Browning (unpublished)
		NI503	n.d.	+	+	+	–	–	n.d.	–	n.d.	n.d.	n.d.	+	Blanco-Urgoiti et al. (1998) and Kerlan et al. (1999)
	$PVY^{N\text{-}Wi}$	SASA-207	n.d.	n.d.	n.d.	n.d.	n.d.	n.d.	n.d.	–	n.d.	n.d.	+	n.d.	Barker et al. (2009)
		Alt	n.d.	+	n.d.	n.d.	–	n.d.	n.d.	–	+	n.d.	n.d.	n.d.	Karasev et al. (2010) and Galvino-Costa et al. (2012)

Serotype-N	**PVY^O**	SASA-61	n.d.	n.d.	n.d.	n.d.	n.d.	n.d.	n.d.	+	n.d.	n.d.	–	n.d.	Barker et al. (2009)
	PVY^N	SCRI-N	n.d.	n.d.	n.d.	n.d.	n.d.	n.d.	n.d.	+	n.d.	n.d.	–	n.d.	Barker et al. (2009)
		605	n.d.	n.d.	n.d.	n.d.	n.d.	n.d.	n.d.	+	n.d.	n.d.	–	n.d.	Browning (unpublished)
		Mont	n.d.	+	n.d.	n.d.	+	n.d.	+	n.d.	–	–	n.d.	n.d.	Lorenzen et al. (2006)
	PVY^Z	L26	+	n.d.	n.d.	n.d.	+	n.d.	+	+	–	n.d.	n.d.	n.d.	Kerlan et al. (2011) and Galvino-Costa et al. (2012)
		unknown	n.d.	n.d.	+	n.d.	n.d.	–	n.d.	n.d.	n.d.	n.d.	n.d.	+	Kerlan et al. (1999)
	PVY^E	MON	+	n.d.	n.d.	n.d.	+	n.d.	+	+	–	n.d.	n.d.	n.d.	Galvino-Costa et al. (2012)
		BC32	n.d.	n.d.	+	+	n.d.	–	n.d.	n.d.	n.d.	n.d.	n.d.	+	Blanco-Urgoiti et al. (1998) and Kerlan et al. (1999)
		12	n.d.	n.d.	n.d.	n.d.	+	n.d.	n.d.	n.d.	+	n.d.	n.d.	n.d.	Chikh Ali et al. (2007)
	PVY^{NTN}	NZ	n.d.	+	+	n.d.	+	+	n.d.	+	–	–	–	n.d.	Glais (unpublished) and Browning (unpublished)
		SYR	n.d.	+	n.d.	n.d.	–	n.d.	n.d.	n.d.	+	n.d.	n.d.	n.d.	Chikh Ali et al. (2007, 2010)
		HN2	n.d.	n.d.	n.d.	n.d.	n.d.	n.d.	n.d.	–	+	n.d.	n.d.	n.d.	Hu et al. (2009)
	Other	AST	+	n.d.	n.d.	n.d.	+	n.d.	–	–	–	n.d.	n.d.	n.d.	Galvino-Costa et al. (2012)

PVY^N- and *PVY^O*- antibodies, *4C3*, *1F5* and *Mab2* monoclonal antibodies distributed by Agdia, *Y123-1* and *Y38-4* monoclonals from FN3PT/INRA laboratory, *Bioreba-N/SASA-N/Neogen-O* monoclonals produced by Bioreba, SASA and Neogen, respectively, *'+/-'* isolate detected and not detected by the serological reagent, *n.d.* not documented

2.2.1 PVY^{O}, PVY^{C}, PVY$^{N\text{-}Wi}$ Strain Groups

Almost all isolates from these strain groups are recognised by PVY^{O} MAbs. However, a PVY^{O}-O5 isolate, was detected by MAb 1F5 considered to be specific for PVY^{N} (Table 3.2) (Karasev et al. 2010). This misidentification was caused by a substitution of an arginine (R) residue for an glutamine (Q) residue at position 98 in the CP protein, resulting in an epitope reacting with the PVY^{N}- monoclonal antibody 1F5. Another exception has been observed with the Scottish SASA-61 isolate which was classified into the PVY^{O} strain group based on its biological reactions on tobacco and potato plants (Barker et al. 2009). However, this isolate was detected with PVY^{N} monoclonal antibodies, which agrees with its whole genome sequence of N-type (Barker et al. 2009).

2.2.2 PVY^{N}, PVY^{Z}, PVY^{E} Strain Groups

So far, PVY^{Z} and PVY^{E} isolates are of serotype-N (Kerlan et al. 2011; Galvino-Costa et al. 2012). Previous studies have identified PVY^{Z} and PVY^{E} isolates that were serotype-O (Blanco-Urgoiti et al. 1998; Kerlan et al. 1999). Unfortunately, these isolates have been lost, so their genome could not be analysed to confirm their nature. To date, the PVY^{E} strain group consists of two isolates: PVY-MON and PVY-12. Unlike PVY-MON, PVY-12 was detected by Mab2 (monoclonal anti-PVY^{O}), and also by 1F5 (monoclonal anti-PVY^{N}) due to the substitution of the amino acid glutamic acid with glycine at position 29 of the coat protein, which switched an epitope from PVY^{N} to PVY^{O} (Chikh Ali et al. 2007).

2.2.3 PVY^{NTN} Strain Group

The majority of PVY^{NTN} isolates classify as serotype-N, but recently serotype-O isolates have been reported. For example, Syrian (SYR-I, SYR-II, SYR-III) and Chinese (PVY-HN2) isolates, initially classified as new variant groups within the PVY^{N} strain group, shared the same biological properties as PVY^{NTN} isolates but displayed different recombination patterns with genomic segments of different origins (Hu et al. 2009; Chikh Ali et al. 2010). Specifically, these PVY^{NTN} isolates reacted with PVY^{O}-MAbs because they possessed an O-type CP gene (Hu et al. 2009; Chikh Ali et al. 2010). However, even if these isolates are different at the genomic level, we propose that these isolates remain classified as PVY^{NTN} strain because of their biological characteristics. They induce PTNRD on susceptible potato cultivars, and their genome structure is similar to that of PVY^{NTN} isolates. The difference in serological reaction compared with representative PVY^{NTN} isolates should be considered as a characteristic of these isolates, but not as a sufficiently different one to justify their assignment to another group.

2.2.4 Limitations of Serotyping

While serological typing is a very important diagnostic tool and may be useful for PVY detection, the previous sections highlighted its limitations for PVY identification and classification. As previously described, serological typing and CP antigenicity are not always correlated with biological typing (PVY^{O} / $PVY^{N\text{-}Wi}$; PVY^{O}/ SASA-61). Scientists are fully aware of the limitations of the serological classification but serotyping remains an important tool for characterising PVY isolates. In this context, the use of antibodies in routine antibody-based detection methods for identifying specific isolates can sometimes be misleading, as indicated by Karasev et al. (2010) with the potential caveat of misidentification (as observed for PVY^{O}-O5 isolates). Such examples illustrate that although monoclonal antibodies are a powerful tool for the diagnosis of PVY, their performance in relation to the identification of PVY isolate must be thoroughly assessed.

Monoclonal antibodies produced during the last 30 years might recognise different individual epitopes that are not necessarily specific to a particular PVY strain group because classification into strain groups is based on PVY pathogenicity (Table 3.1) (Singh et al. 2008). The basis of this diversity in serotype is the result of nonsynonymous nucleotide changes occurring through mutations-recombination events that alter the amino acid composition and, in some cases, the structure of the CP. Consequently, the absence of a complete validation of any antibodies (MAbs and PAbs) with a recognised range of PVY reference isolates is a limitation for the use of serotyping for PVY classification as some PVY isolates might not be recognised by some Mabs (Dhar and Singh 1997) or can be misidentified (Chikh Ali et al. 2007; Karasev et al. 2010). To circumvent this inherent issue, full genome sequencing and phylogenetic studies of PVY are recommended before assigning PVY isolates to a given genotype.

2.3 Molecular Characterisation of PVY Isolates: Definition of Phylogenetic Groups and Recombination Analysis

In the 1980s, molecular methods were developed initially to sequence the CP genes of PVY isolates and then whole genomes. This involved very labour-intensive cloning and sequencing of short DNA sequences, and the first complete RNA sequence of a PVY isolate was described by Robaglia et al. (1989). Subsequently, the development of the reverse transcriptase-polymerase chain reaction (RT-PCR) to amplify fragments of the RNA genome of PVY and improved Sanger sequencing technologies, have led to the sequencing of many PVY isolates, either partially, for example P1, HC-Pro, CP or the whole genome (approx. 9700 bp). This has resulted in the genetic characterisation of numerous isolates from all over the world. So far, 256 whole and 2383 partial sequences of PVY genome have been deposited in the GenBank public database. In Table 3.3, a selection of whole-genome sequences of

Table 3.3 Representative isolates of PVY strain groups with the GenBank accession numbers for the whole genome sequences. Isolates in bold correspond to those used in the phylogenetic analysis (Fig. 3.6)

Strain group	Isolate	GenBank accession No	Country of origin
PVY^{o}	**0139**	**U09509**	**Canada**
PVY^{o}	Oz	EF026074	USA
PVY^{o}	**SASA-61**	**AJ585198**	**UK**
PVY^{o}	**SASA-110**	**AJ585195**	**UK**
PVY^{o}	**SCRI-O**	**AJ585196**	**UK**
PVY^{C}	**Adgen**	**AJ890348**	**France**
PVY^{C}	PRI-509	EU563512	Netherlands
PVY^{-N}	**Mont**	**AY884983**	**USA**
PVY^{N}	**N605**	**X97895**	**Switzerland**
PVY^{N}	Nicola	AJ890346	Germany
PVY^{N}	**N-Jg**	**AY166867**	**USA**
PVY^{NTN}	Ditta	AJ890344	Poland
PVY^{NTN}	HN1	HQ631374	China
PVY^{NTN}	**HN2**	**GQ200836**	**China**
PVY^{NTN}	HR1	FJ204166	USA
PVY^{NTN}	**NZ**	**AM268435**	**New Zealand**
PVY^{NTN}	**NE-11**	**DQ157180**	**USA**
PVY^{NTN}	**SYR-NB-16**	**AB270705**	**Syria**
PVY^{NTN}	**SYR-II-2-8**	**AB461451**	**Syria**
Other	**AST**	**JF928460**	**Brazil**
Other	**H**	**M95491**	**Hungary**
Other	NIB	AJ585342	Slovenia
$PVY^{N\text{-}Wi}$	L56	AY745492	Canada
$PVY^{N\text{-}Wi}$	Mb112	AY745491	Canada
$PVY^{N\text{-}Wi}$	**SASA-207**	**AJ584851**	**UK**
$PVY^{N\text{-}Wi}$	Wilga5	AJ890350	Germany
PVY^{E}	**PVY-MON**	**JF928458**	**Brazil**
PVY^{Z}	**L26**	**FJ204165**	**USA**

isolates representing various strain groups with their Genbank accession numbers is shown as a guide for selecting reference sequences useful for characterisation of isolates.

A comparison of the whole genome sequences of two PVY^{N} isolates (N605, Mont), one PVY^{NTN} (NZ) and two PVY^{O} isolates (O139, SASA-110) indicates that these three strain groups share 86% of their nucleotide identity, whereas isolates within each strain group share 92 to 94% of their nucleotide identity. When this analysis was performed across the genome, greater nucleotide variability was found in the 5'NTR region and in P1 gene. The nucleotide identity for PVY^{O} and PVY^{N} isolates in these two regions was 67 and 72% respectively, compared with an average of 90% for other genes (Fig. 3.4).

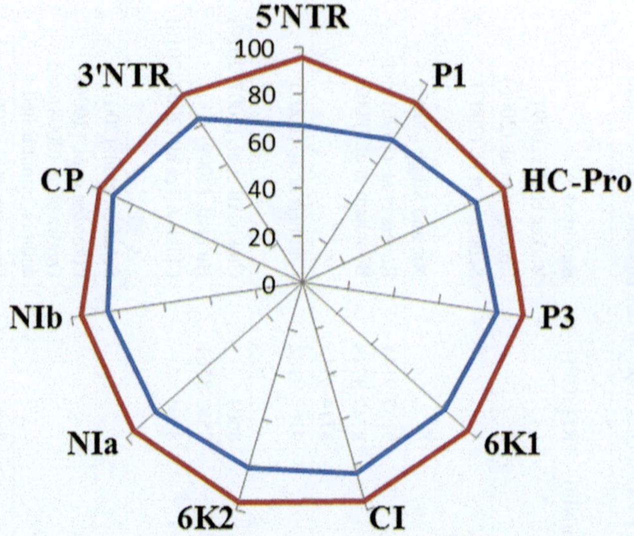

Fig. 3.4 Schematic representation of the percentage of nucleotide homology for each gene and nontranslated parts of PVY genome for various isolates of PVY^{O} and PVY^{N} strain groups; the percentages were obtained by comparing nucleotide sequences of two PVY^{N} isolates (N605, Mont) and one PVY^{NTN} (NZ) and two PVY^{O} (O139, SASA-110) (Table 3.3). The average homologies within the three strain groups (PVY^{O}, PVY^{N}, PVY^{NTN}) are represented by the red line; and those between the two strains is in blue. 5'NTR, P1, HC-Pro, P3, 6K1, CI, 6K2, NIa, NIb, CP and 3'NTR correspond to the different genes and nontranslated regions of PVY genome

Based on this polymorphism, 5'NTR and P1 regions were initially used as taxonomic criteria to classify PVY isolates. Some researchers reported a separation of PVY isolates by nucleotide sequence analysis of the 5'-terminal region of PVY into three main groups: the non-necrotic potato PVY group (PVY^{O}, PVY^{C}), the necrotic (PVY^{N}, PVY^{NTN}) group and the non-potato PVY group (van der Vlugt et al. 1993; Marie-Jeanne Tordo et al. 1995; Nie and Singh 2002; Lorenzen et al. 2006). More detailed descriptions of the PVY genome have been created by sequence analysis and phylogenetic studies of the increasingly large number of PVY nucleotide sequences (complete or partial) available in public databases. This has highlighted the extensive molecular variability of PVY isolates (Table 3.4; Fig. 3.5).

2.3.1 PVY^{O} Strain Group

Until 2010, isolates of PVY^{O} group were considered to be relatively homogeneous. The serological identification of a PVY^{O} isolate (PVY^{O}-O5) in the USA as PVY^{N}, (Karasev et al. 2010) prompted further interest in examining the molecular diversity of PVY^{O} isolates in the USA (Karasev et al. 2011) and in Japan (Ogawa et al. 2012). The complete genome sequence analysis of a large number of PVY^{O} isolates from

Table 3.4 Details of the various genomic features described for PVY strain groups and variants, associated with the nucleotide positions of reported recombination breakpoints; RJsup: corresponds to supplementary recombination junctions present in NIa or NIb regions; nucleotides are numbered according to the whole sequence of PVY^{N}-605 isolate (accession number X97895) (Jakab et al. 1997). Absence of a recombination junction is indicated by /. The white and grey highlighted sections correspond to genomic structures resulting from recombination event(s) between isolates of PVY^{N} and PVY^{O} strain groups, or only between isolates of PVY^{N} strain group, respectively. GenBank accession numbers of the main PVY isolates listed in this table are given in Table 3.3

Strain group	Isolate	RJ1 (P1)	RJ2 (HC-Pro/P3)	RJ3 (6K2/VPg)	RJsup (NIa/NIb)	RJ4 (CP)	Reported by
O	SASA-61	/	2665	/	/	/	Barker et al. (2009)
N	Mont	/	/	/	/	/	Lorenzen et al. (2006)
	SCRI-N	/	/	/	7921-7928/ 8335-8347	/	Schubert et al. (2007)
	N-Jg	/	2665	/	/	/	Nie and Singh (2002)
NTN	T-13	/	2414-2418	5809-5837	/	9137-9144	Revers et al. (1996) and Boonham et al. (2002)
	RU					8747-8748	
	VN					8714-8715	
	12-94, 34/01	499-500	2392-2393	5809-5837	/	9163-9182	Schubert et al. (2007)
	GR99		2158-2177				
	SYR-I	/	2414-2418	5809-5837	/	8604	Chikh Ali et al. (2007, 2010)
	HN2		2509-2522	5863-5867		8552-8561	Hu et al. (2009)
	SYR-II	499-500	2394	5809-5837	/	8604	Chikh Ali et al. (2007, 2010)
	SYR-III	691-694	*ns*	*ns*			
	NZ	/	/	/	/	/	Schubert et al. (2007)
	Tu660, RRA-1, Nicola, NTNOK105	/	2665	/	/	/	Lorenzen et al. (2006), Ogawa et al. (2008) and authors, unpublished
	NE11	/	2002-2015/ 2666-2696	/	/	8853	Lorenzen et al. (2008) and Chikh Ali et al. (2016)
Other	AST, H, 423.3, NIB	/	2414-2418	5809-5837	/	9163-9182	Glais et al. (2002), Lorenzen et al. (2006) and Galvino-Costa et al. (2012)

N-Wi	LW, B11, Isol5	499-500	2392-2393	/	/	/	Glais et al. (2002) and Schubert et al. (2007)
	CF_YL21		2414-2418				Chang et al. (2015)
	MAF-VOY		2320-2328				Galvino-Costa et al. (2012)
	SP17, OR-1, SASA-207, Alt, L56, PN10A	/	2412	/	/	/	Glais et al. (2002), Piche et al. (2004), Lorenzen et al. (2006, 2008), Schubert et al. (2007) and Barker et al. (2009)
	156vari	499-500	2395-2396	5809-5837	6700-6720	/	Schubert et al. (2007)
	156				8564-8572		
E	MON	/	2414-2418	5809-5837	/	/	Galvino-Costa et al. (2012)
	12	499-500				9163-9182	Chikh Ali et al. (2008)
Z	L26	/	2414-2418	5809-5837	/	9163-9182	Kerlan et al. (2011)

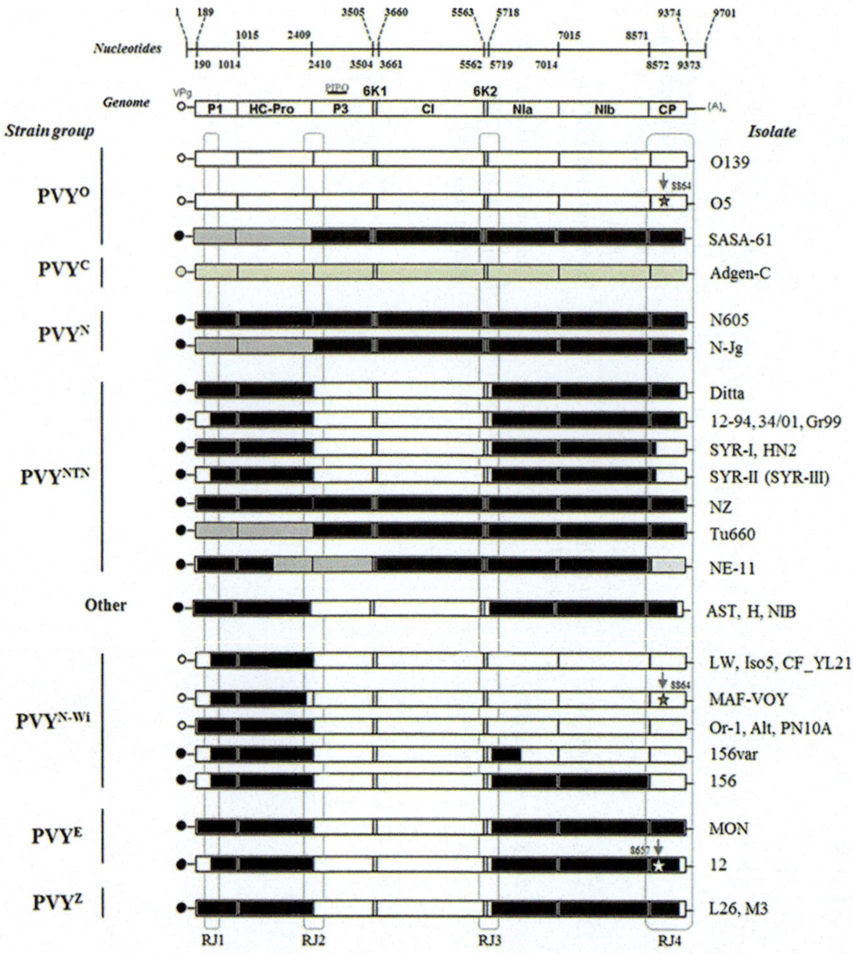

Fig. 3.5 Schematic representation of the molecular diversity of PVY strain groups and typical isolates; white/brown/black/grey/light grey colours correspond to PVY^{O}, PVY^{C}, European-PVY^{N}, North America-PVY^{N} and unknown nucleotide sequence. Arrows indicate the position of a single mutation in the coat protein gene leading to a misidentification of a PVY isolate by ELISA (the numbering is based on the PVY^{N}-605 nucleotide sequence-accession number X97895) (Jakab et al. 1997). White and black stars report a nucleotide of PVY^{O}- or PVY^{N}-type, respectively. RJ1 to RJ4: main recombination junctions describe in PVY genomes (according to Karasev and Gray 2013a). GenBank accession numbers of the main PVY isolates listed in this illustration are given in Table 3.3

these two countries revealed a high percentage (97–99.6%) of similarity (authors, unpublished). Further genome analysis using RDP3 software (Martin et al. 2015) to detect recombination events, revealed that these PVY^O isolates were composed of up to six distinct subgroups of different origins (Karasev et al. 2011; Ogawa et al. 2012). This analysis indicates that PVY^O is a more complex strain group than originally thought. A graphical representation of the genome of PVY^O isolates and PVY^O-O5 isolate is presented in Fig. 3.5. The relevance of these molecular variants within the PVY^O strain group on PVY evolution will be discussed later. In addition, isolate SASA-61 has been found to have the biological properties similar to a PVY^O isolate in spite of having an N-type genome identical to N-Jg and Tu660 isolates, with one recombination event (Fig. 3.6).

2.3.2 PVY^N Strain Group

Two main genetic molecular subgroups have been identified by nucleotide sequence comparison and phylogenetic analysis (Fig.3.5). The first one (initially termed the 'A' pattern or European PVY^N isolate EU-PVY^N) corresponds to PVY^N isolates like PVY^N-605, with a genome that shows no evidence of a recombination event (Glais et al. 2004). One exception has been noted by Schubert et al. (2007) with isolate SCRI-N which exhibits a PVY^O nucleotide sequence at position 7921-8347 (genome not presented in Fig. 3.5). The second subgroup (the 'B' pattern or called North American PVY^N isolate NA-PVY^N) is composed of PVY^N isolates resulting from recombination between PVY^N isolates of different origins (PVY isolate N-Jg). Sequence analyses based on the 5'NTR and P1 region revealed a clear separation between EU-PVY^N and NA-PVY^N isolates. The similarity of nucleotide identity for these two subgroups for the first 2665 nucleotides of their genome was only 90% (Nie and Singh 2002; Lorenzen et al. 2006) but, upstream of this recombination site, the homology was around 99% for the two subgroups. Although the differentiation 'EU' and 'NA' based on the geographic origin of the isolates seemed justified in the early 2000s, this was discontinued because EU-PVY^N isolates (Mont, N-Jg) were found in the USA too. Similarly, PVY^N isolates with a molecular genotype almost identical to that of NA-PVY^N isolates were also found in Europe as well as in Japan, (*e.g.* isolates SASA-61 and NTND6) (Ogawa et al. 2008). Nevertheless, PVY^N isolates are still being reported as EU- or NA- in the literature for convenience, although this is no longer indicative of the origin of an isolate (Lorenzen et al. 2006).

2.3.3 PVY^{NTN} Strain Group

The serious loss of marketable tuber yield in potato crops affected by PTNRD led to numerous efforts to characterise the PVY^{NTN} genome so as to develop specific diagnostic tools. Since the first description of this potato disease in Hungary in 1984 (Beczner et al. 1984), 34 partial and 19 complete genomes of isolates of this strain group have been sequenced. Initially, all PVY isolates collected in Europe from

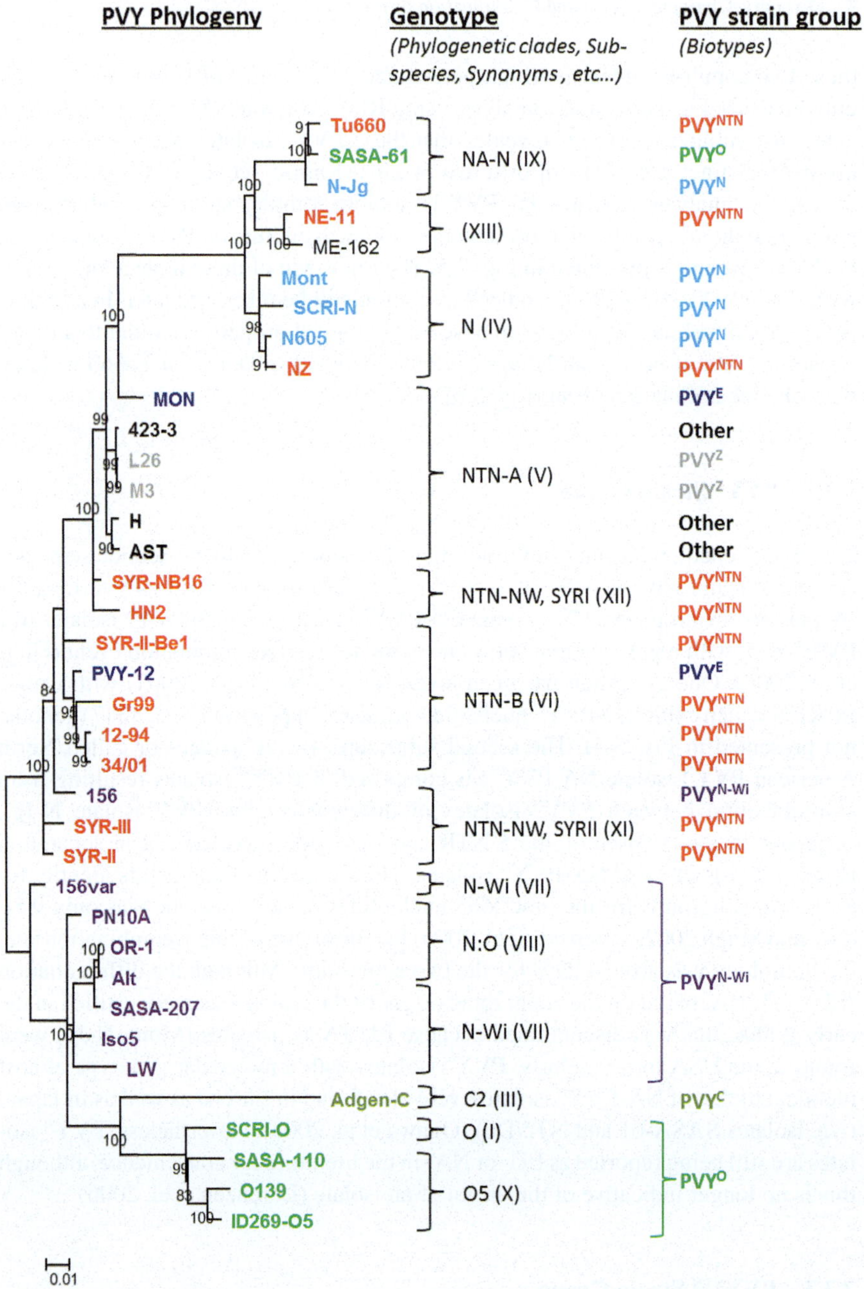

Fig. 3.6 Maximum likelihood phylogenetic trees of a selection of 37 PVY isolates (adapted from Kehoe and Jones 2015); alignments were generated in MEGA6 using CLUSTALW (branches bootstrapped with 1000 replications). The 13 clades representing PVY genotypes are indicated in Roman numbers as previously reported (Kehoe and Jones 2015) except the C1 (II) clade. Branch lengths indicate the number of substitutions per site. Genotypes are listed as genotypic groups (Kehoe and Jones 2015) together with synonyms. When the strain type (biotype) is known, the strain group is colour coded. 'Other' defines a new strain group in which isolates have distinct biological features. GenBank accession numbers of the main PVY isolates listed in this illustration are shown in Table III

potato tubers with symptoms of PTNRD were identified as recombinant isolates containing genomic segments of PVY^{O} and PVY^{N} (Revers et al. 1996; Glais et al. 1998 and Glais et al. 2002; Boonham et al. 2002). This enabled NTN-specific PCR-RFLP or RT-PCR diagnostic methods to be developed based on the nucleotide polymorphism reported in the P1 region (Glais et al. 1996; Weidemann and Maiss 1996; Weilguny and Singh 1998).

Whole sequence analyses on PVY^{NTN} isolates from many regions of the world defined four PVY^{NTN} subgroups resulting from recombination events between genome segments of PVY^{N} and PVY^{O} isolates (Revers et al. 1996; Glais et al. 1998 and Glais et al. 2002; Boonham et al. 2002). Within these four subgroups, the main recombination pattern consisted of three recombination sites (RJ2, RJ3, RJ4). The first occurred at the C-terminal part of HC-Pro gene between bp 2414–2418, the second at the N-terminal region of VPg between nt 5809–5837 (Glais et al. 1998 and Glais et al. 2002), and the third at the C-terminal region of CP encoding region between nt 9170–9183 (Revers et al. 1996; Boonham et al. 2002). Sequence analyses of CP of a collection of European PVY^{NTN} isolates from the UK, Slovenia, Denmark and Hungary, revealed that 70% of these isolates exhibited this recombinant profile (Revers et al. 1996; Boonham et al. 2002). The remaining PVY^{NTN} isolates differed only in that the breakpoint in CP region occurred between nt 8714 and 9144 and occurred less frequently in PVY^{NTN} populations (Revers et al. 1996; Boonham et al. 2002). Nevertheless, all these subgroups have the same pattern (Fig. 3.5).

Sequencing the whole genome of PVY isolates obtained from extensive sampling of infected potato crops identified other recombinant patterns. A second pattern observed in PVY^{NTN} isolates consisted of four recombination sites (RJ1, RJ2, RJ3, RJ4) (Table 3.4). In addition to the breakpoints observed in HC-Pro, VPg and CP regions, a fourth was identified in P1 region, located at nt 499–500 (Schubert et al. 2007).

Recently, the whole nucleotide sequence analysis of Syrian and Chinese PVY isolates highlighted additional recombinant patterns in the P1 and CP regions (Hu et al. 2009; Chikh Ali et al. 2010). To date, these recombinant patterns have been identified only in isolates from Syria (isolates SYR-NB-16, SYR-I, SYR-II, the latter called $PVY^{NTN\text{-}NW}$) and from China (PVY-HN2) (Hu et al. 2009; Chikh Ali et al. 2010). The Syrian isolates have been initially classified by Chikh Ali et al. (2010) as variants closely related to the PVY^{NTN} and $PVY^{N\text{-}Wi}$ strain groups because they have similar properties. The isolates reacted to serotype O, induced veinal necrosis on tobacco and the last 1000 nucleotides of their genome CP-3'NTR regions were similar to Syrian $PVY^{N\text{-}Wi}$ isolates (99.4% homologous). However, they also induced PTNRD on susceptible potato cultivars and had a genomic structure similar to that of some PVY^{NTN} isolates. Therefore based on these two last features we propose that these isolates remain classified as PVY^{NTN} isolates. The differences from typical PVY^{NTN} isolates (serological and $PVY^{N\text{-}Wi}$-type sequence at the 3'end) should be considered as peculiarities, but not as evidence of belonging to a different strain group.

As well as recombinant isolates, non-recombinant PVY^{NTN} isolates have also been described. A full-length genomic sequence of a PVY isolate originating from New Zealand (PVY^{NTN}-NZ) from potato tubers with PTNRD symptoms (Fig. 3.1) revealed a genome closely related (around 99% homologous) to EU-PVY^{N} isolates (Glais et al. 2004). Non-recombinant PVY^{NTN} isolates were found initially in North America (Nie and Singh 2003) and were termed NA-PVY^{NTN} isolates to distinguish them from European recombinant PVY^{NTN} isolates (EU-PVY^{NTN}) (Weidemann and Maiss 1996). Additionally, whole-genome sequencing of a range of isolates demonstrated that NA-PVY^{NTN} and EU-PVY^{NTN} molecular variants of PVY^{NTN} isolates were present both in the USA and in Europe (Lorenzen et al. 2006). In addition, NA-PVY^{NTN} isolates have also been identified in Asia (Ogawa et al. 2008). Therefore, as stated earlier for PVY^{N}, geographical classification of PVY (Lorenzen et al. 2006) is no longer relevant as it appears that EU and North American PVY isolates can be found in several potato-growing areas worldwide.

Other PVY^{NTN} isolates described in the USA, UK, Germany and Japan, initially considered to be non-recombinants (Nie and Singh 2003; Lorenzen et al. 2006; Ogawa et al. 2008), were found to have originated from recombination events between North American and European PVY^{N} isolates (Schubert et al. 2007; Ogawa et al. 2008). Phylogenetic analysis of different parts of the genome of Tu660, RRA-1, Nicola, SASA-61 and NTNOK105 isolates revealed the presence of a single break point located around nt 2665 in the HC-Pro region which defined a North American PVY^{N} nucleotide sequence downstream of this recombinant site (*ca.* 99% of nucleotide homology for N-Jg isolate compared with *ca.* 90% for PVY^{N}-605 isolate), and a European PVY^{N} nucleotide sequence upstream this point (*ca.* 98% homology for N-Jg and PVY^{N}-605 isolates) (Schubert et al. 2007; Ogawa et al. 2008) (Table 3.4; Fig. 3.5). The genomic structure for the PVY-NE11 isolate (Lorenzen et al. 2008), although exhibiting the same biological properties as isolates of PVY^{NTN} strain, was found to contain three unusual recombination events (Lorenzen et al. 2008).

2.3.4 $PVY^{N\text{-}Wi}$ Strain Group

Two main genomic patterns have currently been identified based on recombinations between PVY^{O} and PVY^{N} isolates. The first, exemplified by isolate LW (Fig. 3.5), has two recombination junctions located in P1 region at bp 499–500 and in the C-terminal part of HC-Pro gene at a position nt 2400. These are the same or very similar recombination sites to those reported for PVY^{NTN} (Table 3.4) (Glais et al. 2002; Piche et al. 2004; Lorenzen et al. 2006; Schubert et al. 2007; Barker et al. 2009). In addition to recombination events, mutations have apparently enhanced the diversity of these isolates, as illustrated by the $PVY^{N\text{-}Wi}$ isolate (MAF-VOY) (Galvino-Costa et al. 2012), which has been serologically misidentified as an PVY^{N} isolate because of a single mutation in the capsid gene at the position 8864. A second pattern within the $PVY^{N\text{-}Wi}$ strain group, previously classified as $PVY^{N:O}$, displayed only one recombination break point located, as previously, in the C-terminal part of HC-Pro gene (Table IV) (Glais et al. 2002; Schubert et al. 2007; Chang et al. 2015). In addition to these two main genomic patterns, isolate 156var displays four recombination events in its genome, that is, two recombination events in P1 and HC-Pro, a third located in 6K2/VPg region (between bp 5809-5837) and a fourth present either in VPg gene at nt 6700–6720 for the isolate 156var, or in the C-terminal part of NIb gene at nt 8564–8572 for the isolate 156 (Schubert et al. 2007).

2.3.5 PVY^E Strain Group

The PVY-MON isolate displays a genome with two recombination junctions (RJ2, RJ3) located at the same positions in HC-Pro/P3 region between nt 2414–2418 and 6K2/VPg region between nt 5809–5837, as previously described for PVY^{NTN} isolates. The other PVY^E isolate (PVY-12) has four recombination junctions (RJ1, RJ2, RJ3, RJ4) at the same positions as with PVY^{NTN}. Despite the apparent genetic relationship between PVY-12 and other PVY^{NTN} isolates, they differ in their biological reactions (see Sect. 2.1.5). One or several unknown mutations elsewhere in the PVY-12 genome could explain these differences, highlighting the extreme complexity of genetic and biological relationships.

2.3.6 PVY^Z Strain Group

Molecular characterisation has been conducted on only two isolates. The first isolate PVY-M3 was identified in Mexico (Robles-Hernandez et al. 2010) and the second, PVY-L26, in the USA (Kerlan et al. 2011). Other isolates have been lost or not retained after biological and serological analyses were completed (Jones 1990; Kerlan et al. 1999). Although PVY^Z isolates cause different biological reactions in tobacco and potato cultivars carrying the Nz_{tbr} resistance gene compared with isolates of PVY^{NTN}, they have identical genomes. Isolates PVY-M3 and L26 carry the three classical recombination junctions RJ2, RJ3 and RJ4 at positions similar to some EU-PVY^{NTN} isolates (Table 3.4; Fig. 3.5) (Robles-Hernandez et al. 2010; Kerlan et al. 2011). PVY-L26 isolate has been classified as a PVY^Z-NTN variant (Kerlan et al. 2011). However, we propose that this nomenclature should be avoided and that isolates of this strain should be classified as PVY^Z (Kerlan et al. 2011).

2.4 Inter- and Intrastrain Genetic Diversity of PVY: Phylogenetic and Strain Groups Relationships

This classification of potato PVY species into eight strain groups (*i.e.* seven defined strain groups and an additional strain group of PVY isolates of distinct biological properties) masks an even more complex variability in PVY. This classification into strain groups is based on the biological properties of PVY isolates. However, this does not take account of the prevalence of recombination events and mutations of PVY genomic regions occurring in PVY populations although these often do not result in significant biological changes in isolates' reactions.

Whole genome sequencing and phylogenetic studies have defined a large number of PVY genotypes that separate into at least 13 different subgroups (numbered from I to XIII) (Kehoe and Jones 2015) (Fig. 3.6). This shows that intrastrain genomic variability is very significant. It is also becoming increasingly apparent that strain groups often do not resolve within a specific genetic group, suggesting that the genetic determinants of PVY pathogenicity may be different for each PVY subspecies.

2.4.1 PVY^{N} Strain Group

Genetic diversity in this strain group is exemplified in two genetic patterns based on the nucleotide-sequence homology in the first 2665 nucleotides of their genome and occurred in recombinant and nonrecombinant PVY^{N} isolates (Fig. 3.5) (Nie and Singh 2002; Lorenzen et al. 2006). This suggests that these two populations result from different evolutionary processes.

2.4.2 PVY^{O} Strain Group

Since 2011, considerable molecular heterogeneity has been found in this strain group. Full-genome sequencing and phylogenetic analyses of 44 isolates from North America revealed that this strain group comprised three distinct molecular patterns PVY^{O}-O1, PVY^{O}-O2 and PVY^{O}-O3, in addition to the PVY^{O}-O5 clade (Karasev et al. 2011). A similar clustering has also been found in a Japanese collection of PVY^{O} isolates, in which two new molecular profiles were detected (Ogawa et al. 2012). Based on molecular analyses of the entire genome and a study of three genomic regions (nucleotides 700/2000; 2600/5800; 6270/7500), the isolates of the PVY^{O} strain group can be divided into six molecular patterns exhibiting an average percentage of nucleotide homology of 97% between each other: O1 present in UK, USA and Japan, O2 and O3 found in USA and Canada, O5 present in USA and Canada, and O-J1 and O-J2 present only in Japan. The molecular data suggests that these populations might come from different origins (Ogawa et al. 2012). Moreover, in PVY^{O} and PVY$^{N\text{-}Wi}$ isolates (irrespective of their clustering), a comparison of the nucleotide sequence of the PVY^{O} genomic segment from nucleotide 1 to 500 revealed an identity close to 90% between isolates. This suggests the presence of an unidentified seventh PVY^{O} population which must be the parental genome of PVY$^{N\text{-}Wi}$ isolates for this region (Glais et al. 2002; Karasev et al. 2011).

2.4.3 PVY$^{N\text{-}Wi}$ and PVY^{NTN} Strain Groups

As described earlier, molecular data showed that PVY$^{N\text{-}Wi}$ strain group consisted of two subgroups based on the presence of one or two recombination junctions. In addition, intrastrain diversity has been also reported (Karasev et al. 2011). Phylogenetic analyses of the PVY genomic region of nucleotide 2406–5821 of a collection of recombinant PVY isolates revealed that the PVY^{O} parental donor of this segment was not the same for the two populations. A PVY$^{N\text{-}Wi}$ subgroup (*e.g.* isolate LW) had a PVY^{O} sequence close to PVY^{O}-O2, whereas PVY$^{N\text{-}Wi}$ subgroup (*e.g.* isolate Or-1) acquired an O-type segment from an unknown, ancestral PVY^{O} isolate, which differed from the six PVY^{O} subgroups described above. In addition, recombinant PVY^{NTN} isolates, which also have an O-type sequence at the same region (Fig. 3.5) are closely related to PVY$^{N\text{-}Wi}$ subgroup (Or-1), suggesting that these PVY$^{N\text{-}Wi}$ isolates could be a precursor of recombinant PVY^{NTN} isolates (Karasev et al. 2011).

3 Relationships Between Genomic Traits and Pathogenicity

PVY is one of the plant viruses that has evolved rapidly since it was first recorded. The plasticity of its genome is one of the reasons why it is able to evolve and adapt to new and changing environments and biotic/abiotic constraints. This evolution has been often accompanied by isolates able to produce more severe symptoms in tobacco and potato plants and to overcome resistance genes bred into potato cultivars. Any modification of PVY genome in terms of pathogenicity and virulence is correlated with minor or major genetic changes, for example mutations and recombination events. Modifications in the PVY genome may not necessarily be associated with changes in the biological reactions attributable to infection by an isolate because mutations may not impact on the viral amino acid sequence (synonymous mutations), may affect gene(s) not linked to pathogenicity, or simply the isolate has yet to evolve to overcome corresponding resistance genes within the host plant. Since the late 1990s, numerous researchers have focused their efforts on the identification of the viral molecular determinants responsible for pathogenicity to get a better understanding of PVY–plant interactions and, hence, resistance management. Furthermore, these biological traits constitute some of the criteria used for classifying PVY.

3.1 Tobacco Veinal Necrosis

As previously described, the first criterion used to differentiate between isolates of PVY^{N} and PVY^{O} strain groups was their ability to induce necrosis in tobacco leaves. An initial study of representative isolates of PVY^{O}, PVY^{N}, PVY^{NTN} and PVY$^{N\text{-}Wi}$

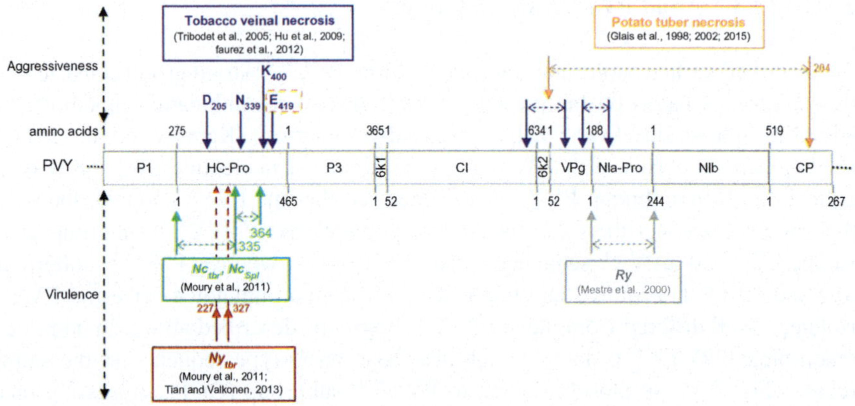

Fig. 3.7 Schematic representation of the PVY genome with the location of viral molecular determinants or genomic regions proven or suspected to be involved in PVY pathogenicity; factors responsible for the expression of veinal necrosis in tobacco leaves are represented in blue; those involved in the Potato Tuber Necrotic Ringspot Disease (PTNRD) are in orange. Genomic regions demonstrated to control the plant resistance (Hypersensitive Response-*N*, Extreme Resistance-*R*) are shown in green (Nc_{tbr}), in red (Ny_{tbr}) or in grey (*Ry*). Amino acid scale is presented according to Thole et al. (1993)

strain groups compared their biological properties and assessed the genome by RFLP. This showed that the molecular determinant(s) involved in the expression of the necrotic reaction were probably located between the C-terminal part of the P1 gene and the N-terminal region of the P3 gene (Glais et al. 1998 and Glais et al. 2002). A reverse genetics approach involving the construction of $PVY^{N/O}$ chimeras and mutants resulting from genomic exchanges between the infectious clone $PVY^{N}605$ and the reference isolate PVY^{O}-139 showed that two amino acids, K400 and E419, in the C-terminal part of the HC-Pro protein were involved in the necrosis of tobacco leaves (Tribodet et al. 2005) (Fig. 3.7). A third amino acid, N339 in the HC-Pro protein and two regions from the N-terminal part of CI protein to the C-terminal part of NIa-Pro protein were also shown to be linked to the necrotic reaction (Faurez et al. 2012). However, the systematic sequencing of a large number of PVY isolates from potato crops showed that some of them, such as L26 (Hu et al. 2009), LW (Schubert et al. 2007) and SASA-61 (Barker et al. 2009), were not able to induce veinal necrosis in tobacco leaves despite the presence of NKE residues in the HC-Pro protein. These three amino acids are now regarded as necessary, but not sufficient, to induce veinal necrosis in tobacco, and other determinants need be present elsewhere in the PVY genome. The aspartic acid residue (D) at position 205 in the HC-Pro protein was also suspected to be a potential candidate, because it was the only amino acid which differed among three necrotic isolates and the nonnecrotic isolate L26 (Hu et al. 2009). However, the alignment of the amino acid sequence of this region in both necrotic and non-necrotic isolates showed the presence of a D residue at position 205, suggesting that this amino acid could not be associated with the expression of tobacco veinal necrosis for all PVY^{N} isolates, but only for the L26 isolate. These studies have highlighted the complexity of PVY-tobacco interactions and suggest that other, still unknown, residues in and outside the HC-Pro region could be involved in the necrotic reaction.

3.2 *Potato Tuber Necrotic Ringspot Disease (PTNRD)*

In potato, a comparison of the biological properties and molecular characteristics of isolates causing PTNRD indicated the involvement of NIa, NIb and/or CP regions in the expression of this disease (Glais et al. 1998; Glais et al. 2002) (Fig. 3.7). The same reverse genetics approach used for the tobacco veinal necrosis study demonstrated that the replacement of glutamic acid by aspartic acid at position 419 (E419 to D419) in the HC-Pro protein alone was sufficient to change the necrotic isolate PVY^{N}-605 into a non-necrotic isolate (Glais et al. 2015). This first molecular determinant for tuber necrosis has proven to be crucial in the expression of this reaction, but not sufficient, because the residue E419 was also present in non-necrotic isolates (Tribodet et al. 2005). This molecular determinant has been validated for the four susceptible cultivars Béa, Hermes, Monalisa and Nicola. However, given the diversity of the genetic background of potato and the complexity already noted in PVY–tobacco interactions, it can be reasonably speculated that other molecular determinants must be involved in PTNRD, and could differ depending on cultivar/PVY species combinations.

3.3 *Increased Virulence*

In addition to their ability to induce severe necrotic symptoms in tobacco leaves or in potato tubers, many identified PVY variants have been found to be more virulent, *i.e.* able to overcome the resistance present in some potato cultivars. For instance, several studies reported that PVY^{NTN} isolates were able to infect some potato cultivars that had been considered to be resistant to PVY^{N} (Beczner et al. 1984; Le Romancer et al. 1994). Moreover, isolates of PVY^{N}, PVY^{Z}, PVY^{E}, PVY^{NTN} and PVY$^{N\text{-}Wi}$ strain groups, were able to overcome the hypersensitive resistance (HR) conferred by Nc_{tbr} and Ny_{tbr} genes to infection by PVY^{C} and PVY^{O} isolates respectively (Jones 1990; Kerlan et al. 1999). Mestre et al. (2000) demonstrated the requirement of the PVY NIa proteinase for *Ry*-mediated resistance. However, in spite of the ability of strains of PVY to evolve and adapt to their environment, *Ry* genes that confer extreme resistance to PVY in potato have not yet been reported to be broken by any PVY isolate.

To obtain a better understanding of the various resistance reactions, studies were undertaken to identify PVY avirulence factors eliciting the HR reaction in potato cultivars carrying Nc_{tbr}, Nc_{spi} or Ny_{tbr} genes (Moury et al. 2011; Tian and Valkonen 2013). A collection of chimeras between cDNA clones of PVY^{N}-605 and

PVY^{C} SON41 isolates revealed that two regions in the HC-Pro cistron formed the molecular basis of the HR reaction mediated by *Nc* resistance genes originally from *Solanum sparsipilum* or from *S. tuberosum* to PVY^{C} infection. The first region, corresponding to codons 335 to 364 of the HC-Pro, was proven to be responsible for the expression of local, necrotic lesions in inoculated potato leaves; the second, encompassing the C-terminal part of HC-Pro up to codon 334, was defined as being involved in resistance to the systemic spread of PVY^{C} isolates within a plant (Moury et al. 2011) (Fig. 3.7). Finally, the region of the HC-Pro protein corresponding to residues 227 to 327 was shown to carry the avirulence factors involved in the two components of resistance conferred by the Ny_{tbr} gene to PVY^{O} isolates (*i.e.* necrotic reaction and inhibition of viral spread) (Tian and Valkonen 2013 and Tian and Valkonen 2015).

4 Conclusions and Recommendations

The collation of biological and molecular data about strain groups and variants of PVY over many decades has revealed the considerable variability in PVY and the difficulties that can arise in defining the exact nature of an isolate (Figs. 3.6 and 3.8).

Singh et al. (2008) proposed that descriptions of new isolates of PVY should require both biological assessment and sequencing of CP and whole genome to provide essential phenotypic and genotypic information.

Over the past years, there has been a trend to classify novel PVY isolates as new variants or strain groups, for example PVY-SYR I, PVY-SYR II, PVY-NE11, PVY$^{NTN\text{-}NW}$, or PVY^{Z}-NTN (Chikh Ali et al. 2008; Lorenzen et al. 2008). We feel that this approach is no longer sustainable and propose that an isolate should be classified as a new strain group only if it displays biological properties that significantly differ from those of the known strain groups (PVY^{O}, PVY^{N}, PVY^{E}, PVY^{Z}, PVY^{NTN}, PVY$^{N\text{-}Wi}$) (Singh et al. 2008). Modifications of the genome alone, for example a shift of the recombination junctions or being putatively composed of genomic segments of different subspecies should not justify the assignment of a PVY isolate to a new strain group. It is acknowledged that new strain groups (biotypes) may be defined in the future and that there will be further examples of PVY isolates displaying different types of responses on the currently used reference cultivars (potato and tobacco) (see Table 3.1 'Other', exemplified by the isolate AST).

As stated earlier, the assignment of a PVY isolate to a strain group requires a thorough assessment of its pathogenicity. However, in spite of efforts to harmonise protocols and approaches (Browning et al. 2004), inherent variability in development of symptoms and assessment may be seen between laboratories. This is in part due to the subjective nature of the assay when it comes to analysing symptoms (or the lack of) on indicator plants in response to PVY infection, and that expression of symptoms (HR response or development of PTNRD) is dependent on environmental conditions, explaining the difficulties in harmonising PTNRD assessment in different institutions (Browning et al. 2004). In order to minimise variability, we

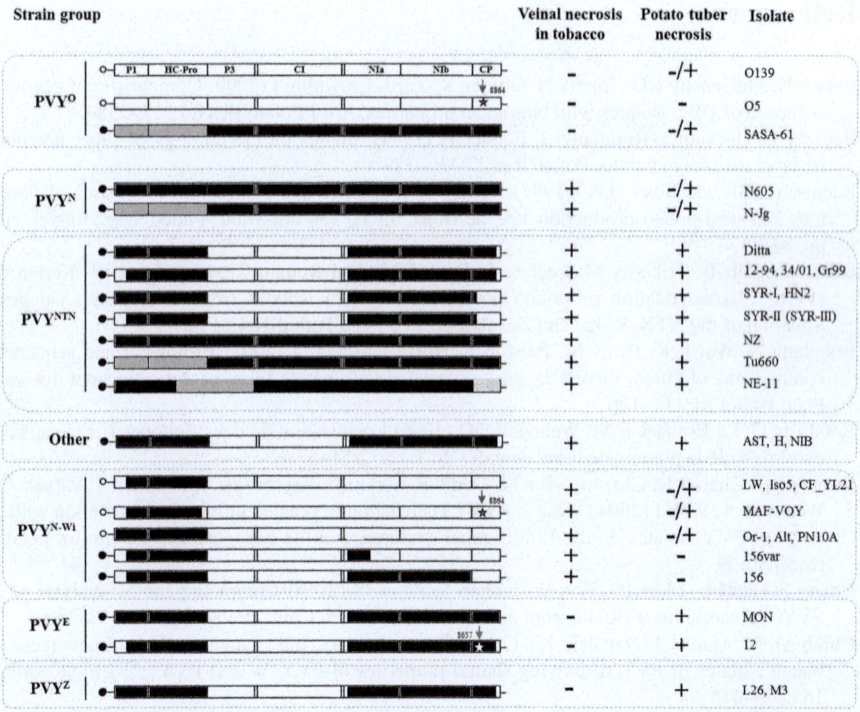

Fig. 3.8 A summary of the molecular and biological serological properties that define the various PVY strain groups and variants; '–', '+': unable or able to induce a particular biological property; '–/+': possible expression of the potato tuber necrosis in optimal experimental conditions or in specific potato cultivars

recommend that the biological responses of a novel PVY isolate should be assessed together with known PVY isolates and with potato cultivars with known characteristics as presented earlier (Table 3.1) (Singh et al. 2008).

While ELISA is an adequate method for detecting PVY and for the monitoring of PVY in crops, it is important to ensure that the antibody(ies) used are sufficiently inclusive to detect all known PVY variants. Whole genome sequencing and phylogenetic studies should be the benchmarks for the accurate definition of PVY genotypes. Such studies should be complemented with suitable biological characterisation to assign PVY isolates to a specific strain group.

While it is necessary to report accurately the molecular diversity of PVY isolates by appropriate phylogenetic and/or recombination patterns studies, it is important to emphasise that the purpose of PVY characterisation is to ultimately evaluate the pathogenicity and impact of a strain/isolate/variant on its hosts. This is a prerequisite to monitoring and managing viruses and their disease efficiently and to sustain food production worldwide.

References

Barker H, McGeachy KD, Toplak N, Gruden K, Zel J, Browning I (2009) Comparison of genome sequence of PVY isolates with biological properties. Am J Potato Res 86:227–238

Beczner L, Horvath J, Romhanyi I, Forster H (1984) Studies on the etiology of tuber necrotic ringspot disease in potato. Potato Res 27:339–352

Beemster ABR, de Bokx, J.A. (1987) Survey of properties and symptoms. In Viruses of potatoes and seed-potato production J.A. de Bokx, J.P.H. van der Want: Pudoc Wageningen, ed. pp. 84–113

Blanco-Urgoiti B, Tribodet M, Leclere S, Ponz F, de San Roman CP, Legorburu FJ, Kerlan C (1998) Characterization of potato potyvirus Y (PVY) isolates from seed potato batches. Situation of the NTN, Wilga and Z isolates. Eur J Plant Pathol 104:811–819

Boonham N, Walsh K, Hims M, Preston S, North J, Barker I (2002) Biological and sequence comparisons of *Potato virus Y* isolates associated with potato tuber necrotic ringspot disease. Plant Pathol 51:117–126

Boudazin G, Le Romancer M, Weimann DO (1995) Localisation de deux épitopes sur la capside du virus Y de la pomme de terre. Annales du Tabac 2:51–57

Browning I, Charlet K, Chrzanowska M, Dedic P, Kerlan C, Kryszczuk A, Schubert J, Varveri C, Werkman A, Wolf I (2004) What is PVY? Thereaction of potato cultivars to inoculation with a range of PVY isolates. In the 12th triennal conference of the European association for potato research

Chang F, Gao FL, Shen JG, Zou WC, Zhao S, Zhan JS (2015) Complete genome analysis of a $PVY^{N\text{-}Wi}$ recombinant isolate from *Solanum tuberosum* in China. Potato Res 58:377–389

Chikh Ali M, Maoka T, Natsuaki KT (2007) The occurrence and characterization of new recombinant isolates of PVY displaying shared properties of $PVY^{N}W$ and PVY^{NTN}. J Phytopathol 155:409–415

Chikh Ali M, Maoka T, Natsuaki KT (2008) Whole genome sequence and characterization of a novel isolate of PVY inducing tuber necrotic ringspot in potato and leaf mosaic in tobacco. J Phytopathol 156:413–418

Chikh Ali M, Maoka T, Natsuaki T, Natsuaki KT (2010) $PVY^{NTN\text{-}NW}$, a novel recombinant strain of *Potato virus Y* predominating in potato fields in Syria. Plant Pathol 59:31–41

Chikh Ali M, Rowley JS, Kuhl J, Gray SM, Karasev AV (2014) Evidence of a monogenic nature of the *Nz* gene conferring resistance against *Potato virus Y* strain Z (PVY^{Z}) in potato. Am J Potato Res 91:649–654

Chikh Ali M, Alruwaili H, Vander Pol D, Karasev AV (2016) Molecular characterization of recombinant strains of *Potato virus Y* from Saudi Arabia. Plant Dis 100:292–297

Cockerham G (1970) Genetical studies on resistance to potato viruses X and Y. Heredity 25:309–348

Davie K (2014) The biodiversity and epidemiology of Potato virus Y (PVY) in Scotland. University of Nottingham, Nottingham

de Bokx JA, Huttinga H (1981) *Potato virus Y*. CMI/AAB Descript Plant viruses 242

Dhar AK, Singh RP (1997) Molecular characterization of coat protein genes of serologically distinct isolates of *Potato virus Y* necrotic strain. Can J Microbiol 43:677–683

Ellis P, Stace-Smith R, Bowler G, Mackenzie DJ (1996) Production of monoclonal antibodies for detection and identification of strains of *Potato virus Y*. Can J Plant Pathol 18:64–70

Faurez F, Baldwin T, Tribodet M, Jacquot E (2012) Identification of new *Potato virus Y* (PVY) molecular determinants for the induction of vein necrosis in tobacco. Mol Plant Pathol 13:948–959

Gago S, Elena SF, Flores R, Sanjuán R (2009) Extremely high mutation rate of a hammerhead viroid. Science 323:1308

Galvino-Costa SBF, Figueira AD, Camargos VV, Geraldino PS, Hu XJ, Nikolaeva OV, Kerlan C, Karasev AV (2012) A novel type of *Potato virus Y* recombinant genome, determined for the genetic strain PVY^{E}. Plant Pathol 61:388–398

Glais L, Kerlan C, Tribodet M, Marie-Jeanne Tordo V, Robaglia C, Astier-Manifacier S (1996) Molecular characterization of potato virus Y^{N} isolates by PCR-RFLP. Eur J Plant Pathol 102:655–662

Glais L, Tribodet M, Gauthier JP, Astier-Manifacier S, Robaglia C, Kerlan C (1998) RFLP mapping of the whole genome of ten viral isolates representative of different biological groups of *Potato virus Y*. Arch Virol 143:2077–2091

Glais L, Tribodet M, Kerlan C (2002) Genomic variability in potato potyvirus Y (PVY): evidence that $PVY^{N}W$ and PVY^{NTN} variants are single to multiple recombinants between PVY^{O} and PVY^{N} isolates. Arch Virol 147:363–378

Glais L, Colombel AS, Tribodet M, Kerlan C (2004) PVY^{N}-605, the reference PVY^{N} isolate, displays a PVY^{NTN} non-recombinant genome. In the 12th Triennal Conference of the European Association for Potato Research. Virology Section Meeting, Le Tronchet, France, p 50

Glais L, Faurez F, Tribodet M, Boulard F, Jacquot E (2015) The amino acid 419 in HC-Pro is involved in the ability of PVY isolate N605 to induce necrotic symptoms on potato tubers. Virus Res 208:110–119

Gray S, De Boer S, Lorenzen J, Karasev A, Whitworth J, Nolte P, Singh R, Boucher A, Xu HM (2010) *Potato virus Y*: an evolving concern for potato crops in the United States and Canada. Plant Dis 94:1384–1397

Gugerli P, Fries P (1983) Characterization of monoclonal antibodies to *Potato virus Y* and their use for virus detection. J Gen Virol 64:2471–2477

Hataya T, Inoue AK, Ohshima K, Shikata E (1994) Characterization and strain identification of *Potato virus Y* isolate nonreactive with monoclonal antibodies specific to the ordinary and necrotic strains. Intervirology 37:12–19

Horvath J (1966) Studies on strains of *Potato virus Y*. 2. Normal strain. Acta Phytopathol Acad Sci Hung 1:334–352

Hu XX, He CZ, Xiao Y, Xiong XY, Nie XZ (2009) Molecular characterization and detection of recombinant isolates of *Potato virus Y* from China. Arch Virol 154:1303–1312

Jakab G, Droz E, Brigneti G, Baulcombe D, Malnoe P (1997) Infectious *in vivo* and *in vitro* transcripts from a full-length cDNA clone of PVY-N605, a Swiss necrotic isolate of *Potato virus Y*. J Gen Virol 78:3141–3145

Jones RAC (1990) Strain group specific and virus specific hypersensitive reactions to infection with potyviruses in potato cultivars. Ann Appl Biol 117:93–105

Karasev AV, Gray SM (2013) Continuous and emerging challenges of *Potato virus Y* in potato. Annu Rev Phytopathol 51:571–586

Karasev AV, Nikolaeva OV, Hu XJ, Sielaff Z, Whitworth J, Lorenzen JH, Gray SM (2010) Serological properties of ordinary and necrotic isolates of *Potato virus Y*: a case study of PVY^{N} misidentification. Am J Potato Res 87:1–9

Karasev AV, Hu XJ, Brown CJ, Kerlan C, Nikolaeva OV, Crosslin JM, Gray SM (2011) Genetic diversity of the ordinary strain of *Potato virus Y* (PVY) and origin of recombinant PVY Strains. Phytopathology 101:778–785

Kehoe MA, Jones RAC (2015) Improving *Potato virus Y* strain nomenclature: lessons from comparing isolates obtained over a 73-year period. Plant Pathol 65:322–333

Keller H, Pomp R, Bakker J, Schots A (2005) Epitope identification and in silico prediction of the specificity of antibodies binding to the coat proteins of *Potato virus Y* strains. Eur J Plant Pathol 111:391–397

Kerlan C, Tribodet M (1996) Are all PVY isolates able to induce potato tuber necrosis ringspot disease? In the 13th triennal conference of the European association for Potato research. Veldhoven, pp 65–66

Kerlan C, Tribodet M, Glais L, Guillet M (1999) Variability of *Potato virus Y* in potato crops in France. J Phytopathol 147:643–651

Kerlan C, Nikolaeva OV, Hu XJ, Meacham T, Gray SM, Karasev AV (2011) Identification of the molecular make-up of the *Potato virus Y* strain PVY^{Z}: genetic typing of PVY^{Z}-NTN. Phytopathology 101:1052–1060

Le Romancer M, Nedellec M (1997) Effect of plant genotype, virus isolate and temperature on the expression of the potato tuber necrotic ringspot disease (PTNRD). Plant Pathol 46:104–111

Le Romancer M, Kerlan C, Nedellec M (1994) Biological characterization of various geographical isolates of *Potato virus Y* inducing superficial necrosis on potato tubers. Plant Pathol 43:138–144

Lorenzen JH, Meacham T, Berger PH, Shiel PJ, Crosslin JM, Hamm PB, Kopp H (2006) Whole genome characterization of *Potato virus Y* isolates collected in the western USA and their comparison to isolates from Europe and Canada. Arch Virol 151:1055–1074

Lorenzen J, Nolte P, Martin D, Pasche JS, Gudmestad NC (2008) NE-11 represents a new strain variant class of *Potato virus Y*. Arch Virol 153:517–525

Marie-Jeanne Tordo V, Chachulska AM, Fakhfakh H, Le Romancer M, Robaglia C, Astier-Manifacier S (1995) Sequence polymorphism in the 5'NTR and in the P1 coding region of *Potato virus Y* genomic RNA. J Gen Virol 76:939–949

Marte M, Bellezza G, Polverari A (1991) Infective behavior and aphid-transmissibility of Italian isolates of *Potato virus Y* in tobacco and peppers. Ann Appl Biol 118:309–317

Martin DP, Murrell B, Golden M, Khoosal A, Muhire B (2015) RDP4: Detection and analysis of recombination patterns in virus genomes. Virus Evolution 1:vev003. doi:10.1093/ve/vev003

McDonald JG, Kristjansson GT (1993) Properties of strains of potato virus Y^N in North America. Plant Dis 77:87–89

Mestre P, Brigneti G, Baulcombe DC (2000) An *Ry*-mediated resistance response in potato requires the intact active site of the NIa proteinase from *Potato virus Y*. Plant J 23:653–661

Moury B, Caromel B, Johansen E, Simon V, Chauvin L, Jacquot E, Kerlan C, Lefebvre V (2011) The helper component proteinase cistron of *Potato virus Y* induces hypersensitivity and resistance in potato genotypes carrying dominant resistance genes on chromosome IV. Mol Plant-Microbe Interact 24:787–797

Nie XZ, Singh RP (2002) A new approach for the simultaneous differentiation of biological and geographical strains of *Potato virus Y* by uniplex and multiplex RT-PCR. J Virol Methods 104:41–54

Nie X, Singh RP (2003) Evolution of North American PVY^{NTN} strain Tu660 from local PVY^N by mutation rather than recombination. Virus Genes 26:39–47

Nikolaeva OV, Roop DJ, Galvino-Costa SBF, Figueira AD, Gray SM, Karasev AV (2012) Epitope mapping for monoclonal antibodies recognizing tuber necrotic isolates of *Potato virus Y*. Am J Potato Res 89:121–128

Ogawa T, Tomitaka Y, Nakagawa A, Ohshima K (2008) Genetic structure of a population of *Potato virus Y* inducing potato tuber necrotic ringspot disease in Japan; comparison with North American and European populations. Virus Res 131:199–212

Ogawa T, Nakagawa A, Hataya T, Ohshima K (2012) The genetic structure of populations of *Potato virus Y* in Japan; based on the analysis of 20 full genomic sequences. J Phytopathol 160:661–673

Ounouna H, Kerlan C, Lafaye P, Loukili MJ, ElGaaied A (2002) Production of monoclonal antibodies against synthetic peptides of the N-terminal region of *Potato virus Y* coat protein and their use in PVY strain differentiation. Plant Pathol 51:487–494

Piche LM, Singh RP, Nie X, Gudmestad NC (2004) Diversity among *Potato virus Y* isolates obtained from potatoes grown in the United States. Phytopathology 94:1368–1375

Ranki H, Hepojoki J, Lankinen H, Tian YP, Rännäli M, Valkonen JPT (2008) Epitopes of the Potato virus Y (PVY) coat protein recognized by monoclonal antibodies. In The 17th triennial conference of European association for Potato research. Brasov, pp 55

Revers F, Le Gall O, Candresse T, Le Romancer M, Dunez J (1996) Frequent occurrence of recombinant potyvirus isolates. J Gen Virol 77:1953–1965

Robaglia C, Durand-Tardif M, Tronchet M, Boudazin G, Astier-Manifacier S, Cassedelbart F (1989) Nucleotide sequence of *Potato virus Y* (N strain) genomic RNA. J Gen Virol 70:935–947

Robles-Hernandez L, Gonzalez-Franco AC, Hernandez-Huerta J, Meacham TM, Nikolaeva OV, Karasev AV (2010) First identification of an unusual recombinant *Potato virus Y* strain in potato in Mexico. Plant Dis 94:1262–1262

Rowley JS, Gray SM, Karasev AV (2015) Screening potato cultivars for new sources of resistance to Potato virus Y. Am J Potato Res 92:38–48. doi:10.1007/s12230-014-9409-5

Schubert J, Fomitcheva V, Sztangret-Wisniewska J (2007) Differentiation of *Potato virus Y* strains using improved sets of diagnostic PCR-primers. J Virol Methods 140:66–74

Shukla DD, Ward CW (1988) Amino acid sequence homology of coat proteins as a basis for identification and classification of the Potyvirus group. J Gen Virol 69:2703–2710

Shukla DD, Strike PM, Tracy SL, Gough KH, Ward CW (1988) The N and C termini of the coat proteins of Potyviruses are surface-located and the N terminus contains the major virus-specific epitopes. J Gen Virol 69:1497–1508

Singh M, Singh RP (1994) A fast-reacting bioassay for the tobacco veinal necrosis strain of *Potato virus Y* (PVY^N). Plant Dis 78:775–778

Singh M, Singh RP (1996) Nucleotide sequence and genome organization of a Canadian isolate of the common strain of *Potato virus Y* (PVY^O). Can J Plant Pathol 18:209–224

Singh RP, Valkonen JPT, Gray SM, Boonham N, Jones RAC, Kerlan C, Schubert J (2008) Discussion paper: The naming of *Potato virus Y* strains infecting potato. Arch Virol 153:1–13

Stobbs LW, Poysa V, Vanschagen JG (1994) Susceptibility of cultivars of tomato and pepper to a necrotic strain of *Potato virus Y*. Can J Plant Pathol 16:43–48

Szajko K, Chrzanowska M, Witek K, Strzelczyk-Zyta D, Zagorska H, Gebhardt C, Hennig J, Marczewski W (2008) The novel gene *Ny-1* on potato chromosome IX confers hypersensitive resistance to *Potato virus Y* and is an alternative to *Ry* genes in potato breeding for PVY resistance. Theor Appl Genet 116:297–303

Szajko K, Strzelczyk-Zyta D, Marczewski W (2014) *Ny-1* and *Ny-2* genes conferring hypersensitive response to *Potato virus Y* (PVY) in cultivated potatoes: mapping and marker-assisted selection validation for PVY resistance in potato breeding. Mol Breed 34:267–271

Thole V, Dalmay T, Burgyan J, Balazs E (1993) Cloning and sequencing of *Potato virus Y* (Hungarian isolate) genomic RNA. Gene 123:149–156

Tian YP, Valkonen JPT (2013) Genetic determinants of *Potato virus Y* required to overcome or trigger hypersensitive resistance to PVY strain group O controlled by the gene *Ny* in potato. Mol Plant-Microbe Interact 26:297–305

Tian YP, Valkonen JPT (2015) Recombination of strain O segments to HCpro-encoding sequence of strain N of *Potato virus Y* modulates necrosis induced in tobacco and in potatoes carrying resistance genes *Ny* or *Nc*. Mol Plant Pathol 16:735–747

Tian YP, Hepojoki J, Ranki H, Lankinen H, Valkonen JPT (2014) Analysis of *Potato virus Y* coat protein epitopes recognized by three commercial monoclonal antibodies. PLoS One 9:1–20

Tomczyńska I, Jupe F, Hein I, Marczewski W, Sliwka J (2014) Hypersensitive response to Potato virus Y in potato cultivar Sárpo Mira is conferred by the Ny-Smira gene located on the long arm of chromosome IX. Mol Breed 34:471–480

Tribodet M, Glais L, Kerlan C, Jacquot E (2005) Characterization of *Potato virus Y* (PVY) molecular determinants involved in the vein necrosis symptom induced by PVY^N isolates in infected *Nicotiana tabacum* cv. Xanthi. J Gen Virol 86:2101–2105

Tromas N, Elena SF (2010) The rate and spectrum of spontaneous mutations in a plant RNA virus. Genetics 185:983–989

Valkonen JPT (1997) Novel resistances to potyviruses in tuber-bearingpotato species, and temperaturesensitive expression of hypersensitiveresistance to Potato virus Y. Ann Appl Biol 130:91–104

van der Vlugt RAA, Leunissen J, Goldbach R (1993) Taxonomic relationships between distinct *Potato virus Y* isolates based on detailed comparisons of the viral coat proteins and 3' non-translated regions. Arch Virol 131:361–375

Vuento M, Paananen K, Vihinenranta M, Kurppa A (1993) Characterization of antigenic epitopes of *Potato virus Y*. Biochim Biophys Acta 1162:155–160

Weidemann HL, Maiss E (1996) Detection of the potato tuber necrotic ringspot strain of Potato virus Y (PVY) by reverse transcription and immunocapture polymerase chain reaction. Zeitschrift fur Pflanzenkrankheiten und Pflanzenschutz-J Plant Dis Protect 103:337–345

Weilguny H, Singh RP (1998) Separation of Slovenian isolates of PVY^{NTN} from the North American isolates of PVY^{N} by a 3-primer PCR. J Virol Methods 71:57–68

Chapter 4
Evolution and Origin of PVY

Dirk U. Bellstedt, Laurent Glais, Kim Davie, and Christophe Lacomme

Abstract The mechanisms by which *Potato virus Y* variants are generated and selected are still unclear. Spontaneous mutations generated by uncorrected replication error and recombination events between viral isolates during co-infection of plant cells are the main likely source of genetic diversity. This high level of diversity generation is essential for virus evolution and survival in different environments. Different PVY strain groups have appeared over time: firstly, non-recombinant PVY^{C}, PVY^{O} and PVY^{N} strains and, more recently, recombinant $PVY^{N\text{-}Wi}$ and PVY^{NTN} strains with novel biological characteristics and the ability to cause potato tuber necrotic ringspot disease (PTNRD). Increased fitness of the recombinant strains appears to have enabled them to replace the non-recombinant variants in most potato growing areas of the world. Partial sequencing of PVY genome (P1, HC-Pro, CP, recombinant junctions) and whole genome sequencing has shown that non-recombinant and recombinant variants are present in potatoes and other plant hosts. Phylogenetic analyses have been applied to document changes in viral isolates and to establish the relationships between different viral isolates. Traditional phylogenetic analysis was, however, developed for bifurcating phylogenies and not for analysing recombination, as a result of which this presents challenges to these analyses which will be outlined in this chapter. The ongoing worldwide studies on PVY characterisation suggest that new variants with distinct biological properties are likely to be uncovered in the future. Emerging technologies such as next generation sequencing will provide valuable insights into PVY population dynamics and evolution in future.

Keywords *Potato virus Y* • Recombination • Phylogeny • Evolution

---

D.U. Bellstedt (✉)
Department of Biochemistry, University of Stellenbosch, 7602 Stellenbosch, South Africa
e-mail: dub@sun.ac.za

L. Glais
FN3PT/RD3PT, 43–45 rue de Naples, F-75008 Paris, France

IGEPP, Agrocampus Ouest, INRA, Université de Rennes 1,
F-35650 Le Rheu, France

K. Davie • C. Lacomme
Science and Advice for Scottish Agriculture (SASA),
1 Roddinglaw Road, Edinburgh EH22 4HR, UK

C. Lacomme et al. (eds.), *Potato virus Y: biodiversity, pathogenicity, epidemiology and management*, DOI 10.1007/978-3-319-58860-5_4

1 Introduction

Potato virus Y (PVY) has remained a consistent and serious threat to solanaceous crops such as tobacco, peppers, tomatoes and, particularly, potatoes despite the use of a range of control measures. For potatoes, control measures include the production of initial seed potato material free of virus, the application of strict tolerances for virus diseases in certification schemes, testing potato tubers for PVY in order to discard infected stocks, applying measures to control the vectors of PVY and breeding for PVY resistance. Due to its ability to adapt to new plant hosts and aphid vectors over short periods of time, PVY has become a major pathogen of crops worldwide (Scholthof et al. 2011). The genetic variation of PVY is generated by mutation and recombination, both of which occur at faster rates than in eukaryotes and most prokaryotes. These changes occur by replication of the viral genome by the virus-encoded RNA polymerase enzyme that introduces errors at the rate of 1 in 10^3–10^4 base pairs per generation and by recombination between viral isolates during co-infection of plant cells. This genetic variation is structured by evolutionary processes such as natural selection, genetic drift and gene flow (Roosinck 2003; Moya et al. 2004; Rubio et al. 2013). The high genetic variability of the PVY genome provides means of overcoming endogenous resistance genes naturally occurring or introgressed into potato cultivars, changes in the environment or control/cultural measures. Recombinant $PVY^{N\text{-}Wi}$ and PVY^{NTN} isolates appear to have replaced non-recombinant PVY^{C}, PVY^{O} and PVY^{N} isolates in most potato growing areas of the world, suggesting that the former are fitter than the latter. Partial sequencing of PVY genome (P1, HC-Pro, CP, recombinant junctions) and whole genome sequencing has, however, shown that many non-recombinant and recombinant isolates are present in potatoes and other plant hosts (see Chaps. 3 and 9). Further changes in PVY genomes and the symptoms induced by PVY in hosts can, therefore, be expected on an ongoing basis. In this chapter, an overview will be given of how evolution has shaped the genetic diversity of PVY since its discovery in 1930, and the consequences of this genomic plasticity on its pathogenicity.

2 Origin and Worldwide Spread of PVY

2.1 The Origin of PVY

In South America, the Incas, who were the first farmers to grow the potato, appear to have recognised the presence of viral diseases affecting potatoes and therefore grew their potatoes intermittently at higher altitudes where the colder temperatures inhibited the spread of viruses by aphids and the build-up of viral diseases (Bertschinger 1992). It is likely that PVY was introduced into Europe, North America and Asia in the fifteenth and sixteenth centuries on infected host plants from South America. It is impossible to determine on which of its hosts, that is, potato, tobacco, tomato and pepper, PVY was introduced into cultivated crops in these areas. Because PVY cannot be transmitted via seeds of any of its hosts, but can be transmitted through vegetatively propagated potato tubers, this suggests that the most probable means of PVY introductions worldwide were via infected potato tubers. The evidence that potyviruses originated in South America (Gibbs and Ohshima 2010; Quenouille et al. 2013 and references cited therein) is supported by the fact that a large majority of PVY strains/subgroups and related potyvirus species show clear phylogenetic relationships with isolates found in plants that are native to South America and suggest that PVY diversity is probably much greater there than anywhere else (Gibbs and Ohshima 2010).

2.2 Emergence of PVY Strain Groups

Historically, the first strain group of PVY that was described and characterised was PVY^C (Salaman 1930), followed by PVY^O (Smith 1931). Subsequently, PVY^N (Orlando and Silberschmidt 1945; Crosslin et al. 2005) was found for the first time in 1935 in tobacco plants growing in the vicinity of a potato crop (Smith and Dennis 1940), and in 1941 and 1942 in potato cultivars originating from Peru and Bolivia (Nobrega and Silberschmidt 1944; Silberschmidt 1960). PVY^C was the dominant strain group from the 1930s to 1950s with a steady increase in the incidence of PVY^O isolates in the 1950s and 1960s. This was followed by an increase in the occurrence of PVY^N isolates in the 1970s and 1980s. A significant change in the pathogenicity of PVY isolates occurred in the 1980s when potato tuber necrotic ringspot disease (PTNRD) associated with PVY infection was found for the first time in Hungary (Beczner et al. 1984). The causal virus was classified as PVY^{NTN} strain group. PTNRD (Beczner et al. 1984) has subsequently been found in most potato growing countries in the last 30 years (Fig. 4.1).

In the 1980s, PTNRD caused the collapse of the Slovenian seed potato industry because the most extensively grown cultivar Igor was very susceptible to infection by PVY^{NTN} and sensitive to PTNRD development (Weilguny and Singh 1998). In the 1990s, isolates of PVY causing PTNRD were also found in America and Asia (McDonald and Kristjansson 1993; McDonald and Singh 1996; Oshima et al. 2000).

As previously mentioned in Chap. 3, isolates of the PVY^{NTN} strain group can be divided into recombinant and non-recombinant isolates. In Europe and in some non-European countries, a large majority of the PVY^{NTN} population is composed of recombinant isolates with three to four recombination sites (Revers et al. 1996; Boonham et al. 2002; Glais et al. 2002; Schubert et al. 2007; Boukhris-Bouhachem et al. 2008; Rigotti et al. 2011; Pourrahim and Farzadfar 2016). However, in New Zealand (Fletcher and Lister 2004) and Japan (Ogawa et al. 2008), recombinant PVY^{NTN} isolates are either absent or very infrequent. Ten years after the first description of PVY^{NTN} isolates in Europe, they became prevalent in the main potato growing

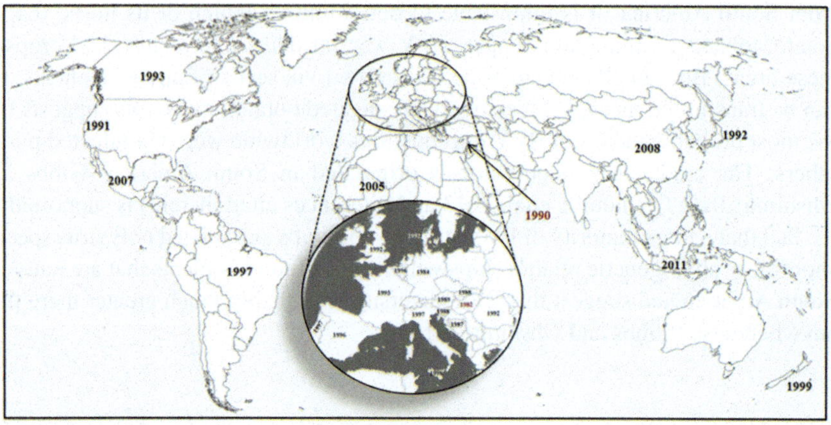

Europe

1982: Hungary (Beczner et al., 1984)
1984: Germany (Radtke, 1984)
1987: Bosnia (Buturovic and Kus, 1989)
1988: Czechoslovakia (Dedic et al., 1988)
Slovenia (Kus, 1990)
1989: Austria (Schiessendoppler, 1990)
1992: Denmark (Nielsen, 1992)
Great Britain (Wright, 1992)
Romania (Weidemann and Maiss, 1996)
1993: France (Le Romancer, 1993)
1996: Belgium (Weidemann and Maiss, 1996)
Spain (Blanco-Urgoiti et al., 1998)
1997: Portugal (Serra and Weidemann, 1997)
Italy (Tomassoli et al., 1998)
1998: Greece (Bem and Varveri, 1999)

Non-Europe

1990: Lebanon (Le Romancer et al., 1991)
1991: California (McDonald and Singh, 1996)
1992: Japan (Oshima et al., 2000)
1993: Canada (McDonald and Singh, 1996)
1997: Brazil (Galvino-Costa et al., 2012)
1999: New Zealand (Fletcher and Lister, 2004)
2005: South Africa (Pieterse, 2005)
2007: Mexico (Hernandez-de la Cruz et al., 2007)
2008: China (Bai et al., 2010)
2011: Indonesia (Damayanti et al., 2014)

Fig. 4.1 Map showing the first reports of PVY^{NTN} isolates identified in various countries worldwide. The dates in red correspond to the first observations of isolates causing tuber necrosis (PTNRD). Not listed are countries where either PTNRD has not been reported or the first report of PTNRD does not correspond to the first emergence of this disease but rather to the time when PVY^{NTN} was "officially recorded"

areas (Table 4.1, Lindner and Billenkamp 2005; Rolland et al. 2008; van der Vlugt et al. 2008; Bahrami Kamangar et al. 2014). In South Africa, Brazil, Japan, Syria and New Zealand, PVY^{NTN} isolates also constitute the majority of the PVY population and, since 2000, have been found to be responsible for 30–70% of all infections by PVY in potato crops (Table 4.1).

Another recombinant strain group is PVY$^{N\text{-}Wi}$, which includes the PVY$^{N\text{-}Wilga}$ isolate identified for the first time in Poland in 1984 (Chrzanowska 1991). Unlike PVY^{NTN}, the worldwide emergence of isolates of the PVY$^{N\text{-}Wi}$ strain group cannot be accurately defined because they do not induce distinctive symptoms in infected potato plants so they may have been present for many years before being first detected. They were largely detected though surveys to assess the prevalence of PVY strain groups and characterisation of their pathogenicity. This has revealed that PVY$^{N\text{-}Wi}$ isolates are present throughout the world. In 1992, isolates similar to PVY$^{N\text{-}Wi}$ were also recorded in Manitoba, Canada (McDonald and Singh 1996), but they differed in the number of recombination junctions (Glais et al. 2002). These recombinant PVY$^{N\text{-}Wi}$ isolates now occur worldwide with a current average incidence

Table 4.1 The occurrence of PVY strain groups and variants in various countries of the world. "/": no information available

Continent/Country		Host	Year	PVY strains and variants relative distribution (%)								Reported by
				O	C	N or NA-N	Z	E	N-Wi	NTN or NA-NTN	Other	
Africa	Kenya	Potato	2009	15	/	29	/	/	29	29	/	Were et al. (2013)
	South Africa	Potato	2009	1	0	14	/	/	35	50	/	Visser and Bellstedt (2009); personal communication
	Tunisia	Potato	2002–06	7	/	11	/	/	/	80	2	Boukhris-Bouhachem et al. (2010)
The Americas	Brazil	Potato	1985–2009	8	/	/	/	/	28	58	6	Galvino-Costa et al. (2012)
			2005–06	4	/	11	/	/	/	85	/	de Ávilal et al. (2009)
			2011–13	8	0	2	/	3	25	61	2	Figueira, personal comunication
		Tobacco	2009	11	/	19	/	/	/	/	65	Lacroix et al. (2011)
	Canada	Potato	2004–06	71	/	0	/	/	27	1	/	Gray et al. (2010)
			2004–12	70	0	0	/	/	28	1	/	Xu (2013)
	USA	Potato	2001–03	18	/	/	/	/	63	18	/	Piche et al. (2004)
			2004–06	64	0	0	/	/	23	5	/	Gray et al. (2010)
Asia	China	Potato	2008	50	0	50	/	/	0	/	/	Bai et al. (2010)
		Tobacco	2010	12	/	8	/	/	64	8	/	Tian et al. (2011)
	Iran	Potato	2005–07	7	0	0	/	/	50	43	/	Hosseini et al. (2011)
			2007–08	77	62	39	/	/	/	9	/	Mousavi et al. (2014)
			2013	23	16	41	/	/	/	19	/	Reza et al. (2016)
	Japan	Potato	1990–2010	50	0	5	/	/	0	45		Ogawa et al. (2012)
	Syria	Potato	2002–04	9	/	13	/	/	19	59	/	Chikh Ali et al. (2007)

(continued)

Table 4.1 (continued)

Continent/Country			Host	Year	PVY strains and variants relative distribution (%)								Reported by
					O	C	N or NA-N	Z	E	N-Wi	NTN or NA-NTN	Other	
Europe	Belgium		Potato	1998	36	/	64	/	/	/	/	/	Rolot and Steyer (2008)
				2005	13	/	87	/	/	/	/	/	
				2010	8	0	3	/	/	8	75	8	Bahrami Kamangar et al. (2014)
	Estonia		Potato	2011–13	13	/	/	/	/	33	35	13	Piret Peterson, personal communication
	France		Potato	1996	83	0	11	0	1	5	/	/	Kerlan et al. (1999)
				2005	28	/	12	/	/	10	50	/	Rolland et al. (2008)
				2014	7	/	10	/	1	9	73	/	Glais, unpublished
			Tobacco	2007	12	/	75	/	/	/	/	13	Lacroix et al. (2010)
	Germany		Potato	1984	66	/	34	/	/	/	/	/	Lindner (2008)
				2002	10	0	5	/	/	22	63	/	Lindner and Billenkamp (2005)
				2009–11	0	0	0	/	/	57	37	/	Lindner et al. (2016)
	Great Britain	England & Wales	Potato	1998–99	50	0	23	/	/	/	23	/	Boonham et al. (2002)
				2010	31*	/	69*	/	/	?	?	/	Fenton et al. (2012)
		Scotland	Potato	2001	24	0	71*	1	/	5	?	/	Browning et al. (2004)
				2010	4	0	6	/	<1	10	80	/	Davie et al. (2012) and Fenton et al. (2012)
				2013	3	0	0	/	/	6	91	/	Lacomme et al. (2014)

Europe	The Netherlands		Potato	1994	50	39	24	/	/	0	0	/	Van der Vlugt, personal communication
				2006	11	0	0	/	/	16	77	/	Van der Vlugt et al. (2008)
				2012	2	0	0	/	/	23	75	/	van den Bovenkamp, personal communication
	Poland		Potato	1980	90	/	0	/	/	10	0	/	Kaczmarek and Wydzialkowska (1996); Zimnoch-Guzowska et al. (2013)
				1994	3	/	0	/	/	94	3	/	
				2000	4	/	0	/	/	90	6	/	
				2008	2	/	0	/	/	32	66	/	
				2010	10	/	0	/	/	62	14	/	
	Spain		Potato	1994	8	/	15	15	/	8	54	/	Blanco-Urgoiti et al. (1998)
	Switzerland		Potato	2003	11	0	2	/	/	6	78	/	Rigotti et al. (2011)
				2008	5	0	7	/	/	17	70	/	
Oceania	New Zealand		Potato	1985	4	/	9	/	/	/	/	/	Fletcher (1989)
				1988–2007	52	0	16	/	/	0	32	/	Fomitcheva et al. (2009)
				2010	5	/	15	/	/	/	/	/	Fletcher (2012)
	Australia		Potato	?	69	31	0	/	/	0	0	/	Kehoe and Jones (2016)

"Other": corresponds to uncharacterised PVY isolates. "*" indicates serological data only available, no information on the molecular type of variants. Values are the relative percentage of various PVY strain groups identified in PVY infected samples and do not represent the overall incidence of PVY in a listed country of crop infection of 10–20% in Europe and other continents (Table 4.1). However, in Poland, Canada, USA and China this strain has become more prevalent, and accounts for almost 60% of PVY infections in potato crops (Kaczmarek and Wydzialkowska 1996; Piche et al. 2004; Gray et al. 2010; Tian et al. 2011; Xu 2013).

The incidence of isolates of PVY^{NTN} and $PVY^{N\text{-}Wi}$ strain groups has increased, in many countries of the world, to the extent that they are now more prevalent than those of PVY^{C}, PVY^{O} and PVY^{N} strain groups (Rolland et al. 2008). The emergence of these strain groups is assumed to have taken place at the same time in different countries, most likely through trade and commercial exchange of seed potatoes. However, while PVY^{NTN} isolates eliciting PTNRD have been found in New Zealand (NZ isolate), recombinant PVY^{NTN} isolates do not appear to be present, perhaps because the import of seed potatoes of new cultivars has not been permitted since the 1970s (Fomitcheva et al. 2009). Similarly, findings of PVY^{NTN} and $PVY^{N\text{-}Wi}$ have only been reported in isolated cases in Australia (Kehoe and Jones 2016) and PVY^{NTN} recombinants are apparently less prevalent than non-recombinants. In contrast, Galvino-Costa et al. (2012) concluded that isolates of PVY^{NTN} and $PVY^{N\text{-}Wi}$ strain groups had been introduced into Brazil in imported seed potatoes and these have now replaced the older PVY^{C}, PVY^{N} and PVY^{O} strain groups. In South Africa, there is currently a high incidence of the PVY^{NTN} and $PVY^{N\text{-}Wi}$ isolates (Visser and Bellstedt 2009), which may be related to the import of new cultivars from Europe after 1990. Although this material was imported as tissue culture material and screened for PVY by ELISA, one hypothesis is that PVY might have been present at very low levels below the detection threshold (Visser and Bellstedt 2009). Isolates of the PVY^{O}-O5 subgroup of the PVY^{O} strain group were initially found in 1997 in Canada (Manitoba, New Brunswick) and USA (Idaho) (Ellis et al. 1997; Karasev et al. 2010). Isolates of the PVY^{O}-O5 subgroup have become increasingly prevalent in the USA and Canada accounting for 5–9% of isolates of the PVY^{O} strain groups (Gray et al. 2010).

3 Genetic Variability and Evolutionary Dynamics of PVY

3.1 Mutation

As an organism utilising RNA for the storage of its genetic information, PVY uses an RNA-dependent RNA polymerase (RdRp) encoded by its own genome for replication (Chap. 1). In contrast to the DNA-dependent DNA polymerases, this enzyme does not have any proofreading activity during synthesis of the new RNA strand (Friedberg et al. 1995), so spontaneous mutations can be introduced at the rate of 1 in a 1000 up to 100,000 bases during the viral replication process (Drake 1993; Domingo and Holland 1997). For a genome the size of PVY (10 kb), this equates to a rate of 0.1–1 mutation per genome per generation/replication cycle (Tromas and Elena 2010). Mutation is the introduction of differences between nucleotides incorporated into the newly synthesised RNA strand during the replication process (Garcia-Arenal et al. 2003). This phenomenon includes substitutions, deletions and insertions. Malpica et al. (2002) analysed the mutational spectrum of the RNA of *Tobacco mosaic virus* (TMV) by focusing on the viral movement protein and found that 69% of the mutations were deletions or insertions involving at least three bases while 31% were substitutions. Previously, Bujarski (1999) had shown that the most common transition was the substitution of one purine by another or one pyrimidine by another compared to transversions in which a purine residue is substituted by a pyrimidine residue. Given that most RNA viruses including PVY reproduce rapidly inside the plant cell, this could lead to the rapid accumulation of large populations of viral quasispecies within a short period after infection, which may harbour considerable variability and consequently enable a virus to adapt to new circumstances (Novella et al. 1995).

3.2 Recombination

Recombination is a process that results in the complete replacement of sections of the genome between two closely related molecules and is, therefore, termed homologous recombination. In contrast, heterogeneous recombination occurs between two genomes sharing a low percentage of homology when two distantly related viruses or organisms combine parts of their genomes. This mechanism involves the switching of the RNA-dependent RNA polymerase (RdRp) from the mother strand to another during replication (Lai 1992). This template switching is found when the RdRp complex stops the synthesis of the nascent strand, unhooks from the initial template and settles on a new one where it pursues the replication on this new template. In the case of homogeneous recombination, it has been reported that in *Brome mosaic virus* (BMV) the recombination site is located close to an AU-rich region known to be a recombination hot spot promoting the RdRp detachment (Nagy and Bujarski 1997). In addition, the frequency of this recombination is enhanced by the presence upstream of this region of a GC-rich region or a region with almost 60% of AU residues (Nagy and Bujarski 1997). Moreover, in *Turnip crinkle virus* (TCV), a secondary structure in the acceptor strand is the cause of the RdRp switching (Nagy and Simon 1997; Nagy et al. 1998). In addition to mutation, recombination plays an important role in PVY evolution. Chare and Holmes (2006) conducted a survey to establish how frequently recombination occurred in many members of the Potyviridae including PVY, and found that one third of the viruses showed evidence of recombination. CP gene sequence alignments showed 17% recombination but whole genome sequence alignments showed 44% recombination. PVY was reported to be one of the species showing recombination. In Chap. 3, an overview was given of the genomes of PVY strain groups and variants, emphasising that many isolates of PVY have evolved through recombination (Chap. 3, Fig. 3.5). Just as mutations, recombination may also have negative evolutionary consequences, and these variants should also be selected against/removed through purifying selection. The reverse would apply for advantageous changes as described above. The generation of a recombinant virus requires an exchange of genetic material between at least two parental genomes. A study undertaken on 560 infected samples containing various combinations of potyviruses (two to three viruses, in non-transgenic or transgenic plants harbouring viral genomes transgenes) did not generate any recombinant potyviral sequences (Dietrich et al. 2007). This suggests that recombination events between potyviruses in a mixed infection are likely to be extremely rare and might not fully explain the "emergence" of recombinant variants (Dietrich et al. 2007).

3.3 Natural Selection

Natural selection is a directional process by which the fittest variants in a given environment will increase their frequency in the population and, contrastingly, less fit variants will not prevail (Schneider and Roossinck 2001). Changes in the PVY genome might either be advantageous or deleterious for a variant, with the latter type occurring more frequently. Natural selection will not favour deleterious changes and such variants would tend not to survive because of a failure to compete with the inherent fitness of the parent or the increased fitness (selective advantage) of only a very limited type of specific mutant progeny (Hughes 2009). An analysis of the variability of viral proteins has shown that high mutation rates did not lead to increased variation in viral proteins or the genome, and instead that the proteins and genomes of RNA viruses were remarkably stable (Garcia-Arenal et al. 2001). Cuevas et al. (2012) assessed the proportion of non-synonymous to synonymous mutations in non-recombinant PVY isolates and found that purifying selection played an important role in PVY evolution.

3.4 Genetic Drift

Genetic drift is caused by random selection of sequence variants during population bottlenecks (Moya et al. 2004). While mutation and recombination are strictly dependent on the nature of the virus genome and mode of replication, natural selection and genetic drift are tightly associated to the biology of the virus (Rubio et al. 2013). Genetic drift occurs during all aspects of the life cycle of a virus and is affected by (i) virus cell-to-cell (local) and systemic (phloem-mediated) movement within a plant, (ii) transmission by vectors between plants, (iii) interaction with co-infecting viruses, (iv) infection of different hosts (*i.e.* host range) and (v) long distance migration (gene flow) among distinct geographical areas. All these events are potential population bottlenecks contributing to the decrease of genetic diversity albeit to different extents (Moya et al. 2004). Genetic drift is highly dependent on population sizes and if population sizes are small, then random mutations may have a negative effect on the survival of a virus. Hughes (2009) examined viruses of the Potyviridae family (including PVY) and found the effective population sizes of 12 viruses to be in the order of 10^4, which is remarkably small considering that the number of virus particles in a plant cell can be very large, in the range of 10^{11}–10^{12} (Malpica et al. 2002). However, when a virus is exposed to the previously mentioned bottlenecks, then a loss of fitness may occur (Muller's ratchet theory, Novella et al. 1995). Recent estimations of population sizes of PVY during transmission have been as small as 0.5–3.2 virus particles per aphid (Moury et al. 2007) and this would support the fact that PVY populations potentially experience extreme bottlenecks. High aphid numbers of different aphid species capable of transmitting PVY, as is the case in countries with higher average temperatures (see Chap. 6), would alleviate this bottleneck. The relatively small genome of PVY is another factor that needs to be considered in relation to mutation and viral genetic drift. The individual proteins encoded in the PVY genome each perform several functions. This means that most mutations in these coding regions are more likely to have negative than positive effects and will be subject to purifying selection.

However, the molecular mechanisms underlying PVY evolution still remain to be elucidated. The actual contribution of PVY recombination *in vivo* and the underlying molecular mechanisms in generating PVY genetic diversity are still a matter of debate. The question arises whether the mechanisms generating genetic diversity (mutation, recombination) are an "ongoing process" or whether other evolutionary mechanisms (natural selection, genetic drift) are the main factors in shaping PVY evolution.

4 Monitoring the Evolution of PVY

Phylogenetic analysis has been applied to document the changes in viral populations and to establish the relationships between them. In this section, the use and applicability of phylogenetic analysis methods for the identification and characterisation of PVY isolates and PVY evolution will be reviewed.

4.1 Methods Used to Perform Phylogenetic Analyses

Methods of phylogenetic analysis are regularly employed to characterise PVY isolates and to determine the evolutionary relationships between the various PVY strain groups and variants. The methods have proved so effective that almost all publications on PVY strain group identification and evolution studies involve such analyses. Phylogenetic analysis has an extensive theoretical base which cannot be reviewed exhaustively here. Suggested references such as Lemey et al. (2009) are available to provide a basic background to phylogenetic analysis and its basic assumptions. Recent reviews on informatics tools are also available to study virus evolution (Duffy 2016).

The three basic methods of phylogenetic analysis available for RNA, DNA and amino acid sequence data are distance, parsimony and likelihood analyses. PVY phylogeny has been commonly generated by distance analysis to define molecular subgroups (genotype) of PVY isolates (Kehoe and Jones 2016). The method is primarily used because it is fast. However, it suffers from shortcomings in its assumptions and, consequently, should only be used in specific applications. Parsimony analysis is still accepted in phylogenetic analysis, but is computationally more complex and requires specialist interpretation. Maximum Likelihood (ML) analysis is currently the most widely accepted method in phylogenetic analysis to define molecular subgroups of PVY (Kehoe and Jones 2016), but requires long run times as it is computationally intensive. These constraints have largely been overcome with the development of high speed computers, and especially portals for conducting any analyses (Stamatakis 2006). Likelihood analyses such as Maximum Likelihood and Bayesian statistics are the most frequently used methods in phylogenetic analysis in general at present and should be used for phylogenetic analysis of PVY sequences more frequently as they are currently viewed to be the most robust methods for phylogenetic analysis. BEAST (Bayesian Evolutionary Analysis by Sampling Trees, Drummond et al. 2012) employs Bayesian statistics to analyse sequence data and generate dated phylogenies which can be used to establish when isolates or clades of isolates had originated.

4.2 *Phylogeny and Evolutionary Pathways of PVY*

In phylogenetic analyses, a very important consideration in assessing the phylogenetic trees generated by an analysis is whether they are resolved and supported statistically. Statistical reliability of nodes is most often expressed as bootstrap support or posterior probability values which need to be above 80% or 0.95 respectively before clades can be viewed to be supported. Relationships without statistically supported nodes should be viewed as unreliable. In addition, a fundamental consideration in phylogenetic theory is that it was developed for analysing bifurcating evolutionary pathways such as those occurring in non-recombinant organisms. Recombination, as identified in PVY, is likely to be generated by associating parts of genes or genomes into new combinations, and the standard methods of phylogenetic analysis cannot take this into account. This leads to conflict in phylogenetic analysis and weakens node support. However, there are ways of overcoming these conflicts and using phylogenetic analysis in studies of recombinant PVY isolates. In their assessment of recombination in viruses by phylogenetic analysis, Chare and Holmes (2006) divided the viral genomes into various genomic segments before analysing each segment separately. This overcomes the conflict caused by recombination because the genomic segments used were those in which recombination did not occur. This approach showed that segments of recombinants grouped differently depending on which segment was analysed, thereby illustrating that recombinants appear in different clades from one tree to the next. This effect is referred to as "incongruence", and results in incongruent trees. Testing for incongruence between the derived phylogenies can be conducted using Sawyer's test (Sawyer 1989).

This approach of subdividing whole genome sequences of PVY into segments between specific recombination junctions and performing a phylogenetic analysis on each segment to indicate incongruence has now become a common practice to identify potential parental lineage of PVY recombinants. However, the identification of recombination boundaries prior to performing a phylogenetic analysis on sequence alignments of PVY genome segments or groups of PVY genes or whole genome sequences is an important prerequisite. Defining recombination junctions can be performed using recombination analysis software such as RDP (Martin et al. 2010). In spite of performing recombination analyses, it is nevertheless often not possible to accurately define them especially for new or unknown recombinants.

Due to the inherent difficulty of mapping recombinant junctions accurately and the assumption that parental genomes can be unequivocally identified, many phylogenetic studies have not excluded isolates that contain recombinant segments prior to phylogenetic analysis. Therefore, while such analyses are useful for defining genotypes (molecular subgroups, see Chap. 3), one of the limitations is that they do not truly reflect the evolutionary histories of some isolates.

Once segments of genomes containing recombinant regions have been excluded, then a phylogenetic analysis should give more accurate information about the relatedness of the sequences in that matrix; in other words, about the evolutionary history of each segment of a virus. Recently, recombinants have been carefully removed and phylogenetic analyses performed to characterise and infer relationships of newly identified, non-recombinant isolates (Ogawa et al. 2008; Moury 2010; Ibaba and Gubba 2012; Janzac et al. 2015) or with the ultimate objective of analysing biogeographic patterns (Cuevas et al. 2012).

4.3 *The Evolution History of PVY: A Hypothesis*

4.3.1 Origin of PVY Strain Groups

From the documented history of the evolution of PVY in Europe, a succession of PVY^{C}, PVY^{O} and PVY^{N} strain dominance followed after 1930 until the 1980s, when the first recombinant strain, PVY^{NTN}, was described (Beczner et al. 1984). Since 1990, many "new" recombinants have been reported. As stated previously, the question arose whether recombination is indeed a relatively recent event driving PVY diversity in a range of environmental conditions worldwide or if the identification of PVY recombinants is essentially circumstantial resulting from the relatively recent availability of full-genome sequences of PVY variants allowing a more accurate genome analysis on a larger scale. In early studies, reference was already made to the possibility that recombination occurs between PVY isolates, for example, that recombination occurred between PVY^{C} and PVY^{O} (Watson 1960).

The first estimation of the dates for the evolution of non-recombinant and recombinant isolates of PVY was made by phylogenetic analysis after identifying recombination breakpoints with RDP software, and splitting of whole genomes into segments to generate a dated BEAST phylogenetic tree (Visser et al. 2012) (see Fig. 4.2a). This analysis showed that the split between the PVY^{O} and PVY^{N} lineages occurred about 350 years ago (~1600s). The first introduction of potatoes to Spain

A

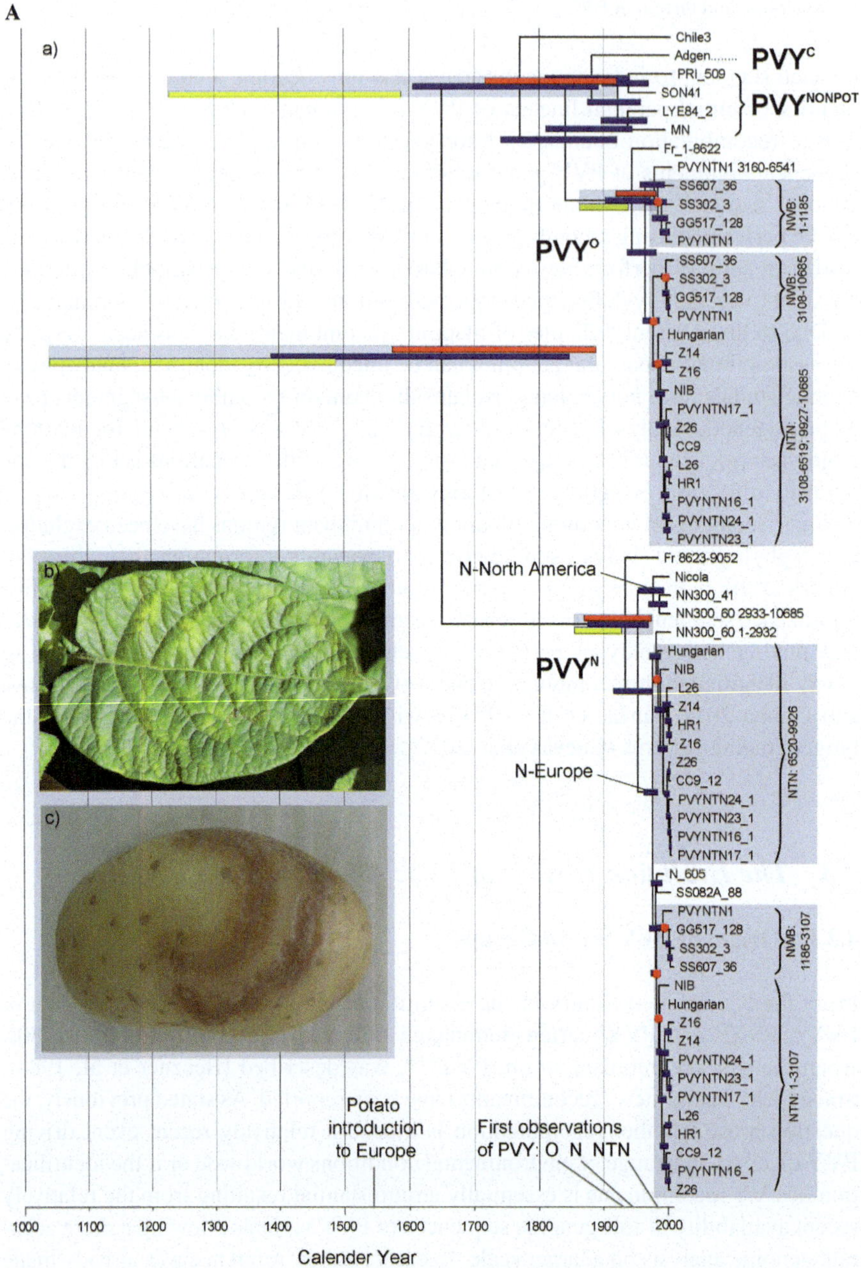

Fig. 4.2 Evolutionary pathways of PVY. (**a**) The recent, recombinant origins of *Potato virus Y* genomes. (a) The maximum clade credibility tree from the BEAST LN relaxed clock analysis is shown with error bars representing the 95% HPD of node ages according to LN (*blue*), EX (*red*) and SC (*yellow*) models. Recombinant strains such as pathogenic PVY^{NTN} and PVY$^{NW\text{-}B}$ are represented as multiple taxa, each representing a subset of the alignment (as indicated) with distinct phylogenetic signal. Topology constrained nodes are indicated with *red dots*. Inserts: (b) potato leaf showing mosaic patterns and (c) tuber displaying potato tuber necrotic ring disease (as published in Visser et al. 2012).

B

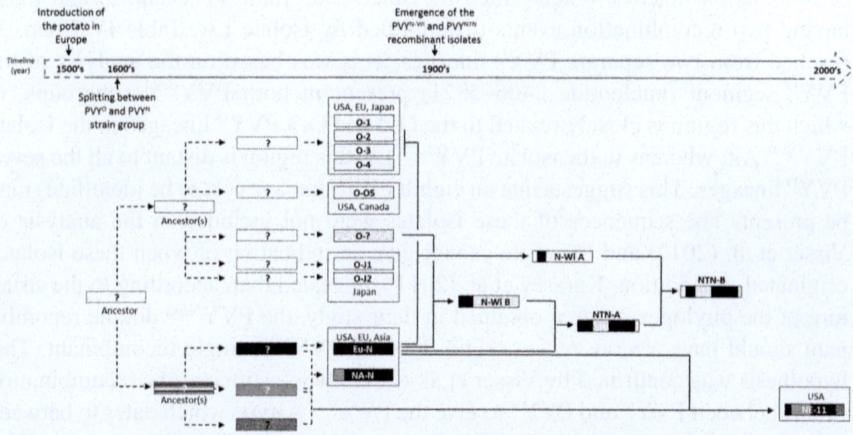

Fig. 4.2 (continued) (**b**). Schematic representation of the evolutionary relationships between PVY populations. The *black dotted lines* correspond to the evolution of hypothetical PVY^{O} and PVY^{N} ancestors. "?" corresponds to unknown ancestors. "O-1, O-2, O-3, O-4, O-O5, O-J1, O-J2" correspond to PVY^{O} subpopulations described in some countries. "O-?" is a hypothetical ancestral PVY^{O} subpopulation. "Eu-N" referred to European non-recombinant PVY^{N} isolates. "NA-N" referred to North American PVY^{N} isolates. "N-Wi A" corresponds to PVY$^{N\text{-}Wi}$ isolate resulting from two recombination events. "N-Wi B" corresponds to PVY$^{N\text{-}Wi}$ isolate with one recombination point. "NTN-A" and "NTN-B" referred to PVY^{NTN} isolates with three and four recombination junctions, respectively. More details about PVY genome diversity are shown in Chap. 3, Fig. 3.5 (Adapted from Karasev et al. (2011) and Ogawa et al. (2012))

date from around 1540 to 1565 and to Great Britain by 1565 (Brücher 1975). This implies that the PVY^{O} and PVY^{N} strain groups originated during modern potato cultivation (Visser et al. 2012). The current hypothesis is that after this splitting, PVY^{O} and PVY^{N} ancestors evidently evolved over time to produce the PVY strain groups currently described (Visser et al. 2012) (Fig. 4.2a).

Phylogenetic analyses applied to PVY polyprotein sequences or on partial genomic sequences revealed that isolates of PVY^{O} strain group did not originate from a unique ancestor but from multiple ancestors, due to the high amino acid intra-strain heterogeneity observed in PVY^{O} isolates (up to 3%). These can be grouped into subpopulations (O-1, O-2, O-3, O-4, O-O5, O-J1, O-J2) of which some are present on all continents or while others are only present in the USA or Japan (Karasev et al. 2011; Ogawa et al. 2012). However, assigning a PVY^{O} subpopulation to only one continent is only based on the current available literature. This might not reflect the reality because no large-scale surveys for these isolates have been undertaken in other countries. In the same way, two subgroups were also reported within the PVY^{N} strain group: one being non-recombinant (so-called European PVY^{N} isolates) and the other a recombinant (referred as North American NA-PVY^{N} isolates) derived from two PVY^{N} populations of different origins (Fig. 4.2b). Visser et al. (2012) dated this split to between 150 and 30 years ago

(Fig. 4.2a). Karasev et al. (2011) suggested that PVY^{N-Wi} isolates displaying one recombination junction (exemplified by isolate Alt, Table IV Chap. 3) and those having two recombination events (exemplified by isolate Lw, Table IV Chap. 3) resulted from two separate PVY^{O} lineages. This was based on the analysis of the PVY^{O} segment (nucleotide 2406–5821) present in both PVY^{N-Wi} subgroups, in which this region is closely related to the O-2 and O-3 PVY^{O} lineages in the isolate PVY^{N-Wi} Alt, whereas in the isolate PVY^{N-Wi} Lw this region is distant to all the seven PVY^{O} lineages. This suggests that an eighth PVY^{O} lineage (yet to be identified) may be present. The sequences of these isolates were not included in the analysis of Visser et al. (2012) and therefore cannot give an indication of when these isolates originated. In addition, Karasev et al. (2011) suggested that, according to the structure of the phylogenetic tree obtained in their study, the PVY^{N-Wi} double recombinant should have a more recent origin than the PVY^{N-Wi} single recombinant. This hypothesis was confirmed by Visser et al. (2012) who estimated the recombination events between PVY^{O} and PVY^{N} to give the PVY^{N-W} A-type which dates to between 48 and 20 years ago, and to PVY^{N-W} B-type to between 47 and 6 years ago. For PVY^{NTN} isolates, the analysis of the same PVY^{O} segment suggests that this region is closely related to that of PVY^{N-Wi} showing two recombination junctions. This suggests that PVY^{NTN} has evolved from additional recombination events from this subgroup (Karasev et al. 2011). Visser et al. (2012) dated the origin of the PVY^{NTN} double recombinant to between 47 and 19 years ago. All these estimations on the chronology of recombination events are in agreement with their first descriptions and support the hypothesis that recombination events in PVY are relatively recent occurrences.

From these phylogenetic studies, it is suggested that the ancestor of the PVY^{O} and PVY^{N} strains was introduced to mainland Europe and evolved relatively recently to generate the current strain groups. A separate PVY lineage that predates the "non-South American" strains of PVY is exemplified by the recombinant Brazilian isolate PVY-1054 found in tobacco plants (Janzac et al. 2015). Together with the identification of the PVY "Chilean" strain (Moury 2010), this provides further evidence of the high genetic diversity of PVY that occurs in its original environment in South America, and that additional lineages/strain groups of PVY have yet to be described.

4.3.2 Plant Virus Co-evolution: Relationship Between PVY Host Origin and Phylogenetic Groups

The evolutionary history of PVY has been shaped by its interaction with its host-vector-environment (*i.e.* selective advantages to overcome all potential population bottlenecks) and its biogeography. In terms of biogeography, the evolutionary history of PVY should be considered in two geographical areas. The first area is the Andes region of South America, where it is likely that PVY originated (Gibbs and Ohshima 2010) and many of its host including wild solanaceous plants occur naturally. The second area is the countries into which the hosts of PVY have been introduced. Recent hypotheses on the co-evolution of PVY and its hosts are based on the

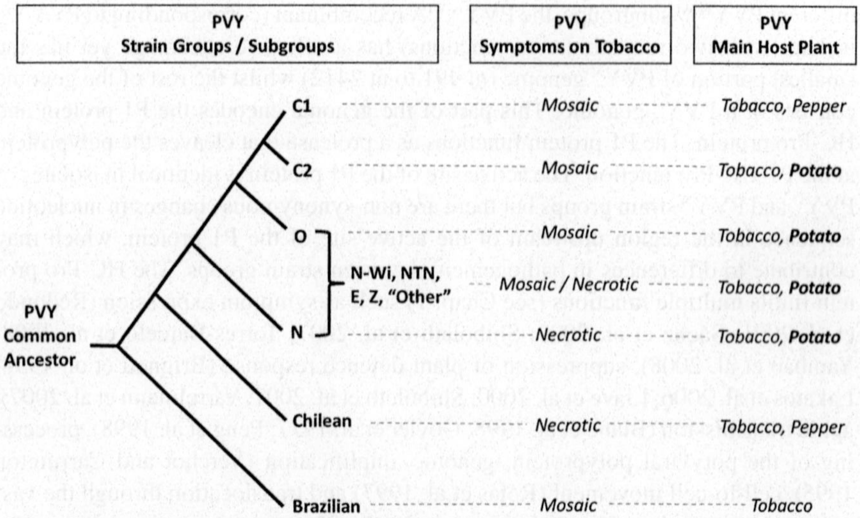

Fig. 4.3 Hypothetical PVY evolutionary lineages. Simplified PVY phylogeny and host range support evolutionary scenarios of potato and pepper infectivity, and of tobacco vein necrotic properties of PVY (adapted from Moury 2010; Janzac et al. 2015). Based on whole genome alignments, the Brazilian and Chilean clades have been considered to predate the N, O and C clades. During its evolution from a common ancestor, PVY has had to adapt to different hosts, eliciting different types of symptoms on some hosts such as tobacco. The main host species are presented on the right-hand side. The "recently evolved" recombinant PVY strains groups (from parental O and N genomes) identified in potato are indicated

reconstruction of ancestral traits and phylogenetic analysis (Fig. 4.3). Indeed, a correlation between phylogenetic classification and some host range properties can be observed for PVY groups N, O and C and their ability to infect (or not) tobacco, pepper, tomato or potato (Moury 2010; Quenouille et al. 2013; Janzac et al. 2015). It has been hypothesised that PVY infected tobacco and pepper, and that it subsequently evolved to be able to infect potato (Moury 2010; Quenouille et al. 2013). Janzac et al. (2015) concluded that the most parsimonious hypothesis for the ancestral state of PVY was to be adapted only to tobacco requiring only four "evolutionary steps" to adapt to different observed currently for the PVY N, O and C groups. This aspect is discussed further in Chap. 9.

4.4 *What Are the Selective Advantages That PVY$^{N\text{-}Wi}$ and PVY^{NTN} Possess Compared to Others?*

In order to assess the possible fitness advantages of the PVY$^{N\text{-}Wi}$ and PVY^{NTN} strain groups in host plants, an analysis of the regions of their recombinant genome may give insight into which genes or regions of genes confer such advantages. Of the different $PVY^{N\text{-}Wi}$ subgroups, the $PVY^{N\text{-}Wi}$A recombinant (corresponding to $PVY^{N\text{-}Wi}$ isolates with two recombination junctions) has a selective advantage yet has the smallest portion of PVY^{N} genome (nt 491 to nt 2412) whilst the rest of the genome consists of a PVY^{O} genome. This part of the genome encodes the P1 protein and HC-Pro protein. The P1 protein functions as a protease that cleaves the polyprotein at the P1-HC-Pro junction. The active site of the P1 protein is identical in isolates of PVY^{O} and PVY^{N} strain groups but there are non-synonymous changes in nucleotide sequence in the region upstream of the active site of the P1 protein, which may contribute to differences in pathogenicity between strain groups. The HC-Pro protein fulfils multiple functions (see Chap. 1) such as symptom expression (Redondo et al. 2001; Sáenz et al. 2002; Shiboleth et al. 2007; Torres-Barcelo et al. 2008; Yambao et al. 2008), suppression of plant defence response (Brigneti et al. 1998; Lakatos et al. 2006; Llave et al. 2000; Shiboleth et al. 2007; Varrelmann et al. 2007), aphid transmission (Blanc et al. 1998; Govier et al. 1977; Peng et al. 1998), processing of the potyviral polyprotein, genome amplification (Verchot and Carrington 1995), cell-to-cell movement (Rojas et al. 1997) and translocation through the vascular system of the plant (Sáenz et al. 2002). Therefore, mutations within the HC-Pro may alter PVY fitness. HC-Pro has been shown to be the avirulence factor for Nc_{tbr} and Ny_{tbr} resistance genes conferring resistance to the PVY^{C} and PVY^{O} strain groups (Moury et al. 2011). The incorporation of these resistance genes into most potato cultivars is thought to be responsible for the selection and prevalence of $PVY^{N\text{-}Wi}$ and PVY^{NTN} variants (Moury et al. 2011). In the C-terminal part of the HC-Pro from PVY^{O}, PVY^{C} and PVY^{N} strain groups, Moury et al. (2011) identified seven differences in amino acids (from aa 339 to 419) of which they concluded that the I363V change conferred resistance by the *Nc* gene. In a related study, Tian and Valkonen (2013) found that thc central region of the HC-Pro (from aa 227 to 327) played an important role in *Ny* recognition. Protein modelling studies of this region suggest that K269R and R270K substitutions determined the differences in the structure of HC-Pro between the two strains. Previous studies identified different regions of the HC-Pro as pathogenicity determinants of HC-Pro conferring *Ny* resistance (Moury et al. 2011; Tian and Valkonen 2013). Breeding potato cultivars with resistance conferred by the *Nc* and *Ny* genes has affected the extent to which the older PVY^{C} and PVY^{O} strains have been able to infect potato crops. As the acreage of potato cultivars with these resistance genes expanded and, as a result, the number of plants susceptible to these strain groups was reduced, the PVY^{C} strain became rare and the prevalence of PVY^{O} strain declined (Quenouille et al. 2013). However, isolates of the PVY^{NTN} and $PVY^{N\text{-}Wi}$ strain groups can overcome this resistance resulting in an increased prevalence over the older strain groups because of their improved fitness and ability to infect normally resistant cultivars. The assessment of the relative efficiency factors (REF) of various aphid species in transmitting isolates of PVY^{N}, $PVY^{N\text{-}Wi}$ and PVY^{NTN} strain groups revealed that not only *Myzus persicae* but also other aphid species might transmit $PVY^{N\text{-}Wi}$ and PVY^{NTN} more efficiently than other strain groups (Verbeek et al. 2010). These results support the fact that the HC-Pro region plays a key role in determining PVY pathogenicity and may confer selective advantages to some PVY variants.

Most PVY^{NTN} recombinants are composed of a PVY^{N} fragment spanning from nt 5825 to nt 9180. This part of the genome encodes the VPg, NIa-Pro and NIb genes and the first 605 bp of the CP. Potyvirus VPg proteins cap the 5′ end of the viral RNA and recruit eIF4E isoforms to promote replication and translation of the viral genome (Chap. 1). The NIa-Pro mediates the polyprotein cleavage and movement function, while the NIb protein fulfils multiple roles in pathogenicity and replication (Chap. 1). Using the dN/dS method, Moury and Simon (2011) examined the first 605 bp of the CP and found that two positions in the protein (aa 25 and aa 68) gave a selective advantage to PVY in a host-dependent manner in potato and tobacco but not in pepper or tomato. The CP interacts with HC-Pro to promote binding to aphid mouthparts during aphid transmission and interacts with NIb to stimulate genome amplification (Chap. 1). Based on this evidence it is proposed that the first 605 bp of PVY^{NTN} CP (nt 8575– nt 9180) harbours essential genetic determinants that might contribute to the selection of these PVY variants.

5 Conclusion

As for all obligate parasites, PVY is subjected to an ongoing process of natural selection and co-evolution by its hosts. While the mechanisms driving PVY diversity remain to be elucidated, ongoing mutation, recombination and evolutionary forces provide means to ensure its survival in its host plant. PVY has been extremely effective in evolving and adapting to changing environments. These adaptive capacities offer selective advantages to variants such as overcoming host resistance mechanisms, efficiently spreading in the host plant and being transmitted from one plant to another by many aphid species. The identification of new PVY lineages in different host species gives important insights into PVY evolution which is likely to have started with the advent of potato domestication in Peru (~ 8000 years ago), persisted and spread with cultivation, trade, changes of environment and hosts. More exhaustive surveys of PVY diversity in its natural habitat in cultivated and wild host species together with the use of next generation sequencing platforms will unveil new information on the diversity of PVY and its evolutionary pathways, as for related potyviruses and other virus species in future.

References

Bahrami Kamangar S, Smagghe G, Maes M, De Jonghe K (2014) *Potato virus Y* (PVY) strains in Belgian seed potatoes and first molecular detection of the N-Wi strain. J Plant Dis Prot 121:10–19

Bai Y, Nie X, Li X, Gao Y, Fan G, Zhang W, Geng H, Liu W (2010) The seed potato inspection and virus research in China. Am J Potato Res 87:112

Beczner L, Horvath J, Romhanyi I, Förster H (1984) Studies on the etiology of tuber necrotic ringspot disease in potato. Potato Res 27:339–352

Bem F, Varveri C (1999) First report of occurrence of potato tuber necrotic ringspot disease in Greece. Plant Dis 83:488

Bertschinger L (1992) Modeling of potato virus pathosystems by means of quantitative epidemiology: an exemplary case based on virus degeneration studies in Peru. PhD thesis. Swiss Federal Institute of Technology, Zurich

Blanc S, Ammar ED, Garcia-Lampasona S, Dolja VV, Llave C, Baker J, Pirone TP (1998) Mutations in the potyvirus helper component protein: effects on interactions with virions and aphid stylets. J Gen Virol 79:3119–3122

Blanco-Urgoiti B, Tribodet M, Leclere S, Ponz F, de San Roman CP, Legorburu FJ, Kerlan C (1998) Characterization of potato potyvirus Y (PVY) isolates from seed potato batches. Situation of the NTN, Wilga and Z isolates. Eur J Plant Pathol 104:811–819

Boonham N, Walsh K, Hims M, Preston S, North J, Barker I (2002) Biological and sequence comparisons of *Potato virus Y* isolates associated with potato tuber necrotic ringspot disease. Plant Pathol 51:117–126

Boukhris-Bouhachem S, Khamassy N, Glais L, Kerlan C (2008) Occurrence in Tunisia of potato tuber necrotic ringspot disease (PTNRD) caused by variant PVY^{NTN} of *Potato virus Y*. Plant Pathol 57:388

Boukhris-Bouhachem S, Djilani-Khouadja F, Fakhfakh H, Glais L, Tribodet M, Kerlan C (2010) Incidence and characterization of *Potato virus Y* in seed potatoes in Tunisia. Potato Res 53:151–166

Brigneti G, Voinnet O, Li WX, Ji LH, Ding SW, Baulcombe DC (1998) Viral pathogenicity determinants are suppressors of transgene silencing in *Nicotiana benthamiana*. EMBO J 17:6739–6746

Browning I, Fox A, Holmes R, Mulholland L Darling D (2004) Variability in *Potato virus Y* in seed potato crops in Scotland. In: The 12th European Association of Potato research Virology section meeting, Rennes, France. p 4

Brücher H (1975) Domestikation und Migration von *Solanum tuberosum* L. Genet Resour Crop Evol 23:11–74

Bujarski JJ (1999) Molecular basis of genetic variability in RNA viruses. In: Mandahar CL (ed) Molecular biology of plant viruses. Kluwer Academic Publishers, Norwell, pp 121–141

Buturovic D Kus M (1989) The occurrence of potato tuber ring necrotic diseases in Yugoslavia. In: The 6th Triennial Conference of European Association for Potato Research, Virology Section Meeting, Bologna, Italy, p 6

Chare ER, Holmes EC (2006) A phylogenetic survey of recombination frequency in plant RNA viruses. Arch Virol 151:933–946

Chikh Ali M, Maoka T, Natsuaki KT (2007) The occurrence and characterization of new recombinant isolates of PVY displaying shared properties of $PVY^{N}W$ and PVY^{NTN}. J Phytopathol 155:409–415

Chrzanowska M (1991) New isolates of the necrotic strain of potato virus Y (PVY^{N}) found recently in Poland. Potato Res 34:179–182

Crosslin M, Hamm PB, Shiel PJ, Hane DC, Brown CR, Berger PH (2005) Serological and molecular detection of tobacco veinal necrosis isolates of *Potato virus Y* (PVY^{N}) from potatoes grown in the western United States. Am J Potato Res 82:263–269

Cuevas JM, Delaunay A, Visser JC, Bellstedt DU, Jacquot E, Elena SF (2012) Phylogeography and molecular evolution of *Potato virus Y*. PLoS One 7:e37853

Damayanti TA, Alabi OJ, Hidayat SH, Crosslin JM, Naidu RA (2014) First report of *Potato virus Y* in potato in West Java, Indonesia. Plant Dis 98:287

Davie K, Lacomme C, Dickinson M (2012) The biodiversity and epidemiology of *Potato virus Y* (PVY) in Scotland. Crop Protection in Northern Britain, Dundee, pp 255–260

de Ávilal AC, de Melo PE, Leite LR, Inoue-Nagata AK (2009) Ocorrência de vírus em batata em sete estados do Brasil. Hortic Bras 27:490–497

Dedic P, Kanak E Nohejl J (1988) Unusual strain of PVY with necroses on potato tubers. In: The 11th Czechoslovak Plant protection conference, Nitra, Czechoslovakia, pp 179–180

Dietrich C, Miller J, McKenzie G, Palkovics L, Balasz E, Palukaitis P, Maiss E (2007) No recombination detected in artificial potyvirus mixed infections and between potyvirus derived transgenes and heterologous challenging potyviruses. Environ Biosaf Res 6:207–218

Domingo E, Holland JJ (1997) A virus mutations and fitness for survival. Annu Rev Microbiol 51:151–178

Drake JW (1993) Rates of spontaneous mutation among RNA viruses. Proc Natl Acad Sci USA 90:4171–4175

Drummond AJ, Suchard MA, Xie D, Rambaut A (2012) Bayesian phylogenetics with BEAUti and the BEAST 1.7. Mol Biol Evol 29:1969–1973

Duffy S (2016) Viral informatics: tools for understanding the evolution of biology's most varied genomes. In: Weaver SC, Denison M, Roosinck M, Vignuzzi M (eds) Virus evolution: current research and future directions. Caister Academic Press, Wymondham, pp 29–60

Ellis P, Stace-Smith R, de Villiers G (1997) Identification and geographic distribution of serotypes of *Potato virus Y*. Plant Dis 81:481–484

Fenton B, Lacomme C, Collins L, Fox A, Pickup J, Evans A, Northing P, Anderson E (2012) Aphids and virus transmission in seed crops. Potato council and agriculture and horticulture development board (AHDB, UK) project R428 final report. http://www.potato.org.uk/sites/default/files/publication_upload/R428_Final%20Report.pdf

Fletcher JD (1989) Potato virus Y^N – host range and incidence in seed potato crops in New Zealand. N Z J Crop Hortic Sci 17:259–263

Fletcher JD (2012) A virus survey of New Zealand fresh, process and seed potato crops during 2010–2011. N Z Plant Prot 65:197–203

Fletcher JD, Lister RA (2004) PVY^{NTN} in New Zealand potato crops. In: The 12th Triennial Conference of the European Association for Potato Research, Virology Section Meeting, Le Tronchet, France. p 55

Friedberg EC, Walker GC, Siede W (1995) DNA repair and mutagenesis. In: DNA repair and mutagenesis. ASM Press, Washington, DC, p 698

Fomitcheva VW, Fletcher JD, Schubert J (2009) *Potato virus Y* strain spectrum in New Zealand – absence of recombinant N:O strains. J Phytopathol 157:507–510

Galvino-Costa SBF, Figueira AR, Rabelo-Filho FAC, Moraes FHR, Nikolaeva OV, Karasev AV (2012) Molecular typing of *Potato virus Y* isolates from Brazil reveals a diverse set of recombinant strains. Plant Dis 96:1451–1458

Garcia-Arenal F, Fraile A, Malpica JM (2001) Variability and genetic structure of plant virus populations. Annu Rev Phytopathol 39:157–186

Garcia-Arenal F, Fraile A, Malpica JM (2003) Variation and genetic structure of plant virus populations. Int Microbiol 6:225–232

Gibbs A, Ohshima K (2010) Potyviruses and the digital revolution. Annu Rev Phytopathol 48:205–223

Glais L, Tribodet M, Kerlan C (2002) Genomic variability in *Potato potyvirus Y* (PVY): evidence that PVY^NW and PVY^{NTN} variants are single to multiple recombinants between PVY^O and PVY^N isolates. Arch Virol 147:363–378

Govier DA, Kassanis B, Pirone TP (1977) Partial purification and characterization of *Potato virus Y* helper component. Virology 78:306–314

Gray S, De Boer S, Lorenzen J, Karasev A, Whitworth J, Nolte P, Singh R, Boucher A, Xu HM (2010) *Potato virus Y*: an evolving concern for potato crops in the United States and Canada. Plant Dis 94:1384–1397

Hernandez-de la Cruz M, Gomez-Leyva JF, Lopez-Muraira IG, Dimas-Estrada MS, Andrade-Gonzalez I (2007) Deteccion serologica y molecular del virus PVY^N y su variante PVY^{NTN} en papa (*Solanum tuberosum* L.) y hospedantes alternos en Tapalpa, Mexico. Revista Mexicana de Fitopatologia 25:167–172

Hosseini A, Massumi H, Heydarnejad J, Pour AH, Varsani A (2011) Characterisation of *Potato virus Y* isolates from Iran. Virus Genes 42:128–140

Hughes AL (2009) Small effective population sizes and rare non-synonymous variants in potyviruses. Virology 393:127–134

Ibaba JD, Gubba A (2012) Phylogenetic studies of selected isolates of *Potato virus Y* (PVY) infecting vegetable crops in KwaZulu-Natal, Republic of South Africa, South African. J Plant Soil 29(2):117–120

Janzac B, Willemsen A, Cuevas JM, Glais L, Tribodet M, Verrier JL, Elena SF, Jacquot E (2015) Brazilian *Potato virus Y* isolates identified as members of a new clade facilitate the reconstruction of evolutionary traits within this species. Plant Pathol 64:799–807

Kaczmarek U, Wydzialkowska J (1996) The spread of PVY strains in potato plants in Poland. In: The 13th Triennial conference of the European association for Potato research, Veldhoven, The Netherlands, pp 304–305

Karasev AV, Nikolaeva OV, Hu XJ, Sielaff Z, Whitworth J, Lorenzen JH, Gray SM (2010) Serological properties of ordinary and necrotic isolates of *Potato virus Y*: a case study of PVY^N misidentification. Am J Potato Res 87:1–9

Karasev AV, Hu XJ, Brown CJ, Kerlan C, Nikolaeva OV, Crosslin JM, Gray SM (2011) Genetic diversity of the ordinary strain of *Potato virus Y* (PVY) and origin of recombinant PVY Strains. Phytopathology 101:778–785

Kehoe MA, Jones RAC (2016) Improving *Potato virus Y* strain nomenclature: lessons from comparing isolates obtained over a 73-year period. Plant Pathol 65:322–333

Kerlan C, Tribodet M, Glais L, Guillet M (1999) Variability of *Potato virus Y* in potato crops in France. J Phytopathol 147:643–651

Kus M (1990) Potato tuber ring necrosis disease (PTNRD) in Slovenia. In the 11th Triennial conference of the European association for Potato research, Edinburgh, UK, pp 196

Lacomme C, Davie K, Holmes R, Pickup J (2014) PVY^N prevalence in potato crops: impact of strain competition and differential ability to overcome plant resistance mechanisms. In: Proceedings of crop protection in Northern Britain, Dundee, UK, pp 203–208

Lacroix C, Glais L, Kerlan C, Verrier J-L, Jacquot E (2010) Biological characterization of French Potato virus Y (PVY) isolates collected from PVY-susceptible or -resistant tobacco plants possessing the recessive resistance gene va. Plant Pathol 59(6):1133–1143

Lacroix C, Glais L, Verrier JL, Charlier C, Lorencetti C, Jacquot E (2011) Impact of tobacco recessive resistance gene *va* on biological properties of Brazilian *Potato virus Y* (PVY) isolates. Plant Pathol 60:1048–1054

Lai MM (1992) RNA recombination in animal and plant viruses. Microbiol Rev 56:61–79

Lakatos L, Csorba T, Pantaleo V, Chapman EJ, Carrington JC, Liu YP, Dolja VV, Calvino LF, Lopez-Moya JJ, Burgyan J (2006) Small RNA binding is a common strategy to suppress RNA silencing by several viral suppressors. EMBO J 25:2768–2780

Lemey P, Salemi M, Van Damme M (2009) The phylogenetic handbook. Cambridge University Press, Cambridge, 723 pages

Le Romancer M (1993) La maladie des nécroses annulaires superficielles des tubercules de la pomme de terre. Identification et caractérisation de l'agent pathogène: le variant YNTN du Virus Y de la pomme de terre. Université de Rennes, France

Lindner K (2008) PVY strains in Germany – the period between 1984 and 2006. In: The 13th Triennial conference of the European association for Potato research, Virology Section Meeting, Coylumbridge, Scotland

Lindner K, Billenkamp N (2005) Changes in the spectrum of PVY strains and strain groups: could this be a cause for the increase of virus susceptibility of potato and tobacco varieties? Nachrichtenblatt des Deutschen Pflanzenschutzdienstes 57:245–253

Lindner K, Trautwein F, Kellermann A, Bauch G (2016) *Potato virus Y* (PVY) in seed potato certification. J Plant Dis Prot 122:109–119

Llave C, Kasschau KD, Carrington JC (2000) Virus-encoded suppressor of posttranscriptional gene silencing targets a maintenance step in the silencing pathway. Proc Natl Acad Sci U S A 97:13 401–13 406

Malpica JM, Fraile A, Moreno I, Obies CI, Drake JW, Garcia-Arenal F (2002) The rate and character of spontaneous mutation in an RNA virus. Genetics 162:1505–1511

Martin DP, Lemey P, Lott M, Moulton V, Posada D, Lefeuvre P (2010) RDP3: a flexible and fast computer program for analyzing recombination. Bioinformatics 26:2462–2463

McDonald JG, Kristjansson GT (1993) Properties of strains of potato virus Y^{N} in North America. Plant Dis 77:87–89

McDonald JG, Singh RP (1996) Response of potato cultivars to North American isolates of PVY^{NTN}. Am Potato J 73:317–323

Moury B (2010) A new lineage sheds light on the evolutionary history of *Potato virus Y*. Mol Plant Pathol 11:161–168

Moury B, Simon V (2011) dN/dS-based methods detect positive selection linked to trade-offs between different fitness traits in the coat protein of *Potato virus Y*. Mol Biol Evol 28:2707–2717

Moury B, Fabre F, Senoussi R (2007) Estimation of the number of virus particles transmitted by an insect vector. Proc Natl Acad Sci U S A 104:17891–17896

Moury B, Caromel B, Johansen E, Simon V, Chauvin L, Jacquot E, Kerlan C, Lefebvre V (2011) The helper component proteinase cistron of *Potato virus Y* induces hypersensitivity and resistance in potato genotypes carrying dominant resistance genes on chromosome IV. Mol Plant-Microbe Interact 24:787–797

Mousavi L, Mozafari J, Rakhshandehroo F, Chahardehi AM, Mousavi M (2014) Distribution and prevalence of *Potato virus Y* isolates obtained from potatoes grown in the Iran by RT-PCR. J Biol Agric Healthcare 4:45–53

Moya A, Holmes EC, González-Candelas F (2004) The population genetics and evolutionary epidemiology of RNA viruses. Nat Rev Microbiol 2:279–288

Nagy PD, Bujarski JJ (1997) Engineering of homologous recombination hotspots with AU-rich sequences in *Brome mosaic virus*. J Virol 71:3799–3810

Nagy PD, Simon AE (1997) New insights into the mechanisms of RNA recombination. Virology 235:1–9

Nagy PD, Zhang C, Simon AE (1998) Dissecting RNA recombination *in vitro*: role of RNA sequences and the viral replicase. EMBO J 17:2392–2403

Nielsen SL (1992) Ringplet nekrosesyge i kartoffelknolde. Tidsskr Pl Avl 86:201–202

Nobrega NR, Silberschmidt KM (1944) Sobre una provavel variante do virus "Y" da batatinha (Solanum virus 2, Orton) que tem a peculiaridade de provocar necroses em plantas de fumo. Arquivos do Instituto Biologico 15:307–330

Novella IS, Duarte EA, Elena SF, Moya A, Domingo E, Holland JJ (1995) Exponential increases of RNA virus fitness during large population transmissions. Proc Natl Acad Sci U S A 92:5841–5844

Ogawa T, Tomitaka Y, Nakagawa A, Ohshima K (2008) Genetic structure of a population of *Potato virus Y* inducing potato tuber necrotic ringspot disease in Japan; comparison with North American and European populations. Virus Res 131:199–212

Ogawa T, Nakagawa A, Hataya T, Ohshima K (2012) The genetic structure of populations of *Potato virus Y* in Japan; based on the analysis of 20 full genomic sequences. J Phytopathol 160:661–673

Orlando A, Silberschmidt K (1945) Estudos sobre a transmissao da doenca de Solanaceas "Necroses das Nervuras", por afıdios, e algumas relacoes entre esse virus e o seu principle inseto vector. Arquivos do Instituto Biologico 6:133–152

Oshima K, Sako K, Hiraishi C, Nakagawa A, Matsuo K, Ogawa T, Shikata E, Sako N (2000) Potato tuber necrotic ringspot disease occurring in Japan: its association with *Potato virus Y* necrotic strain. Plant Dis 84:1109–1115

Peng YH, Kadoury D, Gal-On A, Huet H, Wang Y, Raccah B (1998) Mutations in the HC-Pro gene of *Zucchini yellow mosaic potyvirus*: effects on aphid transmission and binding to purified virions. J Gen Virol 79:897–904

Piche LM, Singh RP, Nie X, Gudmestad NC (2004) Diversity among *Potato virus Y* isolates obtained from potatoes grown in the United States. Phytopathology 94:1368–1375

Pieterse B (2005) A new variant of PVY in South Africa. Available at http://www.potatoes.co.za/home.asp?pid=219, Vol 19

Pourrahim R, Farzadfar S (2016) Population analysis of Iranian *Potato virus Y* isolates using complete genome sequence. Plant Pathol J 32:33–46

Quenouille J, Vassilakos N, Moury B (2013) *Potato virus Y*: a major crop pathogen that has provided major insights into the evolution of viral pathogenicity. Mol Plant Pathol 14:439–452

Radtke W (1984) Schwellungen an Kartoffelknollen beobachtet: Ursache unbekannt. Kartoffelbau 35:24–25

Redondo E, Krause-Sakate R, Yang SJ, Lot H, Le Gall O, Candresse T (2001) *Lettuce mosaic virus* pathogenicity determinants in susceptible and tolerant lettuce cultivars map to different regions of the viral genome. Mol Plant-Microbe Interact 14:804–810

Revers F, Le Gall O, Candresse T, Le Romancer M, Dunez J (1996) Frequent occurrence of recombinant potyvirus isolates. J Gen Virol 77:1953–1965

Reza M, Behrouz J, Mahrokh R (2016) Investigation of *Potato virus Y* strains in Guilan and Mazandaran provinces using biological, serological and molecular methods. J Appl Environ Biol Sci 6:107–112

Rigotti S, Balmelli C, Gugerli P (2011) Census report of the *Potato virus Y* (PVY) population in Swiss seed potato production in 2003 and 2008. Potato Res 54:105–117

Rojas MR, Zerbini FM, Allison RF, Gilbertson RL, Lucas WJ (1997) Capsid protein and helper component-proteinase function as potyvirus cell-to-cell movement proteins. Virology 237:283–295

Rolland M, Lacroix C, Blanchard A, Baldwin T, Kerlan C, Jacquot E (2008) *Potato virus Y* (PVY): from its discovery to the latest outbreaks. Virologie 12:261–273

Rolot JL, Steyer S (2008) First screening of the different PVY strains in Belgium. In: The13th Triennial conference of the European association for Potato research, Virology Section Meeting, Coylumbridge, Scotland

Roosinck MJ (2003) Plant RNA virus evolution. Curr Opin Microbiol 6:406–409

Rubio L, Guerri J, Moreno P (2013) Genetic variability and evolutionary dynamics of viruses of the family *Closteroviridae*. Front Microbiol 4:1–15

Sáenz P, Salvador B, Simon-Mateo C, Kasschau KD, Carrington JC, García JA (2002) Host-specific involvement of the HC protein in the long-distance movement of potyviruses. J Virol 76:1922–1931

Salaman RN (1930) Virus diseases of potato: streak. Nature 126:241

Sawyer S (1989) Statistical tests for detecting gene conversion. Mol Biol Evol 6:526–538

Schiessendoppler, E. (1990). PVY as causal agent of tuber necrotic ring disease. In: The 11th Triennial conference of the European association for Potato research, Virology Section Meeting, Edinburgh, UK, pp 194–195

Scholthof KG, Adkins S, Czosnek H, Palukaitis P, Jacquot E, Hohn T, Horn B, Saunders K, Candresse T, Ahlquist P, Hemenway C, Foster GD (2011) Top 10 plant viruses in molecular plant pathology. Mol Plant Pathol 12:938–954

Schubert J, Fomitcheva V, Sztangret-Wisniewska J (2007) Differentiation of *Potato virus Y* strains using improved sets of diagnostic PCR-primers. J Virol Methods 140:66–74

Schneider WL, Roossinck MJ (2001) Genetic diversity in RNA virus quasispecies is controlled by host-virus interactions. J Virol 75:6566–6571. doi:10.1128/JVI.75.14.6566-6571.2001

Serra MC, Weidemann HL (1997) First report of potato tuber necrotic ringspot disease caused by PVY^{NTN} in Portugal. Plant Dis 81:694–694

Shiboleth YM, Haronsky E, Leibman D, Arazi T, Wassenegger M, Whitham SA, Gaba V, Gal-On A (2007) The conserved FRNK box in HC-Pro, a plant viral suppressor of gene silencing, is required for small RNA binding and mediates symptom development. J Virol 81(13):135–148

Silberschmidt KM (1960) Types of *Potato virus Y* necrotic to tobacco: history and recent observation. Am Potato J 37:151–159

Smith KM (1931) Composite nature of certain potato viruses of the mosaic group as revealed by the use of plant indicator. Proc R Soc B 109:251–267

Smith KM, Dennis RWG (1940) Some notes on a suspected variant of Solanum virus 2 (*Potato virus Y*). Ann Appl Biol 27:65–70

Stamatakis A (2006) RAxML-VI-HPC: maximum likelihood-based phylogenetic analyses with thousands of taxa and mixed models. Bioinformatics 22(21):2688–2690

Tian YP, Liu JL, Zhang CL, Liu YY, Wang B, Li XD, Guo ZK, Valkonen JPT (2011) Genetic diversity of *Potato virus Y* infecting tobacco crops in China. Phytopathology 101:377–387

Tian YP, Valkonen JPT (2013) Genetic determinants of *Potato virus Y* required to overcome or trigger hypersensitive resistance to PVY strain group O controlled by the gene *Ny* in potato. Mol Plant-Microbe Interact 26:297–305

Tomassoli L, Lumia V, Cerato C, Ghedini R (1998) Occurrence of potato tuber necrotic ringspot disease (PTNRD) in Italy. Plant Dis 82:350–350

Torres-Barcelo C, Martin S, Daros JA, Elena SF (2008) From hypo- to hypersuppression: effect of amino acid substitutions on the RNA-silencing suppressor activity of the *Tobacco etch potyvirus* HC-Pro. Genetics 180:1039–1049

Tromas N, Elena SF (2010) The rate and spectrum of spontaneous mutations in a plant RNA virus. Genetics 185:983–989

van der Vlugt RAA, Verbeek M, Cuperus C, Piron PGM, de Haan E, van de Bovenkamp GW (2008) Strains of *Potato virus Y* in Dutch seed potato culture. Potato Res 51:191

Varrelmann M, Maiss E, Pilot R, Palkovics L (2007) Use of pentapeptide insertion scanning mutagenesis for functional mapping of the plum pox virus helper component proteinase suppressor of gene silencing. J Gen Virol 88:1005–1015

Verbeek M, Piron PGM, Dullemans AM, Cuperus C, van der Vlugt RAA (2010) Determination of aphid transmission efficiencies for N, NTN and Wilga strains of *Potato virus Y*. Ann Appl Biol 156:39–49

Verchot J, Carrington JC (1995) Evidence that the potyvirus P1 proteinase functions in trans as an accessory factor for genome amplification. J Virol 69:3668–3674

Visser JC, Bellstedt DU (2009) An assessment of molecular variability and recombination patterns in South African isolates of *Potato virus Y*. Arch Virol 154:1891–1900

Visser JC, Bellstedt DU, Pirie MD (2012) The recent recombinant evolution of a major crop pathogen, *Potato virus Y*. PLoS One 7:e50631

Watson MA (1960) Evidence for interaction or genetic recombination between *Potato viruses Y* and C in infected plants. Virology 10:211–232

Weidemann HL, Maiss E (1996) Detection of the potato tuber necrotic ringspot strain of *Potato virus Y* (PVY^{NTN}) by reverse transcription and immunocapture polymerase chain reaction. J Plant Dis Prot 103:337–345

Weilguny H, Singh RP (1998) Separation of Slovenian isolates of PVY^{NTN} from the North American isolates of PVY^{N} by a 3-primer PCR. J Virol Methods 71:57–68

Were HK, Kabira JN, Kinyua ZM, Olubayo FM, Karinga JK, Aura J, Lees AK, Cowan GH, Torrance L (2013) Occurrence and distribution of potato pests and diseases in Kenya. Potato Res 56:325–342

Wright DM (1992) Potato tuber necrotic ring disease. Plant Disease Notice N° 74. Central Science Laboratory, Hatching Green, Herts, UK, 5

Xu H (2013) PVY strain status in Canada. In: The 15th Triennial conference of the European association for Potato research, Virology Section Meeting, Antalya, Turkey, p 5

Yambao MLM, Yagihashi H, Sekiguchi H, Sekiguchi T, Sasaki T, Sato M, Atsumi G, Takahashi Y, Nakahara KS, Uyeda I (2008) Point mutations in helper component protease of *Clover yellow vein virus* are associated with the attenuation of RNA-silencing suppression activity and symptom expression in broad bean. Arch Virol 153:105–115

Zimnoch-Guzowska E, Yin ZM, Chrzanowska M, Flis B (2013) Sources and effectiveness of potato PVY resistance in IHAR's breeding research. Am J Potato Res 90:21–27

Chapter 5
Detection and Diagnosis of PVY

Laurent Glais, Mohamad Chikh Ali, Alexander V. Karasev, Denis Kutnjak, and Christophe Lacomme

Abstract The worldwide prevalence of *Potato virus Y* (PVY) poses a continuous challenge to efficient potato production. The accurate diagnosis of viruses such as PVY is inherently challenging due to the broad biological and genetic diversity of PVY strains that elicit a range of symptoms and diseases in various potato cultivars and related solanaceous species. A wide range of techniques have been developed over a period of 50 years for the detection of PVY. Serological methods such as ELISA, using polyclonal and especially monoclonal antibodies, have been widely used by most diagnostic laboratories, due to their cost effectiveness and capacity to implement for a large number of samples. Over the last decade, PCR-based assays have been routinely used in diagnostic laboratories because of their sensitivity, specificity and their capacity to be automated for high-throughput testing. The objective of this chapter is to provide a brief historical overview of the main diagnostic methods used to detect and identify PVY and to highlight those that are suitable for either research or diagnostic purposes.

L. Glais (✉)
FN3PT/RD3PT, 43-45 rue de Naples, F-75008 Paris, France

IGEPP, Agrocampus Ouest, INRA, Université de Rennes 1, F-35650 Le Rheu, France
e-mail: laurent.glais@fnpppt.fr

M. Chikh Ali • A.V. Karasev
Department of Plant Soil and Entomological Sciences, University of Idaho,
875 Perimeter Drive MS 2339, Moscow, ID 83844-2339, USA

D. Kutnjak
Department of Biotechnology and Systems Biology, National Institute of Biology,
Večna pot 111, 1000 Ljubljana, Slovenia

C. Lacomme
Science and Advice for Scottish Agriculture (SASA),
1 Roddinglaw Road, Edinburgh EH12 9FJ, UK

C. Lacomme et al. (eds.), *Potato virus Y: biodiversity, pathogenicity, epidemiology and management*, DOI 10.1007/978-3-319-58860-5_5

1 Introduction

The accurate diagnosis of viruses such as *Potato virus Y* (PVY) is inherently challenging due to the broad biological and genetic diversity of PVY strains that elicit a range of symptoms and disease in various potato cultivars and related solanaceous species. The worldwide prevalence of PVY poses a continuous challenge to efficient potato production. The United Nations Economic Commission for Europe (UNECE) has developed an international standard for seed potatoes that provides a framework for creating a harmonised quality certification system in order to facilitate the worldwide marketing of seed potatoes. This Seed Potato Standard sets out harmonised quality conditions and standards for the certification and marketing of seed potatoes, including virus (Chap. 7). However, countries may apply more stringent quality standards provided that these are technically justified and applied to their own production. For the purpose of plant protection, a nil tolerance for specific PVY strains may be applied to imported seed potatoes. Therefore, the use of reliable diagnostic tools is essential to detect, identify and, when necessary, characterise various PVY strains which could be present in seed potatoes. While diagnostic methods have evolved over time together with the knowledge of the diversity of PVY, they will continue to be based on the biological, serological and molecular properties of PVY.

Rolland (2008) reported that 93 serological and PCR-based methods for the detection of PVY had been published between 1974 and 2008 while only seven new diagnostic methods have been published since 2009 (Fig. 5.1). The antigenic properties of PVY have always been used for the development of serological methods. Publication of most PCR-based methods occurred between 2000 and 2008, highlighting the recent shift from serological to molecular diagnostic methods. This survey also revealed that 40% of the molecular methods were based on capsid/coat protein (CP) nucleotide sequences (Fig. 5.2) because these display a relatively high degree of heterogeneity among PVY isolates (Rolland 2008).

The objective of this chapter is to provide a brief historical overview of the main diagnostic methods used to detect and identify PVY and to highlight those that are currently being used for research and diagnostic purposes.

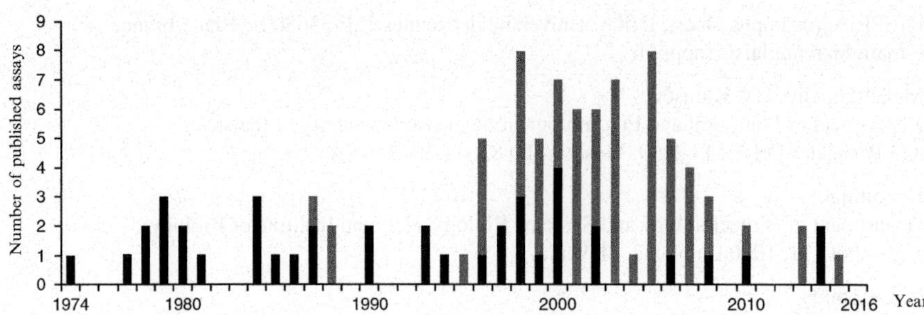

Fig. 5.1 Chronological publication of serological (*black bars*) and molecular (*grey bars*) diagnostic methods for the detection of PVY between1974 and 2016 (Adapted from Rolland 2008)

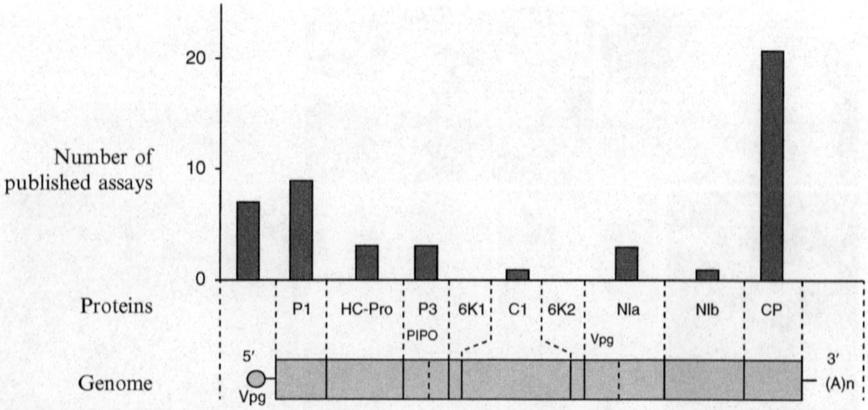

Fig. 5.2 Distribution of PVY molecular assays given their target genomic region (Extracted from Rolland 2008)

2 Biological Characterisation: Early Diagnosis of Potato Viral Diseases and the Definition of PVY Strain

2.1 Visual Observation of Symptoms

In the early days of plant virology, visual symptoms were the main method of recording diseases associated with viruses, including PVY. Until the early 1900s, viral malformations of potato plants were described as 'mosaic', 'crinkle' and 'rugosity' diseases (Smith 1931; MacLeod 1962). Mosaic disease appears as a mottling of paler chlorotic areas on the leaflets without any deformation (Fig. 5.3A). Crinkle disease appears as a mosaic symptom together with a wrinkling of the leaves, especially in the apices of stems (Fig. 5.3B). When infection is transmitted from seed tubers (secondary transmission), affected plants are stunted and severely crinkled (Fig. 5.3C). Similarly, rugosity only develops with secondary infection, with leaves becoming severely wrinkled, mottled and closely bunched together (Fig. 5.3D, E). Lower leaves can become completely necrotic and drop along the stem (Fig. 5.3G) (MacLeod 1962). A wide range of necrotic symptoms can be observed depending on the combination of potato cultivar and PVY isolate. These necrotic symptoms can appear as a darkening of the veins visible under the leaf surface, and may be accompanied by necrosis of the petioles and main stem (Fig. 5.3F, G). In some cases, necrosis on a leaf can form an 'oak leaf' pattern (Fig. 5.3H). Well defined, necrotic, circular spots can also develop in response to PVY infection (Fig. 5.3J, K). The development of dark necrotic rings on chlorotic leaves (so-called "Maritta" symptom) (Kerlan et al. 1985) was first observed in the potato cultivar Maritta (Fig. 5.3I).

Fig. 5.3 Symptoms of PVY infection observed in field conditions on various potato cultivars and breeding lines. (**a**) mild mosaic; (**b**) crinkle symptom with mosaic and leaf deformation induced by primary infection; (**c**) severe crinkle symptom of mosaic, leaf deformation and stunting of the whole plant caused by secondary infection; (**d/e**) rugosity symptoms of mosaic and severe crinkling and shrivelling of virtually all leaves; (**f**): veinal necrosis symptom; (**g**) severe stem and veinal necrosis, and leaf-drop streak (stipple-streak); (**h**) necrosis forming an 'oak leaf' pattern; (**i**) necrotic 'Maritta' symptom; (**j/k**) well-defined necrotic spots caused by primary infection. Additional examples of symptoms caused by PVY can be seen in Chaps. 1 and 3 (Photos are the courtesy of: © GNIS/K. Charlet-Ramage (a–b, d–k); © INRA/C. Kerlan (c))

Historically, specific symptoms have been used as an initial indicator of the causal virus of a disease. However, infected potato plants only express a limited range of symptoms and similar symptoms may be caused by more than one virus species or strain, and by biotic and abiotic stresses *e.g.* climatic conditions, mineral deficiency and herbicide phytotoxicity (Le Hingrat 2011). In addition, infection by some PVY isolates, such as the $PVY^{N\text{-}Wi}$, may be symptomless or may induce only very mild symptoms on potato plants. Infected plants may, therefore, not be detected by visual inspection (Chrzanowska 1991). Moreover, the genotype of a potato cultivar, the developmental stage of a plant, type of infection (primary or secondary), time of primary infection (early or late), number of virus species/strains infecting a plant (single or multiple) and environmental conditions can also affect the expression of viral symptoms by a plant. Although visual observation of symptoms may not be reliable in enabling infected plants to be detected in a crop, it can provide an indication of those crops at risk of viral infection that may require further testing to establish the incidence of infection and the causal virus, if this is required. However, the terminology of foliar symptoms (mosaic, crinkle, rugosity, etc.) may still be relevant for research purposes to document the pathogenicity of a PVY isolate in specific conditions. In certification schemes including UNECE Seed Potato Standard, quality standards for viral diseases in seed potato crops have been defined by symptom expression on the diseased plant *e.g.* mild mosaic, severe mosaic and leaf roll, but this approach is currently under review.

In compatible potato/PVY interactions (Chap. 2), PVY will normally infect all parts of a plant including tubers in which the virus can survive until the next growing season. Infection by some PVY strains can cause superficial, necrotic ringspots on tubers (Fig. 5.4). This disease is called potato tuber necrotic ringspot disease (PTNRD). PTNRD is characterised by the formation of brown, circular, necrotic rings on the tuber skin (Fig. 5.4A–F). These are often barely visible at harvest, but develop during storage. Initially, PTNRD symptoms develop mainly at the heel end or near eyes (Fig. 5.4A–F). After storage for two to three months, PTNRD symptoms can cover most of the tuber surface (Fig. 5.4G–K). As symptoms develop, the necrotic rings become sunken as a result of tissue collapse (Fig. 5.4J). Occasional cracks (Fig. 5.4N) can also develop and these can provide an entry point for infection by other pathogens such as fungi and bacteria. In most cases, PTNRD symptoms are associated with infection by isolates of PVY^{NTN} strain. However, other PVY strains have been reported as being able to cause PTNRD, depending on cultivar and environmental conditions (Chap. 3, Browning et al. 2004). The presence of PTNRD alone is, therefore, not sufficient to identify the causal PVY strain.

2.2 *Biological Characterisation of PVY Using Indicator Plants*

Historically, the first biological characterisation of potato viruses in a laboratory was done on the basis of the type of symptoms induced in tobacco leaves inoculated mechanically or by aphid transmission. In this way, the viral nature of mosaic and crinkle diseases was discovered in the early 1930s (Smith 1931). It was the first time that PVY had been named and described. The use of indicator host plants *e.g. Chenopodium amaranticolor, Datura stramonium, Solanum dulcamara* and *Nicotiana tabacum* (tobacco) gradually became important for the diagnosis of suspected viral diseases (Fig. 5.5A–C). Based on the nature of symptoms induced in *C. amaranticolor* and *N. tabacum,* PVY isolates can be separated into two biological groups. PVY isolates that induce local lesions on inoculated leaves of *C. amaranticolor* and mosaic on tobacco leaves are classified as being biotype-O of PVY (PVY^{O} strain). PVY isolates that do not elicit disease symptoms on *C. amaranticolor* but cause veinal necrosis on tobacco leaves are classified as being biotype-N of PVY (PVY^{N} strain). PVY^{N} and PVY^{O} strains can also be differentiated using the so-called A6 leaf test, in which detached leaves of the hybrid A6 (*S. demissum* x *S. tuberosum* 'Aquila') are mechanically inoculated with a test sample and the time of

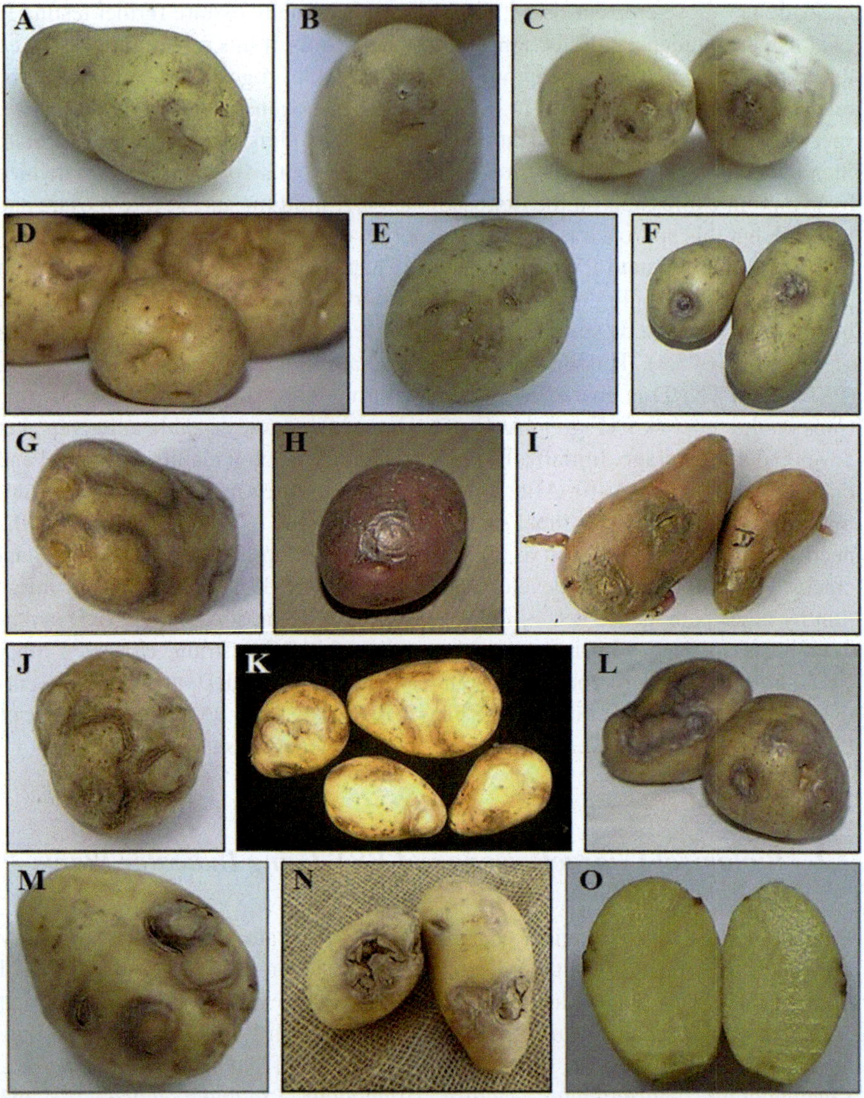

Fig. 5.4 Symptoms of potato tuber necrotic ringspot disease (PTNRD) caused by PVY^{NTN} on tubers of various potato cultivars at harvest or after storage. (**a**)–(**c**)–(**d**)–(**f**)–(**h**)–(**i**)–(**l**)–(**m**): breeding lines; (**b**): cv. Mistral; (**e**): cv. Bea; (**g**): cv. Kaptah-Vandel; (**j**): cv. Lola; (**k**)–(**m**)–(**o**): cv. Monalisa; (**n**): cv. Annabelle. Additional examples of PTNRD symptoms are shown in Chaps. 1 and 3 (Photos are the courtesy of: © GNIS/K. Charlet-Ramage (**a**)–(**l**); © FN3PT-RD3PT/L. Glais (**m**)–(**o**))

expression of local necrosis recorded (de Bokx 1964). Local lesions develop 4 days after inoculation if the isolate is PVY^{O}, whereas 6 days are required if it is a PVY^{N} isolate. This test has been routinely used for a long time in some European countries for PVY diagnosis because it was easy to use, reliable and provided results in a reasonable time (Weidemann 1988; Le Hingrat, personal communication).

Fig. 5.5 Symptoms on indicator plants inoculated with PVY. (**a**) local lesions on leaves of *Chenopodium amaranticolor* inoculated with PVY^O; (**b**) veinal necrosis and deformation of tobacco (*Nicotiana tabacum* cv. Xanthi) leaves infected by a PVY isolate of biotype-N; (**c**) mosaic symptom without deformation of tobacco leaves infected by a PVY isolate of biotype-O; (**d**) severe mosaic symptom in an un-inoculated potato leaf, mainly due to PVY^O; (**e**) veinal necrosis in an un-inoculated potato leaf caused by PVY^O; (**f**) hypersensitive response (HR) resulting in complete necrosis of the apical leaves and stem of a potato plant leading to its death; (**g**) HR resulting in necrosis and dropping of lower and intermediate leaves (leaf-drop streak); (**h/i**) local necrotic spots or green necrotic ringspots (called Maritta symptom) on inoculated potato leaves revealing an HR reaction. Depending on the resistance gene carried by a potato cultivar, symptoms in F, G, H and I could be obtained following infection by PVY^N or PVY^O (Photos: © INRA/C. Kerlan (**a**); © FN3PT/RD3PT/L. Glais (**b**–**i**). (Additional examples of PVY symptoms can be found in Chap. 3)

Nevertheless, the efficiency of these two biological assays for PVY diagnosis declined at the beginning of 1970s when new strains of PVY emerged and caused atypical symptoms on these and other host plants (Horvath 1967; Thompson et al. 1987; McDonald and Singh 1996). The A6 test is no longer used for PVY diagnosis.

As stated earlier, PVY can elicit different types of disease symptoms on the foliage and tubers of potato plants (Fig. 5.4). The severity of symptoms is affected by PVY isolate, virus titre in a plant, the type of infection (*i.e.* primary or secondary) and cultivar (see Chap. 2) (Smith and Dennis 1940; Bawden and Kassanis 1947; Kahn and Monroe 1963; Cockerham 1970; reviewed in Karasev and Gray 2013a). In controlled conditions (*e.g.* glasshouses), the interaction of PVY with a range of potato cultivars carrying various resistance genes to PVY can be used to characterise strains based on whether a hypersensitive response (HR) is induced by a virus after inoculation (Table 5.1). Local, well-defined, necrotic lesions on inoculated leaves are typical of HR (Fig. 5.5). As described in Chap. 2, HR is characterised as localised cell death that is associated with plant resistance to infection by a pathogen. In some cases, this necrosis can extend to other parts of the plant, leading to the dropping of basal and middle leaves and, ultimately, to top necrosis. On non-inoculated leaves, systemic veinal necrosis may also be observed (Fig. 5.5).

Singh et al. (2008) have proposed that strains of PVY should be classified on the basis of the type of symptoms elicited on tobacco plants and the development of HR on potato cultivars carrying the resistance genes Ny_{tbr}, Nc_{tbr}, Nz_{tbr}, *Ny-1* and *Ny-2*. Several strains have been defined: PVY^{O}, PVY^{C}, PVY^{N}, PVY^{Z}, PVY^{E} (Jones 1990; Kerlan et al. 1999; Rolland et al. 2008b; Singh et al. 2008; Szajko et al. 2008, 2012; Galvino-Costa et al. 2012; Karasev and Gray 2013a; Chikh Ali et al. 2014), $PVY^{N\text{-}Wi}$ and PVY^{NTN} (Beczner et al. 1984; Chrzanowska 1994; Nie and Singh 2002a, 2003) (Chap. 3). The ability to elicit PTNRD symptoms on tubers is a necessary condition for classifying a PVY isolate as PVY^{NTN} but not all isolates that cause PTNRD are PVY^{NTN}.

PTNRD assessment can be undertaken by mechanically inoculating the foliage of a potato plant and assessing the incidence and severity of PTNRD on daughter tubers. A disease index can then be calculated based on the proportion of progeny tubers displaying PTNRD and surface area affected by PTNRD lesions (Charlet-Ramage and Kerlan 2005). This test should be undertaken in a greenhouse using a PTNRD-susceptible potato cultivar *e.g.* Yukon Gold, Béa or Nadine as a reference (Le Romancer and Nedellec 1997). Depending on the aggressiveness of an isolate, symptoms may have developed by harvest, but more commonly they appear 1–2 months after storage at 4°C (Le Romancer and Nedellec 1997). As highlighted in Fig. 5.6, PTNRD development is strongly dependent on environmental conditions during plant growth and storage of tubers, the aggressiveness of a PVY isolate and the genetic background of the potato cultivar. While PTNRD-susceptible cultivars have been widely used for PTNRD assessment and PVY strain characterisation, it is also recommended that cultivars 'less susceptible' or 'resistant' to PTNRD are also included in any assessment (Le Romancer and Nedellec 1997) to determine more fully the pathogenicity of an isolate and thus to evaluate potential risks to the potato industry (see Chap. 3).

Although the use of indicator plants enables reliable identification of isolates of PVY, this methodology is cumbersome to implement because it requires greenhouse facilities for the production and maintenance of plants in conditions optimal for symptom expression. Incubation for several months is also necessary for the

Table 5.1 Biological reactions of leaves and tubers of the main indicator plants to infection by PVY strains

PVY strain group (typical isolate)												
			O	**C**	**N**	**Z**	**E**	**N-Wi**	**NTN**	**Other**		
Plant species	**Cultivar**	**Genotype**	(SASA-110)	(Adgen)	(Mont)	(L26)	(MON)	(SASA-207)	(NZ)	(AST)	**Reported by**	
Nicotiana tabacum	Xanthi, White Burley, Samsun		Mo	Mo	VN	Mo	Mo	VN	VN	VN	Chikh Ali et al. (2008), Singh et al. (2008) and Barker et al. (2009)	**On leaves**
Solanum tuberosum	Désirée, Pentland Crown, Alturas	Ny_{tbr}	HR	S	S	S	S	S	S	S	Jones (1990), Kerlan et al. (1999) and Rowley et al. (2015)	
	King Edward, Duke of York (Eersteling), Maris Piper,	Nc_{tbr}	S	HR	S	S	n.d.	S	S	S	Jones (1990), Kerlan et al. (1999) Chikh Ali et al. (2014) and Glais. (unpublished)	
	Pentland Ivory, Maris Bard, Yukon Gold,	Ny_{tbr}, Nc_{tbr}, Nz_{tbr}	HR	HR	S	HR	S	S	S	HR	Jones (1990), Valkonen (1997), Kerlan et al. (1999, 2011), Chikh Ali let al. (2014), Tomczynska et al. (2014) and Rowley et al. (2015)	
	Rywal (Romula)	*Ny-1* (*Ny-2*)	HR	n.d.	HR	n.d.	n.d.	HR	HR	HR	Szajko et al. (2008, 2012, 2014)	
	Béa		o	o	PTNRD*	PTNRD	PTNRD	o	PTNRD	PTNRD	Kerlan and Tribodet (1996), Karasev and Gray (2013) and Glais et al. (2015)	**On tubers**
	Nadine		PTNRD*	o	PTNRD*	n.d.	n.d.	PTNRD*	PTNRD	PTNRD	Browning et al. (2004) and Barker et al. (2009)	
	Yukon Gold		PTNRD*	n.d.	o	PTNRD	PTNRD	PTNRD*	PTNRD	PTNRD	Piche et al. (2004), Hu et al. (2009) Gray et al. (2010) and Galvino-Costa et al. (2012)	

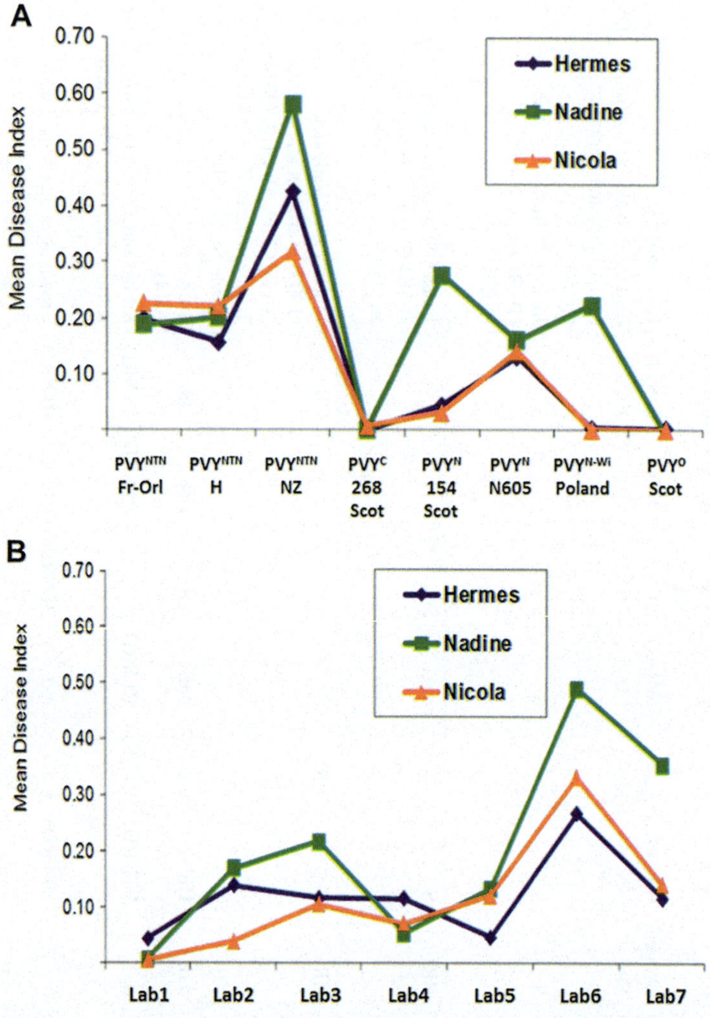

Fig. 5.6 Severity of potato necrotic ringspot disease, expressed as Mean Disease Index (an arbitrary unit), on three cultivars Hermes, Nadine and Nicola; (**a**) in response to infection by eight PVY isolates of various strains (PVY^{O}, PVY^{C}, PVY^{N}, $PVY^{N\text{-}Wi}$, PVY^{NTN}) and (**b**) in laboratories in seven countries (Czech Republic, France, Greece, Hungary, The Netherlands, Poland and the UK) (Adapted from Browning et al. 2004)

expression of symptoms on the leaf or the tuber. This method remains the only tool for classifying PVY isolates into strains but is not suitable for high-throughput detection of PVY.

Detection and characterisation of PVY isolates are generally achieved using serological methods (enzyme-linked immunosorbent assay (ELISA)-based typing using a range of antibodies) and genome sequencing (partial or complete) as described below.

3 Serological Typing

Serological techniques are among the most convenient means of identifying and characterising plant viruses. They involve an interaction between two proteins: the antigen which is a viral coat protein and the antibody which is produced by an animal (rabbit, mouse, goat, guinea pig, etc.) following injection of the viral antigen. PVY is composed of a nucleic acid (RNA) coated by a viral capsid/coat protein (CP) exhibiting immunological properties, which will elicit the host animal's immune system and result in the production of antibodies specific to the capsid protein.

3.1 *Immunoprecipitin*

Soon after the discovery of the antigenic properties of virus particles, some antisera raised against PVY were developed in various institutes throughout the world and used as a serological tool for the detection of PVY by the precipitin reaction in liquid medium or in a double diffusion assay in gels (Bawden and Kassanis 1951; Horvath 1965; de Bokx et al. 1975). The precipitin reaction exploits the fact that virus particles are composed of some subunits of capsid proteins which catch many antibodies resulting in the formation of a visible precipitate. The principle is the same for the double diffusion test except that antigens and antibodies initially separate in different wells of an agarose gel, diffuse towards each other and form a visible line of precipitin if both reactants are related (Fig. 5.7).

These methods were widely used in phytodiagnosis because of their simplicity and speed compared with biological tests, despite requiring a high concentration of virus particles in any infected test tissue. For instance, in France until 1985, the immunoprecipitin method on a glass slide was routinely used for quality control of health of seed potatoes as a complement to the A6 assay (Le Hingrat, personal communication). However, although differentiation between virus species was efficient with this method, it did not enable 'necrotic' and ordinary' strains of PVY to be differentiated (Bawden and Kassanis 1951; Horvath 1965; de Bokx et al. 1975;

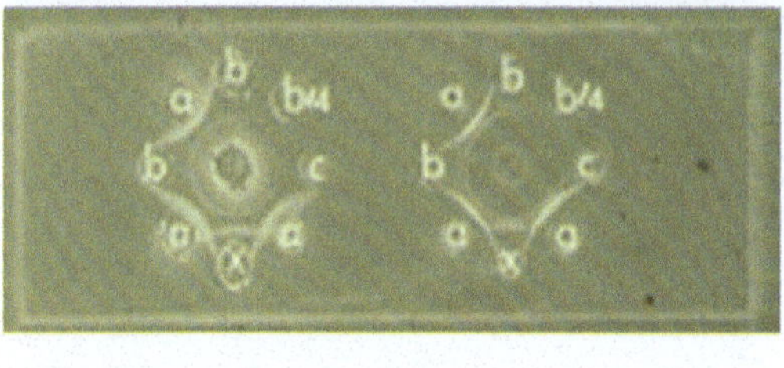

Fig. 5.7 Reaction of double diffusion in agarose gel. The central well contains a specific antibody; peripheral wells contain different infected or non-infected plant saps. The visible precipitin line in front of the 'a' well indicates the presence of a specific antigen (virus) (Photo: © INRA/C. Kerlan)

Rose et al. 1987). This method was later replaced with an innovative and more reliable serological diagnostic method named ELISA.

3.2 Enzyme-Linked Immunosorbent Assay (ELISA)

A 'revolution' in serological detection took place at the end of the 1970s with two technological breakthroughs. In 1975, Köhler and Milstein described the production of monoclonal antibodies by the fusion of 'immortal' mouse myeloma cancer cells and mouse spleen cells infected by an antigen. In 1977, Clark and Adams developed a method for the enzyme labelling of antibodies and their absorption on a solid surface, and named the new diagnostic technique ELISA (enzyme-linked immunosorbent assay). This test is a thousand times more sensitive than the immune-diffusion test, much easier to use, rapid, adapts to large-scale testing, and allows automation and quantification of the target antigen (Clark and Adams 1977). These two scientific advances allowed serological typing to become a powerful tool for PVY detection. Different ELISA formats have been developed, but the most widely used is DAS-ELISA (double antibody sandwich ELISA) which is based on the visualisation of the antigen–antibody interaction through a colorimetric reaction mediated by an enzyme, such as alkaline phosphatase, conjugated to antibody (Fig. 5.8). Briefly, a virus present in a sample is trapped on the surface of the wells of a

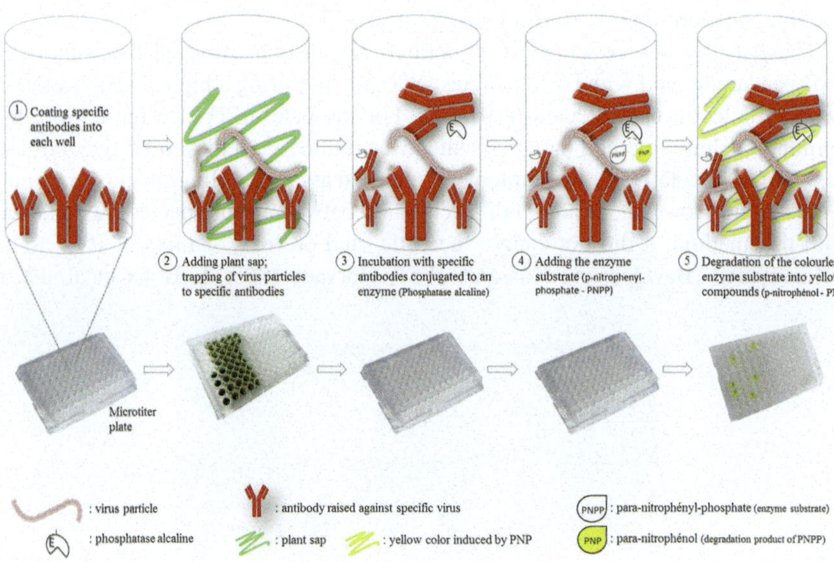

Fig. 5.8 Schematic representation of the different ELISA steps

Table 5.2 Characteristics and origin of various monoclonal antibodies used worldwide for the detection and identification of PVY

Company	Country	Acronym	Specificity	Source
Agdia	USA	4C3	PVY	Ellis et al. (1996)
		1F5	PVY^N	
		Mab2	$PVY^{O/C}$	McDonald and Kristjansson (1993)
Bioreba	Switzerland	PVY-mono	PVY	Gugerli and Fries (1983)
		Bioreba-N	PVY^N	
INGENASA	Spain	13E3	PVY	Sanz et al. (1990)
		1E10	$PVY^{O/C}$	
Neogen Europe Ltd	UK	Neogen-N	PVY^N	Nikolaeva et al. (2012)
		Neogen-O	$PVY^{O/C}$	unpublished
FN3PT (French federation of seed potato growers)	France	Y123	PVY	Kerlan et al. (1987)
		05YN312	PVY^N	Glais (unpublished)
NCSS (National Center for Seed and Seedlings)	Japan	PVY-T	PVY^N	Ohshima et al. (1990)
SASA (Science and Advice for Scottish Agriculture)	UK	SASA-N	PVY^N	Barker et al. (2009)
		SASA-O/C	$PVY^{O/C}$	

microtiter plate previously coated with a specific antibody raised against the virus. The trapped virus is caught in a sandwich by providing the same specific antibody labelled with an enzyme for detection. After washing, enzymatic activity is revealed by the conversion of a suitable colourless enzyme substrate into colour compounds. The optical density is measured with a spectrophotometer (at 405–410 nm) and may reflect the quantity of virus particles present in a sample.

After the technology to produce monoclonal antibodies (MAbs) became available, many laboratories worldwide engaged in a race to develop their own monoclonal antibodies against all PVY isolates, or those specific to PVY^N and/or PVY^O strains: Switzerland (Gugerli and Fries 1983), Scotland, the UK (Rose and Hubbard 1986), France (Kerlan et al. 1987; Ounouna et al. 2002; Glais, unpublished), Japan (Ohshima et al. 1990), Spain (Sanz et al. 1990) and the USA (McDonald and Kristjansson 1993; Ellis et al. 1996). Many of these MAbs became commercially available and have been widely used (Agdia-USA, Bioreba-Switzerland, IGENASA-Spain, Neogen-UK). Others have been produced by private or public laboratories (FN3PT-French Federation of Seed Potato Growers, SASA-Science and Advice for Scottish Agriculture, NCSS-National Center for Seeds and Seedlings) and have been used mainly for scientific research or as part of testing in seed potato certification schemes (Table 5.2).

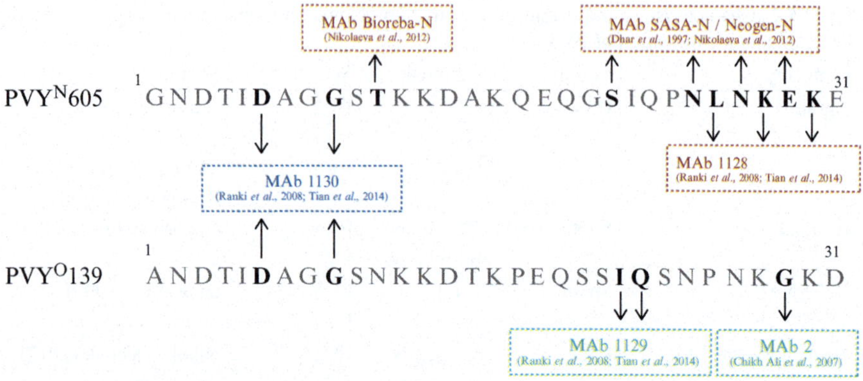

Fig. 5.9 Presentation of amino acids (in bold) in the PVY N-terminal region of the coat protein involved in the interactions with monoclonal antibodies raised against all PVY isolates (in *blue*), or specifically PVY^{N} (in *red*), or PVY^{O} isolates (in *green*). Peptide sequences were based on PVY^{N}605 (Jakab et al. 1997 – accession number X97895) and PVY^{O}139 (Singh and Singh 1996a – accession number U09509)

Some studies based on sequence analysis, alanine scanning and MAb reactivity to synthetic peptides, have identified some amino acids present in N- and C-terminal part of the CP protein involved in epitope structures that are recognised by some monoclonal antibodies specific to PVY^{N} or PVY^{O} isolates (Dhar and Singh 1997; Chikh Ali et al. 2007; Ranki et al. 2008; Nikolaeva et al. 2012; Tian et al. 2014) (Fig. 5.9). Consequently, monoclonal antibodies raised against the same O- or N-serotype population could recognise different binding domains.

However, although these serological tools provided a new dimension to diagnosis, there were some issues because serological typing was not always correlated with biological typing (Chrzanowska 1991). Serological testing specifically differentiates strains within O-serotype (PVY^{O}, PVY$^{N\text{-}Wi}$) or strains within N-serotype (PVY^{N}, PVY^{NTN}, PVY^{Z}, PVY^{E}). The single mutation of one amino acid in the coat protein could lead to an isolate no longer being recognised by a MAb (Dhar and Singh 1997) or being misidentified (Chikh Ali et al. 2007; Karasev et al. 2010) (see Chap. 3 for more details).

Both polyclonal and monoclonal antibodies have been used routinely in ELISA tests for the detection of PVY, particularly in certification schemes. This technique is still the most widely used for PVY diagnostic and detection because of its robustness, practicality, low cost for implementation and operation, and accessibility to every laboratory. To bypass the problem of the specificity of monoclonal antibodies being too great, a recommended procedure is to use a combination of MAbs in cocktail comprising antibodies of different specificities to increase the accuracy and inclusivity of the assay. As discussed in Chap. 3, full genome sequencing and phylogenetic analysis is the recommended method for the characterisation of PVY phylogenetic groups.

3.3 *Methods Derived from ELISA*

3.3.1 Luminex xMAP Technology

The Luminex technology was developed in 1997 (Fulton et al. 1997). First applied in the medical field for the screening of the human cytokines (Kellar and Douglass 2003), it was developed in 2007 for the detection of plant bacteria (Peters et al. 2007). The xMAP technology is a relatively novel method combining an interaction of antigen and antibody, as in ELISA, and a flow cytometer. It uses magnetic or polystyrene microspheres with different internal ratios for two fluorochromes, making them unique. Currently, there are 500 distinct bead sets commercially available. Each bead is coated with a MAb specific to an antigen, incubated with a test sample to which the secondary antibodies conjugated with a reporter fluorochrome are added. The beads in suspension are extracted, individually separated and analysed by the Luminex analyser with two light sources; the first excites the internal dyes of the microspheres and the second excites the reporter fluorochrome conjugated to the secondary antibodies (Fig. 5.10).

Each bead must be detected at least 100 times in order to achieve reliable results (Moalic et al. 2004). This technology has already been applied in the laboratory for the simultaneous detection of three potato viruses PVY, *Potato virus X* (PVX) and

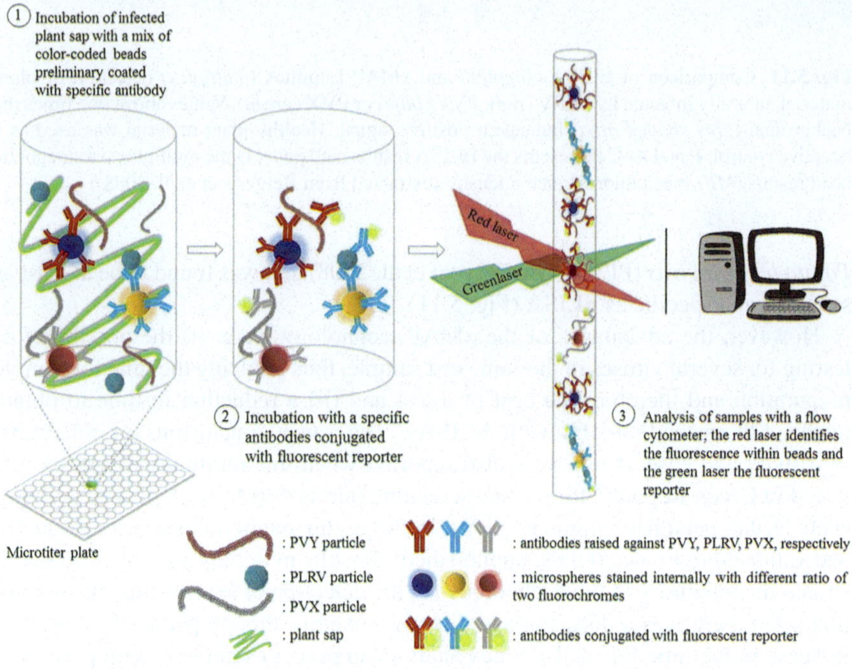

Fig. 5.10 Schematic representation of the method applied for the multiplex detection of *Potato virus Y* (PVY), *Potato leafroll virus* (PLRV) and *Potato virus X* (PVX) by Luminex technology

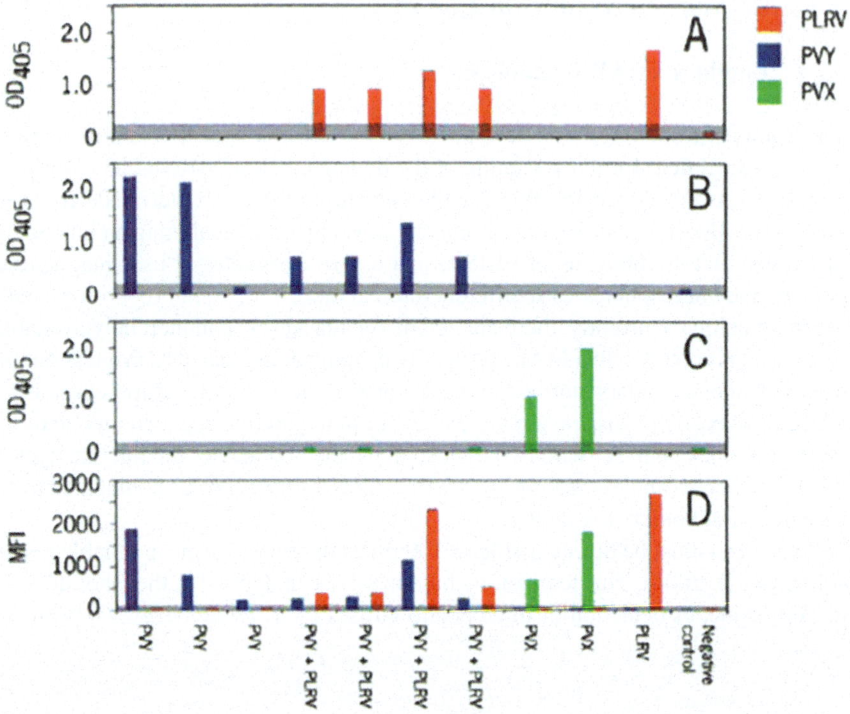

Fig. 5.11 Comparison of ELISA (*simplex*) and xMAP Luminex (*multiplex*) detection of plant material naturally infected by PLRV (*red*), PVY (*blue*) or PVX (*green*). Values above two times the background (*grey shaded area*) indicate a positive signal. Healthy plant material was used as a negative control. Panel A–C represents the ELISA results, and panel D the multiplex paramagnetic bead results (*MFI* mean fluorescence intensity (extracted from Bergervoet et al. 2008))

Potato leafroll virus (PLRV) (Bergervoet et al. 2008) and was found to be at least as sensitive and specific as ELISA (Fig. 5.11).

However, the advantages of the xMAP technology were: (i) the possibility of testing for several viruses in the same test sample, thus reducing the time for sample preparation and therefore the cost of a test and (ii) a reduction in time to obtain results (2h instead of 16h with ELISA). The same conclusions of reliability, sensitivity and ease of use were also reported when this multiplex diagnosis was tested in Czech Republic for routine screening during certification of seed potatoes (Petr Dedic, personal communication). In 2012, this methodology was also tested and validated in France (Glais, unpublished). Results in initial experiments demonstrated the reliability of this technology for the detection of several antigens in multiplex but there was a disadvantage directly linked to this property which was an increase in the time for the Luminex analyser to process samples, in proportion to the number of different beads present in the assay.

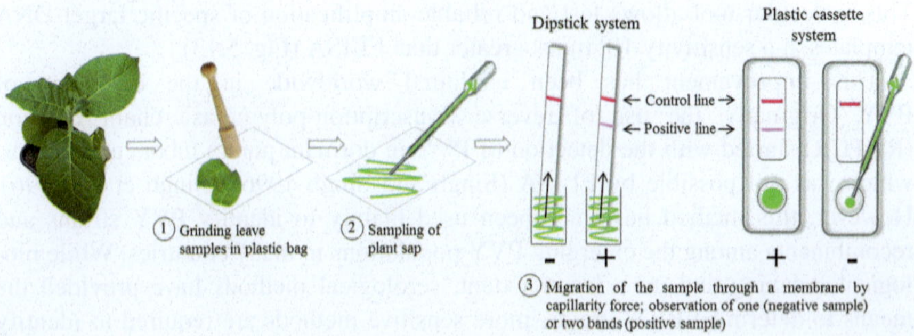

Fig. 5.12 Procedure for PVY detection in the field using lateral flow devices

3.3.2 The Lateral Flow Immunochromatography Technology

This technology uses specific monoclonal or polyclonal antibodies raised against an antigen immobilised on a membrane attached to two devices, either a dipstick or a plastic cassette. For the dipstick, an extremity of a strip is dipped into a sample of plant sap, whereas for the plastic cassette, some drops of sap are applied on to a specific area. The sap migrates up the membrane by capillarity to reach the area where antibodies are captured. If the antigen of interest is present in the sample, a reaction between antigen and antibody is initiated. This complex accumulates and forms a visible band indicating the presence of virus in the sample (Fig. 5.12). In addition to being easy to use, this technology provides results rapidly, in approximately 2 min.

In the UK and USA, this technology was used by seed potato inspectors, and even by growers, as a preliminary test to confirm visual diagnoses of suspect plants. However, in the UK, the seed potato certification authorities require a confirmatory diagnosis of virus-infected plants by an ELISA test of symptomatic leaves. In 2010, a study to assess the reliability of diagnostic field kits was performed by a research team in the USA (Olsen et al. 2011). For 115 samples with or without foliar symptoms of PVY infection, there was a 94–97% correlation of results using this method, depending on the kit used, with those obtained with conventional ELISA and RT-PCR methods. The discrepancies were mainly false negative results. This data demonstrates that lateral flow immunochromatography technology can be reliable but also emphasises the weakness of these kits when viral concentrations in the plant sap are low.

4 Molecular Typing

4.1 Reverse Transcription-Polymerase Chain Reaction

As with the ELISA method, an important milestone occurred in nucleic acid detection when polymerase chain reaction (PCR) technology was developed in the 1980s (Saiki et al. 1988), enabling genetic information on microorganisms to be accessed.

This molecular tool allows fast and reliable amplification of specific target DNA templates at a sensitivity 10^3 times greater than ELISA (Fig. 5.13).

This improvement has been exploited worldwide in the diagnosis of PVY. Originally, the use of reverse transcription-polymerase chain reaction (RT-PCR) started with the detection of PVY in dormant potato tubers and aphids, which was not possible by ELISA (Singh and Singh 1996b; Singh et al. 1996). However, this method has since been used mainly to identify PVY strains and recombinants among the emerging PVY populations in many countries. While biological indexing and, to a lesser extent, serological methods have provided the means to determine PVY strains, more sensitive methods are required to identify differences within strains and more complex recombinant variants of PVY. For example PVY^{NTN} cannot be differentiated from PVY^{N} using MAbs or tobacco indicator plants because they share the same serotype and phenotypic expression in tobacco. Even the ability of an isolate to induce PTNRD was found to vary because all PVY^{N} isolates are able to induce PTNRD symptoms in optimal conditions (Kerlan and Tribodet 1996). Some PVY^{NTN} isolates which cause PTNRD in potato crops in the field may fail to cause the disease under greenhouse conditions (Boonham et al. 2002) and vice versa (Chap. 3). In addition, other PVY strains may induce symptoms of PTNRD of varying severities in susceptible potato cultivars (Piche et al. 2004; Gray et al. 2010). It has been reported that one isolate (EU-12Jp) displaying all the genetic features of PVY^{NTN} do not appear to elicit PTNRD (Chikh Ali et al. 2013) (Chap. 3), which would require a more complete characterisation of reference cultivars prior to assignment to a given strain group. Isolates of the $PVY^{N\text{-}Wi}$ strain have the same serotype as PVY^{O} strain and elicit similar veinal necrosis symptoms on tobacco as PVY^{N}, which implies that they cannot be identified using serological and/or biological methods alone.

Due to the difficulties in differentiating variants within PVY^{NTN} strain using biological and serological methods and the need for a rapid identification method, several RT-PCR assays have been developed to identify and discriminate PVY^{NTN} from other PVY strains. Initially, RT-PCR assays relied on targeting the nucleotide polymorphism of certain parts of the genome, such as the P1 cistron (Rosner and Maslenin 1999, 2001; Weilguny and Singh 1998; Nie and Singh 2002a, b), or even a single nucleotide polymorphism in the CP region (Moravec et al. 2003). Later it was realised that these methods were reliable only for a specific set of isolates and were inconsistent when tested on a wider range of PVY^{NTN} isolates.

Subsequent significant improvement of PVY strain identification was driven by two main factors: firstly, the recent exponential increase of PVY sequences available in public DNA databases which allowed the design of specific and reliable primers; secondly, the development and utilisation of many sequence analysis programmes which led to the discovery of the recombinant nature of PVY^{NTN} and $PVY^{N\text{-}Wi}$ genomes (Glais et al. 1998, 2002), as well as the vast majority of newly characterised PVY recombinant isolates found in potato crops in many countries *e.g.* $PVY^{N\text{-}Wi}$, PVY^{Z}, PVY^{E} and PVY^{NTN} (Lorenzen et al. 2006, 2008; Ogawa et al. 2008; Hu et al. 2009; Chikh Ali et al. 2010; Kerlan et al. 2011; Galvino-Costa et al. 2012) (see Chap. 3). These variants display a huge range of recombination patterns

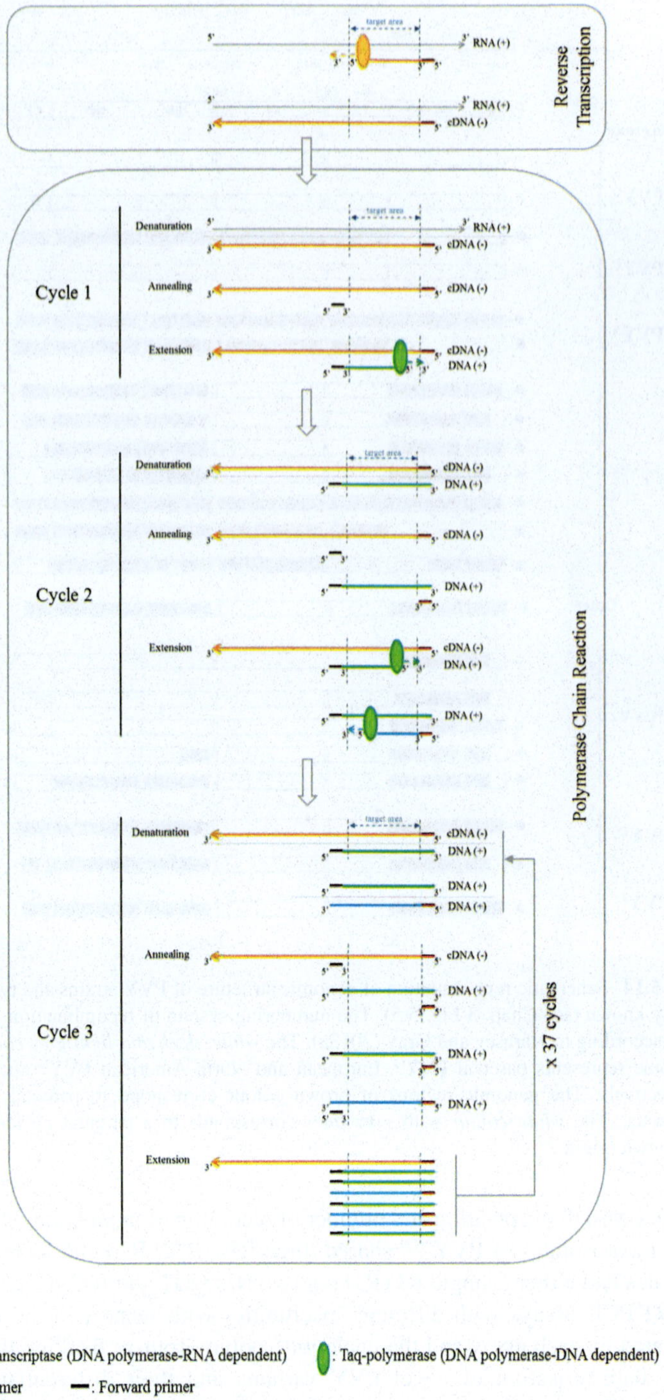

Fig. 5.13 Schematic representation of the RT-PCR method

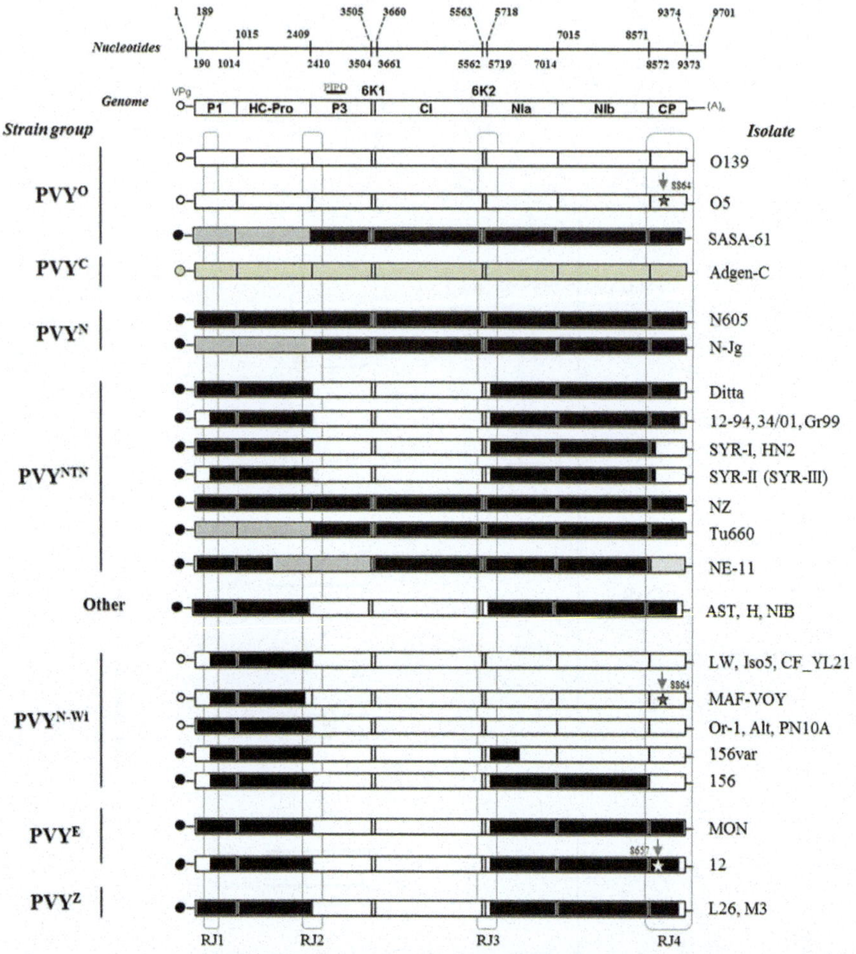

Fig. 5.14 Schematic representation of genomic structure of PVY strains and typical isolates currently known (see Chap. 3 Fig. 5.5). The numbering system of recombination junction (RJ) 1–4 was according to Karasev and Gray (2013a). The *white, dark and light grey colour* code in PVY genome represents parental PVY^{O}, European and North American PVY^{N} nucleotide sequence, respectively. The genomic regions of brown colour correspond to parental PVY^{C} nucleotide sequence. The *white colour* with *grey lines* corresponds to a parental nucleotide sequence of unknown origin

with regard to the position and number of recombination junctions (RJs). For example, most isolates of PVY^{NTN} shared three RJs (RJ2, RJ3 and RJ4) while PVY$^{N\text{-}Wi}$ isolates had either a single RJ (RJ1) or two RJs (RJ1 and RJ2) (Fig. 5.14).

RT-PCR assays with different specificities with respect to the number of RJs targeted by each assay and the nucleotide polymorphism flanking these RJs, led to the characterisation of novel PVY variants and their differentiation into strain groups (Boonham et al. 2002, Nie and Singh 2003, Glais et al. 2005, Schubert et al. 2007, Rigotti and Gugerli (2007)).

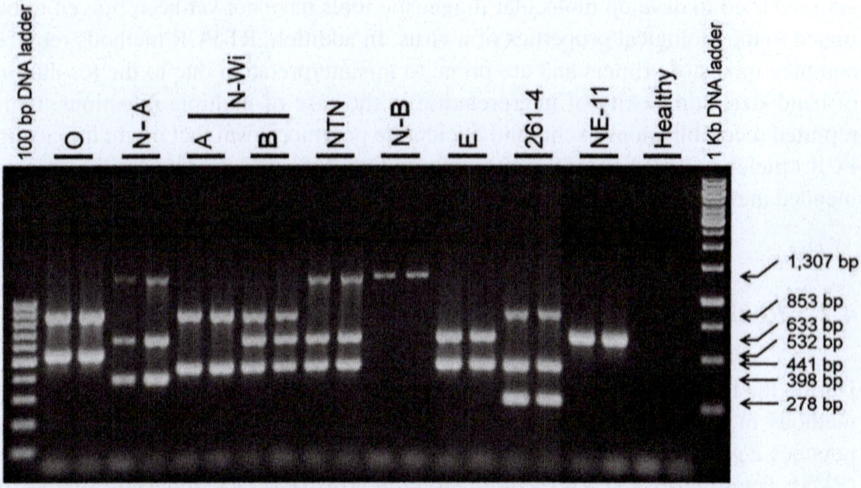

Fig. 5.15 Examples of the identification of reference isolates represent the main strains of PVY using immunocapture-reverse transcription-polymerase chain reaction (IC-RT-PCR) (Chikh Ali et al. 2013) in the potato cv. Russet Norkotah. The results for two reference isolates per strain are shown as labelled *above*. Amplicon's size is indicated with *arrows*.

The multiplex RT-PCR assay is often used to differentiate isolates within a broader range of PVY strains (Lorenzen et al. 2006; Chikh Ali et al. 2010, 2013). This assay targets RJ2 and RJ3 in the PVY genome, which allows the identification of several strains, but is limited in its ability to distinguish between variants of the same strain (Lorenzen et al. 2006). For example, $PVY^{N\text{-}Wi}A$, $PVY^{N\text{-}Wi}B$ and PVY isolate 261-4 could not be differentiated because they had the same amplification pattern (Fig. 5.15).

The multiplex RT-PCR assay was further designed to cover most of the genome and to target five RJs, *i.e.* RJ1, RJ1a, RJ2, RJ3 and RJ3a enabling differentiation among additional PVY strains and variants occurring alone or in combination (Chikh Ali et al. 2010, 2013). Modification of the RT-PCR assay using immunocapture (IC) to bypass the need for RNA extraction has increased its specificity (Karasev et al. 2010) and made it more suitable for large-scale testing (Chikh Ali et al. 2013). In the RT step, a mixture of oligo-dT and hexamer random primers was used to ensure the reverse transcription of all the genomic RNA, including degraded RNAs. In the PCR step, 12 primers were designed based on the most current PVY sequence and strain information and tested on a wide range of PVY isolates from several parts of the world, including the USA, Europe, Japan, Middle East and Brazil. This has enabled the IC-RT-PCR assay to differentiate the 13 PVY strains/recombinants shown in Fig. 5.10 (Chikh Ali et al. 2010, 2013).

While RT-PCR can be useful for studying strain diversity and the population structure of PVY in various potato production areas, it has its limitations. The main drawbacks are that current RT-PCR typing methods have been developed on a limited number of PVY isolates to define putative recombination events identified in a PVY genome, and most of these molecular particularities (mutations, recombinations) used to develop molecular diagnostic tools have not yet been proven to be linked to the biological properties of a virus. In addition, RT-PCR methods require complex mixes of primers and are prone to misinterpretation due to the resolution of band size, complexity of interpretation in the case of multiple infections, non-reported recombinations events and nucleotide polymorphism that might impact on PCR efficiency. Full-genome sequencing and phylogenetic studies are the recommended methods for the characterisation of PVY genomes (Chap. 3).

4.2 Real-Time RT-PCR

During the last decade, real-time RT-PCR has become one of the most widely used methods in pathogen diagnosis and, in particular, in detecting viruses infecting potatoes *e.g. Potato leaf roll virus* (PLRV), *Potato virus A* (PVA), *Potato virus S* (PVS), *Potato virus X* (PVX), *Potato mop top virus* (PMTV), *Tobacco rattle virus* (TRV) and *Tomato spotted wilt virus* (TSWV) in addition to PVY (Schoen et al. 1996; Mumford et al. 2000; Agindotan et al. 2007; Mortimer-Jones et al. 2009; Boonham et al. 2009). The principle of real-time RT-PCR is the detection of amplified fragments using fluorescent reporter dye. As described in the previous section, any PCR of (+)ssRNA viruses *e.g.* PVY requires an initial step of reverse transcription (RT) to convert RNA into cDNA (Fig. 5.13). The PCR assay is then performed using forward and reverse primers suitably designed to amplify a relatively short region of the genome, often between 50 and 200 nt in length (Fig. 5.16). The real-time Taqman® RT-PCR principle relies on the use of a so-called probe that anneals within the amplicon which carries a fluorescent dye at its 5'-end and a quencher at its 3'-end. At the initial stage, both the fluorescent dye and the quencher dye are in close vicinity but there is no fluorescence. During the PCR reaction, both primers and probes anneal to their respective DNA strand, and polymerisation is initiated by the *Taq* polymerase from each forward and reverse primers. Due to the 5'nuclease activity of the *Taq* polymerase, the Taqman probe cleaves during PCR and the fluorescent reporter dye is separated from the quencher dye, resulting in an increase in fluorescence of the reporter. This process occurs during each cycle and accumulation of the PCR product is detected directly by monitoring the increase in fluorescence (Fig. 5.16 upper and lower panels). A sample is generally considered positive when the amount of fluorescence emitted reaches a threshold at a given amplification cycle (Ct), usually below 40 cycles of amplification (Fig. 5.16 lower panel). The amount of viral template in the reaction can be assessed qualitatively and quantitatively.

Real-time RT-PCR has a far greater sensitivity than conventional end-point PCR methods (approx. 100-fold more sensitive) and ELISA (approx. 10,000-fold more sensitive) (Mumford et al. 2000), and is performed in a closed-tube, one-step reaction which does not require subsequent agarose gel electrophoresis for the detection of amplicons. These advantages have led to its increased adoption as the preferred method of diagnosis for PVY. Combinations of multiple fluorescent

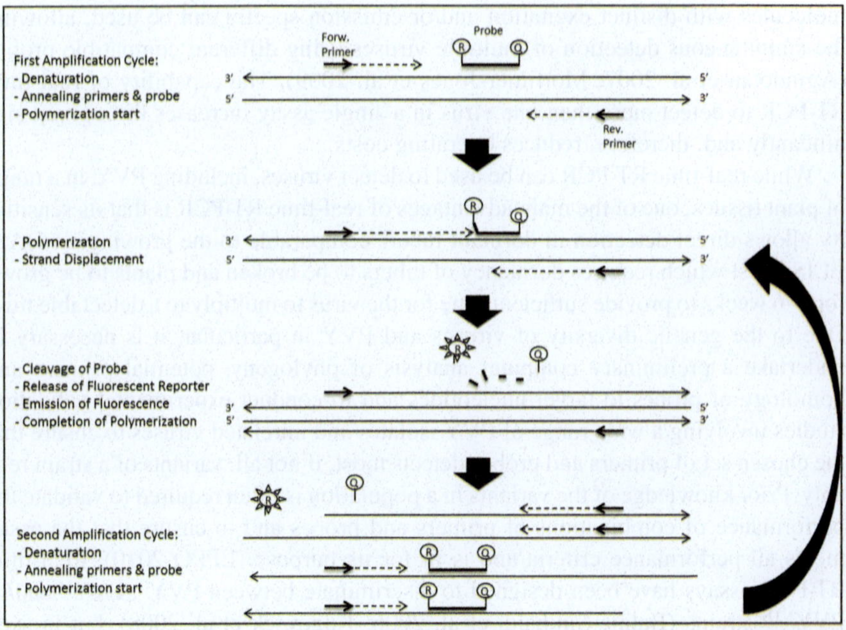

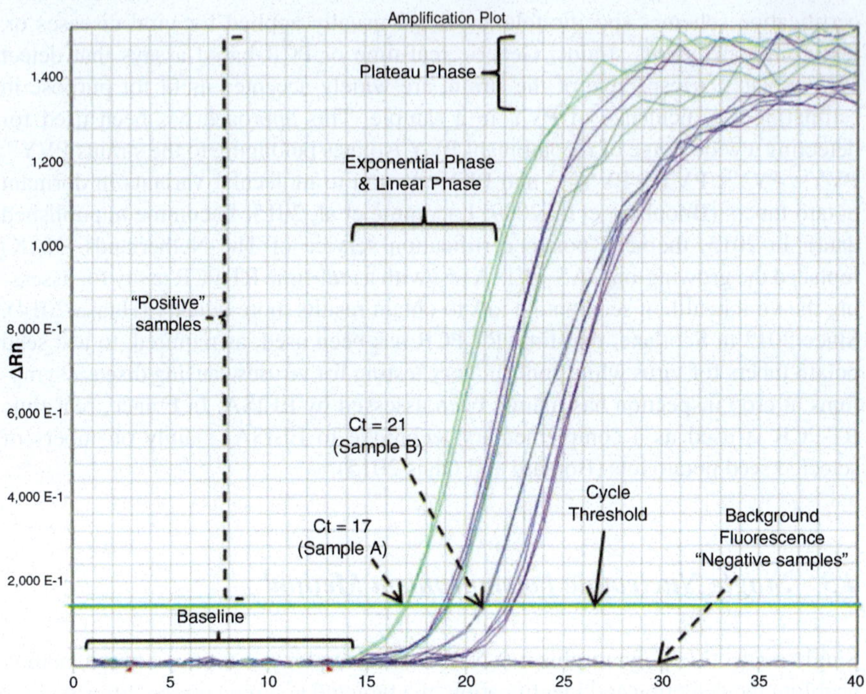

Fig. 5.16 *Upper panel*: Principle of real-time 'Taqman®' PCR with schematic representation of the main steps as described in this section. *R* fluorescent reporter dye, *Q* quencher dye (Adapted from Applied Biosystems, Taqman® Gene Expression Assays Protocol https://tools.thermofisher.com/content/sfs/manuals/cms_041280.pdf. *Lower panel*: typical amplification plots for several samples tested. Cycle threshold (Ct) values for two positive samples are given as example)

molecules with distinct excitation and/or emission spectra can be used, allowing the simultaneous detection of multiple viruses using different compatible probes (Agindotan et al. 2007; Mortimer-Jones et al. 2009). The capability of real-time RT-PCR to detect more than one virus in a single assay increases throughput significantly and, therefore, reduces operating costs.

While real-time RT-PCR can be used to detect viruses, including PVY, in a range of plant tissues, one of the main advantages of real-time RT-PCR is that its sensitivity allows direct detection in dormant tubers comparable to the growing-on DAS-ELISA test which requires dormancy of tubers to be broken and plants to be grown for 4–6 weeks to provide sufficient time for the virus to multiply to a detectable titre. Due to the genetic diversity of viruses and PVY in particular, it is necessary to undertake a preliminary computer analysis of phylogeny, potential primers and homology of probes to target nucleotides and to conduct experimental validation studies involving a wide range of PVY isolates and unrelated viruses to ensure that the chosen set of primers and probes detects most, if not all, variants of a strain reliably. Prior knowledge of the variants in a population is often required to validate the performance of combinations of primers and probes and to ensure that the assay meets all performance criteria and is fit for its purpose (EPPO 2010). Real-time RT-PCR assays have been designed to discriminate between PVY^{O}, PVY^{N} and/or PVY^{NTN} strains (Balme-Sinibaldi et al. 2006; Kogovsek et al. 2008), but in seed certification schemes specific tolerances are usually applied for viral diseases or, less commonly, virus strains. Generic real-time or PCR-based assays that detect PVY isolates, irrespective of the strain, are widely accepted as fit for purpose in estimating the incidence of PVY in a sample. This approach has been used for detecting a wide range of recombinant PVY isolates belonging to the strains PVY^{O}, PVY^{C}, PVY^{E}, PVY^{N}, PVY^{NTN} and $PVY^{N\text{-}Wi}$, and to molecular variants in dormant potato tubers (Boonham et al. 2009; Lacomme et al. 2015; Lacomme unpublished data). In 2015, the seed potato certification service of The Netherlands (NAK) replaced the growing-on DAS-ELISA test with a real-time RT-PCR assay for assessing the viral health of seed potato lots to obtain results more quickly (Hanse 2014). Since 2012 in Scotland, real-time RT-PCR has been used, as required, to test seed potato tubers for virus while confirmatory testing for viruses causing disease symptoms at crop inspection has mainly been assessed by ELISA. In France, real-time RT-PCR is used as a complementary technique to ELISA, mainly on tubers of selected seed potato lots (Boulard and Glais 2015).

4.3 Single Nucleotide Polymorphism Method

Single nucleotide polymorphism (SNaPshot) technology is a fluorescence method enabling the detection and identification of a targeted genomic region differing by a single polymorphic nucleotide in its sequence. This method initially requires PCR amplification with a primer pair surrounding the targeted genomic region where the polymorphic nucleotide is present. After a gel filtration step, a fraction of the

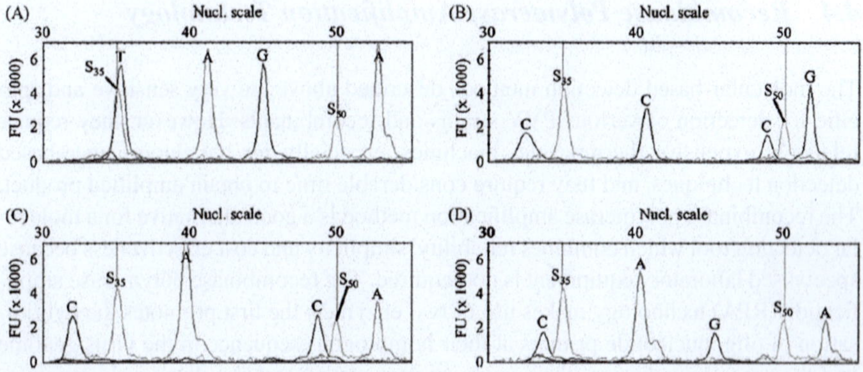

Fig. 5.17 SNaPshot electropherograms obtained for (**a**) PVY^{N}-605, (**b**) PVY^{O}-139, (**c**) PVY^{N}-Wi-P and (**d**) PVY^{NTN}-FrOrl isolates. The nucleotide scale, calculated according to the migration time of the labelled nucleic acid standards (S35 and S50), is denoted. The fluorescence and the calculated length associated with peaks enabled the four-digit code ('TAGA', 'CCCG', 'CACA', 'CAGA') specific to each PVY group to be identified. *FU* fluorescence unit, *Nucl. scale* nucleotide scale (Extracted from Rolland et al. 2008a)

completed PCR reaction is used for a single nucleotide primer extension reaction in the presence of fluorescent-labelled ddNTPs and one or several primers of different lengths, annealing with the specific targeted region immediately upstream of the polymorphic nucleotides of interest. The extension products are size-separated by capillary electrophoresis, allowing them to be identified. The type of the extended nucleotide can then be easily scored because the fluorescence present at the 3'-end of each extended primer is dNTP specific.

This method for PVY typing (Rolland et al. 2008a) is based on the two first nucleotides in the HC-Pro region (A/G_{2213}, A/C_{2271}) that are involved in the production necrosis in tobacco (Tribodet et al. 2005). A multiple assay was developed to identify the PVY^{N}, PVY^{O}, PVY$^{N\text{-}Wi}$ and PVY^{NTN} strains based on the other two single nucleotide polymorphisms (SNPs) positioned in the capsid protein (G/C_{8573}, T/C_{9253}) on either side of the recombination zone reported for the main variants of PVY^{NTN} strain. Four primers of different lengths were designed immediately upstream of these SNPs and used for the extension reaction. At the end of the analysis, a specific code of four letters corresponding to the polymorphism of nucleotides 9259, 2271, 8573, and 2213, was obtained for each PVY strain: 'TAGA' (PVY^{N}), 'CCCG' (PVY^{O}), 'CACA' (PVY$^{N\text{-}Wi}$) and 'CAGA' (PVY^{NTN}) (Fig. 5.17).

The reliability of the SNaPshot method for detecting and discriminating PVY strains and variants was assessed on PVY-infected leaf samples collected from Slovenian potato crops (Rupar et al. 2013). SNaPshot was effective in detecting all PVY isolates which produced necrotic symptoms on tobacco leaves, but was less reliable in assigning recombinant PTNRD-inducing isolates to the PVY^{NTN} strain because consensual PTNRD genetic determinants have yet to be identified. Identifying the genomic determinants of biological traits of PVY such as those responsible for PTNRD using reverse genetic approaches remains an outstanding challenge.

4.4 Recombinase Polymerase Amplification Technology

The molecular-based detection methods described above are very sensitive and specific for detection of various PVY strains and recombinants. However, they require relatively expensive thermocycler machines, especially for the fluorescence-based detection techniques, and may require considerable time to obtain amplified product. The recombinase polymerase amplification method is a good alternative for a molecular detection tool which combines reliability, simplicity and cost effectiveness because specialised laboratory equipment is not required. The recombinase polymerase amplification (RPA) technology makes use of two enzymes: the first promotes the hybridisation of oligonucleotide primers at their homologous sequence in the virus genome and the second synthesises the complementary strand and simultaneously separates the double-stranded DNA molecules. RPA reaction has the advantage that it operates at 37°C for 30–60 min compared with the usual PCR method in which 40 cycles at three different temperatures are usually required for denaturation, hybridisation and elongation. The analysis of the RPA products at the end of the amplification step can be performed by agarose gel electrophoresis or in real time using a fluorescent probe.

This isothermal amplification method has been applied to set up generic and specific RPA tests for discriminating between PVY^{N} and PVY^{O} strains (Glais and Jacquot 2015). To perform these RPA assays, primer pairs were designed in the C-terminal part of HC-Pro region, either surrounding the single nucleotides involved in the PVY necrotic reaction in tobacco (Tribodet et al. 2005; Hu et al. 2009; Faurez et al. 2012) or precisely located on the polymorphic nucleotide A/C_{2271} in HC-Pro cistron where the nucleotide A is specific for PVY^{N} isolates and C for PVY^{O} (Tribodet et al. 2005). Both RPA and ELISA assays produced similar results for strain identification of 12 PVY isolates from different geographical origins.

The efficiency of RPA in identifying and characterising PVY isolates is comparable to the ELISA assay because it differentiates between PVY^{N} and PVY^{O}. This differentiation is currently based on the polymorphism of a viral molecular determinant associated with the necrotic properties of PVY^{N} isolates, so that, unlike with ELISA, isolates of $PVY^{N\text{-}Wi}$ strain can potentially be identified as necrotic with respect to their biological properties. Further development in the mapping of other genetic determinants potentially associated with necrosis induced by different PVY strains will help to refine this approach.

4.5 Loop-Mediated Isothermal Amplification Technology

Loop-mediated isothermal amplification (LAMP) is another relatively novel molecular method allowing a high rate of DNA amplification in a short time with very high specificity and sensitivity without requiring variations in temperature during the process (Notomi et al. 2000). The LAMP method relies on auto-cycling strand displacement DNA synthesis which is conducted at 65°C for less than 60 min in the presence of *Bst* DNA polymerase and a set of four primers (two outers and two inners) annealing specifically to six different regions surrounding the genomic target region. The reaction of LAMP is divided into two steps. The first is relatively short (a few minutes at the beginning) and involves all four primers in the synthesis of the first DNA strand with a double loop structure at both ends. In the second step, only the forward and backward inner primers (FIPs and BIPs) are required for the cycling amplification of the target region. The difference with this new biotechnology is that these two inner primers contain two distinct sequences. At the 3'end, they have a complementary sequence to a specific DNA region allowing, as well as the outer primers, initiation of the beginning of the amplification process and synthesis of the first single-strand DNA. At the 5' end, they contain the same sequence as another specific region present on the DNA template, located just upstream of the target area, which allows the formation of a double loop structure at both ends of the FIP- or BIP-released strand and promotes the self-primed DNA synthesis. The final products correspond to a mixture of stem-loop structures of various sizes corresponding to a succession of sense and antisense of the targeted sequence and 'zigzag'-like structures with multiple loops formed by annealing between alternately inverted repeats of the target sequence in the same strand (Fig. 5.18). Several methods have been

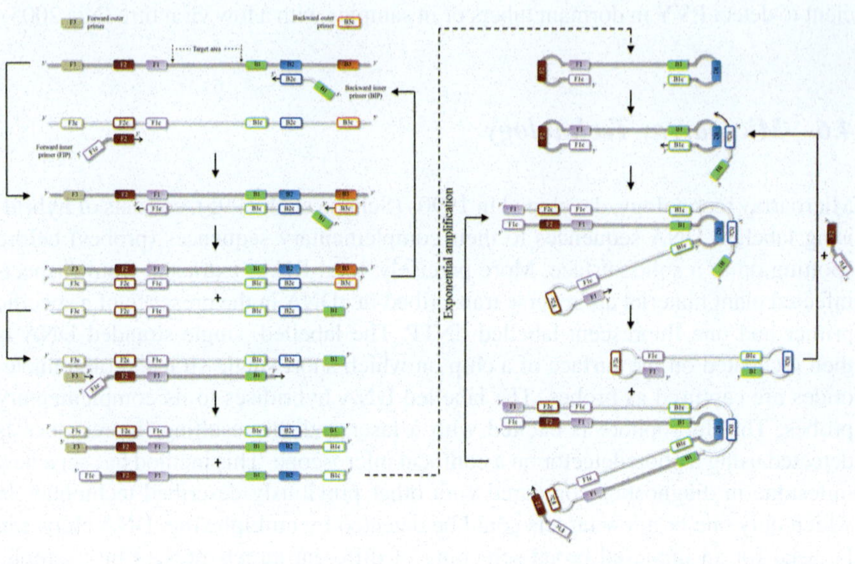

Fig. 5.18 Schematic representation of the process of LAMP (loop-mediated isothermal amplification) reaction (Adapted from New England Biolabs site (https://www.neb.com/applications/dna-amplification-and-pcr/isothermal-amplification). The LAMP procedure requires two steps. The first one leads to the generation of artificial stem loops. Subsequently annealing of BIP primer (or FIP primer) and backward outer primer (or forward outer primer) leads to the synthesis of complementary DNA strands and the releasing of the BIP strand (or FIP strand) which displays at both ends two complementary sequences that self-anneal and form a double loop structure which initiates the exponential amplification in the LAMP reaction. The second step requires FIPs or BIPs; these inner primers hybridise to their specific loop, initiate the synthesis of the complementary strand and release DNA with a new stem-loop structure which is used as a template in the next cycle)

developed for visualisation of the amplified product. A positive LAMP reaction can be detected by visual observation of (i) the turbidity of the reaction mixture caused by the formation of magnesium pyrophosphate in the course of the nucleic acid amplification, (ii) added dye (hydroxynaphthol blue) whose colour pattern changes depending on the presence or absence of the target (iii) an intercalating agent (ethidium bromide, SYBR Green) added in the tube at the end of the reaction and visible under UV light (Mori et al. 2001; Nie 2005; Almasi and Dehabadi 2013). The amplified products can also be checked by electrophoretic analysis of an agarose gel stained with an intercalating molecule, or by a real-time PCR using a FIP primer labelled with fluorophore (Nie 2005; Tanner et al. 2012). In conclusion, the sensitivity and practicality of the LAMP technology for revealing amplified products underline the potential of this method to be a powerful tool in diagnostic laboratories. In the past five years, more than 1000 references can be found on the application of LAMP method to diagnosis.

This isothermal amplification procedure has been successfully applied to the detection of PVY by using the coat protein gene as the target (Nie 2005; Almasi and Dehabadi 2013). A recent comparison of this method with DAS-ELISA and RT-PCR assays showed that LAMP was 100 times more sensitive and three times faster than RT-PCR (Almasi and Dehabadi 2013). However, this sensitivity might not be sufficient to detect PVY in dormant tubers or in samples with a low viral titre (Nie 2005).

4.6 Microarray Technology

Microarray technology, developed in 1990s (Schena et al. 1995), consists of hybridising labelled DNA sequences to their complementary sequences (probes) before spotting on to a solid surface. More precisely, total RNAs extracted from suspect-infected plant material are reverse transcribed in cDNA in the presence of a specific primer and one fluorescent-labelled dNTP. The labelled, single-stranded DNA is then incubated on the surface of a chip on which short single-stranded oligonucleotides are captured as probes. The labelled DNA hybridises to its complementary probes. The fluorophore is excited with a laser and the resulting fluorescence is detected using a photodetector on a confocal microscope. This method has set a new milestone in diagnostics compared with other previously described techniques in which only one or a few targets could be detected by multiplexing. DNA chips can be used for an unbiased broad screening of different targets present in a sample. Tens of thousands of probes can be spotted on one chip. Microarrays were initially used for quantitative gene expression analysis in research laboratories, but over the past 15 years, it has also been designed for the rapid detection of plant pathogens by generic probes (Wei et al. 2009). This method has been developed for the simultaneous detection of at least six major potato viruses: PVY, PLRV, PVX, PVA, PVM and PVS (Boonham et al. 2003; Bystricka et al. 2005; Sip et al. 2010). The effective detection and differentiation of all these viral strains were demonstrated with this technology, but it was no more sensitive than the ELISA assay (Boonham et al.

2003; Sip et al. 2010). Discrimination of viral strains within the same virus species can be achieved, provided the sequence identity between targets is less than 89% (Boonham et al. 2003), and the length of the homologous region within targets does not exceed 30% of the length of the probe (Wei et al. 2009). For these reasons, the DNA chip cannot discriminate between the closely related strains PVY^{O}, PVY^{N} and PVY^{NTN}. While having obvious advantages for the diagnosis of multiple virus species, this method can only be of limited value for high-throughput testing because of the precautions required to ensure the quality and the quantity of the extracted RNA, the cost of the chip and specialist equipment needed.

4.7 Next Generation Sequencing

In the last decade, the advent of high-throughput next generation sequencing (NGS) has made a great impact in nearly every field of life sciences, including plant pathology (Studholme et al. 2011). NGS enables sensitive non-targeted detection of any nucleic acid sequence in a sample. Due to the extensive genetic variability of viruses and consequentially a lack of possible marker genes compared with bacteria and eukaryotes, NGS currently represents the only truly generic molecular method for detecting viruses. Several NGS platforms exist at the moment, employing different solutions for the amplification and sequencing. Nevertheless, a similar workflow is used by most of them; after the initial extraction of nucleic acids from a test sample, sequencing libraries are prepared, nucleic acids are fragmented and specific adaptor sequences are added (ligated) to the short fragments. These enable amplification of the signal and sequencing.

Besides the sequencing itself, there are several other important steps in an NGS analysis workflow (Fig. 5.19). In theory, the sensitivity of NGS analysis is limited by the sequencing depth. Since viral sequences are present in a background of abundant host sequences, removal of host sequences or enrichment for viral sequences would increase the sensitivity of the test and is thus favoured in the sample preparation step. The strategies to achieve this can be different, but the most common are sRNA isolation and sequencing (Kreuze et al. 2009), depletion of ribosomal RNA, (partial) purification of virions and enrichment for double-stranded RNA (Wu et al. 2014). However, other approaches, such as total RNA sequencing (Kehoe and Jones 2016) and sequencing of RNA from purified virions (Kutnjak et al. 2015), have also been successfully employed for characterising PVY isolates.

Because of the large amount of data produced by any of the NGS platforms, efficient analysis pipelines (*i.e.* bioinformatics pipelines) are required to 'fish-out' the viral sequences by employing either de novo assembly approaches or mapping NGS outputs to a database of known viral sequences. Importantly, findings need to be confirmed by other independent techniques (*e.g.* ELISA, PCR, real-time RT-PCR) because of the risk of contamination from several possible sources in the process or other potential experimental artefacts. NGS has been used for the detection and genome characterisation of several new and known potyviruses (Barba et al. 2014), including PVY.

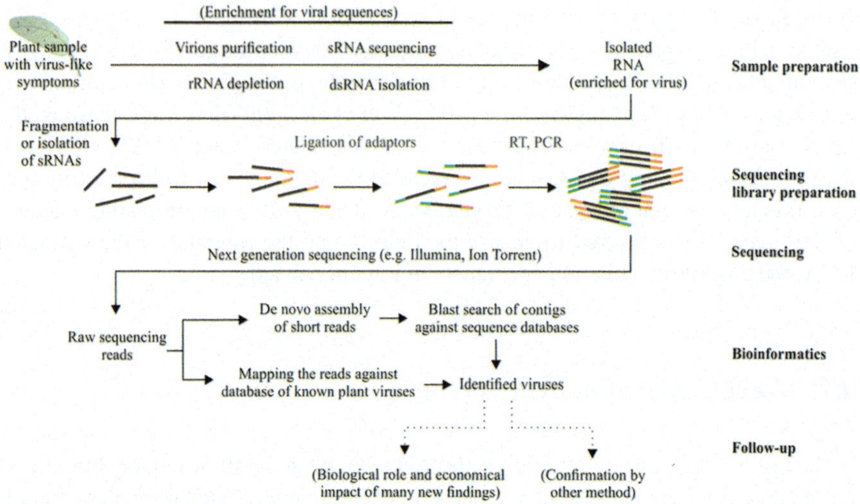

Fig. 5.19 Schematic representation of a possible NGS workflow for detecting plant viruses. The exemplary workflow shows possible steps in NGS-aided analysis of RNA from virus-infected plant, including sample preparation (enrichment for viral sequences), library preparation and sequencing (a simplified library preparation similar to sRNA or directional RNA sequencing approach is shown), bioinformatics and possible follow-up studies

The high-throughput capabilities of NGS also make it especially suitable for studying the diversity and evolution of virus populations (Fabre et al. 2012; Kutnjak et al. 2015). However, the relatively slow turnaround and high cost associated with its use in routine diagnostics limit its application, but it is suitable for studying samples of diseased plants affected by unknown pathogens/viruses, since, in these cases, the use of numerous diagnostic methods and associated staff would result in testing being more expensive than with NGS. NGS is likely to become widely used as the cost of reagents and time required for analysis decrease.

5 Conclusion

A wide range of techniques have been developed over a period of 50 years for the detection of PVY. Some of them are used mainly in research laboratories, and only a small number are used in diagnostic assays to ensure compliance with the sanitary and quality standards for seed potatoes. The criteria that favour the choice of a diagnostic technique are reliability, practicality, suitability for implementation on a large scale and low cost in terms of consumables and/or special equipment. As PVY is a regulated non-quarantine pest, there is currently a lack of formal acceptance of a method for testing for PVY. As a consequence, there are numerous reagents (*i.e.* PVY-specific serological reagents such as polyclonal and monoclonal antibodies) including PCR reagents (oligonuceotides, probes) that are often only used by a limited number of laboratories often without the full knowledge of their performance criteria. As a result, it can be somewhat difficult to agree on the most suitable/reliable technique for the detection of PVY.

While an ELISA test is often used initially to determine the composition of PVY populations by testing symptomatic leaves of growing crops or harvested tubers, partial or full-length sequencing of the PVY genome is the most reliable method to assess the genetic makeup of PVY variants. However, while RT-PCR is still widely used to attempt to characterise recombinant variants, it may not be sufficiently accurate in some cases because of the extensive genomic variability of PVY (see Chap. 3). In the absence of a clear association between genetic characteristics and symptoms elicited by PVY isolates on host plants, characterisation of PVY strains so far can only be determined by bioassays. While a bioassay can be affected by environmental conditions, it remains the only recognised method to assess the pathogenic status of PVY isolates and hence their risk to potato production.

The need to obtain results rapidly and to diagnose a large number of pathogens in a single sample has become increasingly important, making next generation sequencing (NGS) an attractive approach. Although NGS is essentially used in research laboratories for diagnosing microorganisms present in a plant, it is likely to become an indispensable platform for National Plant Protection Organisation and associated laboratories.

References

Agindotan BO, Shiel PJ, Berger PH (2007) Simultaneous detection of potato viruses, PLRV, PVA, PVX and PVY from dormant potato tubers by Taqman ® real-time RT-PCR. J Virol Methods 142:1–9

Almasi MA, Dehabadi SH (2013) Colorimetric immunocapture reverse transcription Loop-mediated isothermal AmPlification assay for rapid detection of the *Potato virus Y*. J Plant Pathol Microbiol 4:188–193

Balme-Sinibaldi V, Tribodet M, Croizat F, Lefeuvre P, Kerlan C, Jacquot E (2006) Improvement of *Potato virus Y* (PVY) detection and quantitation using PVY^N- and PVY^O-specific real-time RT-PCR assays. J Virol Methods 134:261–266

Barba M, Czosnek H, Hadidi A (2014) Historical perspective, development and applications of next-generation sequencing in plant virology. Viruses 6:106–136

Barker H, McGeachy KD, Toplak N, Gruden K, Zel J, Browning I (2009) Comparison of genome sequence of PVY isolates with biological properties. Am J Potato Res 86:227–238

Bawden FC, Kassanis B (1947) The behaviour of some naturally occurring strains of *Potato virus Y*. Ann Appl Biol 34:503–516

Bawden FC, Kassanis B (1951) Serologically related strains of *Potato virus Y* that are not mutually antagonistic in plants. Ann Appl Biol 38:402–410

Beczner L, Horvath J, Romhanyi I, Forster H (1984) Studies on the etiology of tuber necrotic ringspot disease in potato. Potato Res 27:339–352

Bergervoet JHW, Peters J, van Beckhoven J, van den Bovenkamp GW, Jacobson JW, van der Wolf JM (2008) Multiplex microsphere immuno-detection of *Potato virus Y*, *X* and PLRV. J Virol Methods 149(1):63–68

de Bokx JA (1964) Detection of potato virus YN by means of test plants. Doctoral thesis, Wageningen, 84pp
de Bokx JA, Kratchanova B, Maat DZ (1975) Some properties of a deviating strain of *Potato virus Y*. Potato Res 18:38–51
Boonham N, Walsh K, Preston S, North J, Smith P, Barker I (2002) The detection of tuber necrotic isolates of *Potato virus Y*, and the accurate discrimination of PVY^O, PVY^N and PVY^C strains using RT-PCR. J Virol Methods 102:103–112
Boonham N, Walsh K, Smith P, Madagan K, Graham I, Barker I (2003) Detection of potato viruses using microarray technology: towards a generic method for plant viral disease diagnosis. J Virol Methods 108:181–187
Boonham N, Laurenson L, Weekes R, Mumford R (2009) Direct detection of plant viruses in potato tubers using real-time PCR. In: Burns R (ed) Methods in molecular biology, Plant pathology techniques and protocols, vol 508. Humana Press, New York, pp 249–258
Boulard F, Glais L (2015) Détection du virus Y (PVY) directement sur tubercules. La Pomme de Terre Française 602, 56-58.
Browning I, Charlet K, Chrzanowska M, Dedic P, Kerlan C, Kryszczuk A, Schubert J, Varveri C, Werkman A Wolf I (2004) What is PVY^{NTN} ? The reaction of potato cultivars to inoculation with a range of PVY isolates. In the 12th Triennal conference of the European Association for Potato Research, Virology Section Meeting, Le Tronchet, France, pp 51–53
Bystricka D, Lenz O, Mraz I, Piherova L, Kmoch S, Sip M (2005) Oligonucleotide-based microarray: a new improvement in microarray detection of plant viruses. J Virol Methods 128:176–182
Charlet-Ramage K, Kerlan C (2005) Outil de sélection évaluant la sensibilité variétale à la maladie des nécroses annulaires des tubercules causée par le virus Y de la pomme de terre. Le Cahier des Techniques de l'INRA, Méthodes d'appréciation du comportement variétal vis-à-vis des bioagresseurs, N° spécial, pp 203–208
Chikh Ali M, Maoka T, Natsuaki KT (2007) A point mutation changes the serotype of a *Potato virus Y* isolate; genomic determination of the serotype of PVY strains. Virus Genes 35:359–367
Chikh Ali M, Maoka T, Natsuaki KT (2008) Whole genome sequence and characterization of a novel isolate of PVY inducing tuber necrotic ringspot in potato and leaf mosaic in tobacco. J Phytopathol 156:413–418
Chikh Ali M, Maoka T, Natsuaki T, Natsuaki KT (2010) $PVY^{NTN\text{-}NW}$, a novel recombinant strain of *Potato virus Y* predominating in potato fields in Syria. Plant Pathol 59:31–41
Chikh Ali MC, Karasev AV, Furutani N, Taniguchi M, Kano Y, Sato M, Natsuaki T, Maoka T (2013) Occurrence of *Potato virus Y* strain PVY^{NTN} in foundation seed potatoes in Japan, and screening for symptoms in Japanese potato cultivars. Plant Pathol 62(5):1157–1165
Chikh Ali M, Rowley JS, Kuhl J, Gray SM, Karasev AV (2014) Evidence of a monogenic nature of the *Nz* gene conferring resistance against *Potato virus Y* strain Z (PVY^Z) in potato. Am J Potato Res 91(6):649–654
Chrzanowska M (1991) New isolates of the necrotic strain of *Potato virus Y* (PVY^N) found recently in Poland. Potato Res 34(2):179–182
Chrzanowska M (1994) Differentiation of *Potato virus Y* (PVY) isolates. Polish Phytopathol Soc 8:15–20
Clark MF, Adams AN (1977) Characteristics of microplate method of Enzme-Linked ImmunoSorbent Assay for the detection of plant viruses. J Gen Virol 34:475–483
Cockerham G (1970) Genetical studies on resistance to *Potato viruses X* and *Y*. Heredity 25:309–348
Dhar AK, Singh RP (1997) Molecular characterization of coat protein genes of serologically distinct isolates of *Potato virus Y* necrotic strain. Can J Microbiol 43:677–683
Ellis P, Stace-Smith R, Bowler G, Mackenzie DJ (1996) Production of monoclonal antibodies for detection and identification of strains of *Potato virus Y*. Can J Plant Pathol 18(1):64–70
European and Mediterranean Plant Protection Organization (EPPO) (2010) Specific requirements for laboratories preparing accreditation for a plant pest diagnostic activity. Bulletin 40:5–22, OEPP/EPPO PM 7/98

Fabre F, Montarry J, Coville J, Senoussi R, Simon V, Moury B (2012) Modelling the evolutionary dynamics of viruses within their hosts: a case study using high-throughput sequencing. PLoS Pathog 8:1–9

Faurez F, Baldwin T, Tribodet M, Jacquot E (2012) Identification of new *Potato virus Y* (PVY) molecular determinants for the induction of vein necrosis in tobacco. Mol Plant Pathol 13(8):948–959

Fulton RJ, McDade RL, Smith PL, Kienker LJ, Kettman JR (1997) Advanced multiplexed analysis with the FlowMetrix™ system. Clin Chem 43(9):1749–1756

Galvino-Costa SBF, Figueira AD, Camargos VV, Geraldino PS, Hu XJ, Nikolaeva OV, Kerlan C, Karasev AV (2012) A novel type of *Potato virus Y* recombinant genome, determined for the genetic strain PVY^{E}. Plant Pathol 61(2):388–398

Glais L, Jacquot E (2015) Detection and characterization of viral species/sub-species using isothermal Recombinase Polymerase Amplification (RPA) assays. In: Lacomme C (ed) Methods in molecular biology, plant pathology techniques and protocols. Humana Press, New York

Glais L, Tribodet M, Gauthier JP, Astier-Manifacier S, Robaglia C, Kerlan C (1998) RFLP mapping of the whole genome of ten viral isolates representative of different biological groups of *Potato virus Y*. Arch Virol 143(11):2077–2091

Glais L, Tribodet M, Kerlan C (2002) Genomic variability in Potato potyvirus Y (PVY): evidence that PVY^{N}-W and PVY^{NTN} variants are single to multiple recombinants between PVY^{O} and PVY^{N} isolates. Arch Virol 147:363–378

Glais L, Tribodet M, Kerlan C (2005) Specific detection of the PVY^{N}-W variant of *Potato virus Y*. J Virol Methods 125:131–136

Gray S, De Boer S, Lorenzen J, Karasev A, Whitworth J, Nolte P, Singh R, Boucher A, Xu HM (2010) Potato virus Y: an evolving concern for potato crops in the United States and Canada. Plant Dis 94(12):1384–1397

Gugerli P, Fries P (1983) Characterization of monoclonal antibodies to *Potato virus Y* and their use for virus detection. J Gen Virol 64:2471–2477

Hanse L (2014) The new inspections. Potato World, number 1:17–21

Horvath J (1965) Studies on strains of *Potato virus Y*. 2. Strain C. Acta Phytopathol Acad Sci Hung 1:125–138

Horvath J (1967) Studies on strains of *Potato virus Y*. 4. Anomalous strain. Acta Phytopathol Acad Sci Hung 2:195–210

Hu X, Meacham T, Ewing L, Gray SM, Karasev AV (2009) A novel recombinant strain of *Potato virus Y* suggests a new viral genetic determinant of vein necrosis in tobacco. Virus Res 143:68–76

Jakab G, Droz E, Brigneti G, Baulcombe D, Malnoe P (1997) Infectious *in vivo* and *in vitro* transcripts from a full-length cDNA clone of PVY-N605, a Swiss necrotic isolate of *Potato virus Y*. J Gen Virol 78:3141–3145

Jones RAC (1990) Strain group specific and virus specific hypersensitive reactions to infection with potyviruses in potato cultivars. Ann Appl Biol 117:93–105

Kahn RP, Monroe RL (1963) Detection of the tobacco veinal necrosis strain of *Potato virus Y* in *Solanum cardenasii* and *S. andigenum* introduced into the United States. Phytopathology 53:1356–1359

Karasev AV, Gray SM (2013a) Continuous and emerging challenges of *Potato virus Y* in potato. Annu Rev Phytopathol 51:571–586

Karasev AV, Gray SM (2013b) Genetic diversity of *Potato virus Y* complex. Am J Potato Res 90:7–13

Karasev AV, Nikolaeva OV, Hu XJ, Sielaff Z, Whitworth J, Lorenzen JH, Gray SM (2010) Serological properties of ordinary and necrotic isolates of *Potato virus Y*: a case study of PVY^{N} misidentification. Am J Potato Res 87(1):1–9

Kehoe MA, Jones RAC (2016) Improving *Potato virus Y* strain nomenclature: Lessons from comparing isolates obtained over a 73-year period. Plant Pathol 65:322–333

Kellar KL, Douglass JP (2003) Multiplexed microsphere-based flow cytometric immunoassays for human cytokines. J Immunol Methods 279(1-2):277–285

Kerlan C, Tribodet M (1996) Are all PVY^N isolates able to induce potato tuber necrosis ringspot disease ? 13th Triennal conference of the European Association for Potato Research, Veldhoven, The Netherlands, pp 65–66

Kerlan C, Valade C, Guillou M (1985) Production de plants de pommes de terre en France – Différentes formes du virus Y. La Pomme de Terre Française, N 431:323–326

Kerlan C, Robert Y, Perennec P, Guillery E (1987) Survey of the level of infection by PVY^O and control methods developed in France for potato seed production. Potato Res 30(4):651–667

Kerlan C, Tribodet M, Glais L, Guillet M (1999) Variability of *Potato virus Y* in potato crops in France. J Phytopathol 147(11–12):643–651

Kerlan C, Nikolaeva OV, Hu X, Meacham T, Gray SM, Karasev AV (2011) Identification of the molecular make-up of the *Potato virus Y* strain PVY^Z: genetic typing of $PVY^{Z\text{-}NTN}$. Phytopathology 101:1052–1060

Kogovsek P, Gow L, Pompe-Novak M, Gruden K, Foster GD, Boonham N, Ravnikar M (2008) Single-step RT real-time PCR for sensitive detection and discrimination of Potato virus Y isolates. J Virol Methods 149(1):1–11. doi:10.1016/j.jviromet.2008.01.025

Kreuze JF, Perez A, Untiveros M, Quispe D, Fuentes S, Barker I, Simon R (2009) Complete viral genome sequence and discovery of novel viruses by deep sequencing of small RNAs: a generic method for diagnosis, discovery and sequencing of viruses. Virology 388:1–7

Kutnjak D, Rupar M, Gutierrez-Aguirre I, Curk T, Kreuze JF, Ravnikar M (2015) Deep sequencing of virus-derived small interfering RNAs and RNA from viral particles shows highly similar mutational landscapes of a plant virus population. J Virol 89:4760–4769

Lacomme C, Holmes R, Evans F (2015) Molecular and serological methods for the diagnosis of viruses in potato tubers. In: Lacomme C (ed) Plant pathology: techniques and protocols, methods in molecular biology, vol 1302. Springer, New York, pp 161–176. doi:10.1007/978-1-4939-2620-6_13

Le Hingrat Y (2011) A practicle guide to diseases, pests and disorders of the potato, 192 pp, edited by FN3PT, GNIS and Arvalis-Institut du végétal

Le Romancer M, Nedellec M (1997) Effect of plant genotype, virus isolate and temperature on the expression of the potato tuber necrotic ringspot disease (PTNRD). Plant Pathol 46:104–111

Lorenzen JH, Piche LM, Gudmestad NC, Meacham T, Shiel P (2006) A multiplex PCR assay to characterize *Potato virus Y* isolates and identify strain mixtures. Plant Dis 90:935–940

Lorenzen J, Nolte P, Martin D, Pasche J, Gudmestad N (2008) NE-11 represents a new strain variant class of *Potato virus Y*. Arch Virol 153:517–525

MacLeod DJ (1962) In Mosaic and streak viruses of the potato. Edited by Canada Department of Agriculture, research Branch, pp 44–55

McDonald JG, Kristjansson GT (1993) Properties of strains of Potato virus Y^N in North America. Plant Dis 77(1):87–89

McDonald JG, Singh RP (1996) Host range, symptomology, and serology of isolates of potato virus Y (PVY) that share properties with both the PVY^N and PVY^O strain groups. Am Potato J 73(7):309–315

Moalic V, Mercier B, Ferec C (2004) Technologie Luminex: principe, applications et perspectives. Immuno-analyse & Biologie spécialisée 19:181–187

Moravec T, Cerovska N, Boonham N (2003) The detection of recombinant, tuber necrosing isolates of *Potato virus Y* (PVY^{NTN}) using a three- primer PCR based in the coat protein gene. J Virol Methods 109:63–68

Mori Y, Nagamine K, Tomita N, Notomi T (2001) Detection of loop-mediated isothermal amplification reaction by turbidity derived from magnesium pyrophosphate formation. Biochem Biophys Res Commun 289:150–154

Mortimer-Jones SM, Jones MGK, Jones RAC, Thomson G, Dwyer GI (2009) A single tube, quantitative real- time RT-PCR assay that detects four potato viruses simultaneously. J Virol Methods 161:289–296

Mumford RA, Walsh K, Barker I, Boonham N (2000) Detection of *Potato mop top* and *Tobacco rattle virus* using a multiplex real-time fluorescent reverse-transcription polymerase chain reaction assay. Phytopathology 90:448–453

Nie X (2005) Reverse transcription loop-mediated isothermal amplification of DNA for detection of *Potato virus Y*. Plant Dis 89:605–610

Nie X, Singh RP (2002a) A new approach for the simultaneous differentiation of biological and geographical strains of *Potato virus Y* by uniplex and multiplex RT-PCR. J Virol Methods 104(1):41–54

Nie X, Singh RP (2002b) Probable geographical grouping of PVY^{N} and PVY^{NTN} based on sequence variation in P1 and 5-UTR of PVY genome and methods for differentiating North American PVY^{NTN}. J Virol Methods 103:145–156

Nie X, Singh RP (2003) Evolution of North American PVY^{NTN} strain Tu660 from local PVY^{N} by mutation rather than recombination. Virus Genes 26:39–47

Nikolaeva OV, Roop DJ, Galvino-Costa SBF, Figueira AD, Gray SM, Karasev AV (2012) Epitope mapping for monoclonal antibodies recognizing tuber necrotic isolates of *Potato virus Y*. Am J Potato Res 89(2):121–128

Notomi T, Okayama H, Masubuchi H, Yonekawa T, Watanabe K, Amino N, Hase T (2000) Loop-mediated isothermal amplification of DNA. Nucleic Acids Res 28:E63

Ogawa T, Tomitaka Y, Nakagawa A, Ohshima K (2008) Genetic structure of a population of *Potato virus Y* inducing potato tuber necrotic ringspot disease in Japan; comparison with North American and European populations. Virus Res 131:199–212

Ohshima K, Inoue AK, Ishikawa Y, Shikata E, Hagita T (1990) Production and application of monoclonal antibodies specific to ordinary strain and necrotic strain of potato virus Y. Ann Phytopathol Soc Jpn 56(4):508–514

Olsen N, Woodell L, Miller J, Wharton P, Karasev A, Thornton M Nolte P (2011) Using diagnostic test kits in the field and storage. Proceedings of the University of Idaho Winter Commodity Schools, Vol 43

Ounouna H, Kerlan C, Lafaye P, Loukili MJ, ElGaaied A (2002) Production of monoclonal antibodies against synthetic peptides of the N-terminal region of *Potato virus Y* coat protein and their use in PVY strain differentiation. Plant Pathol 51(4):487–494

Peters J, Sledz W, Bergervoet JHW, van der Wolf JM (2007) An enrichment microsphere immunoassay for the detection of Pectobacterium atrosepticum and Dickeya dianthicola in potato tuber extracts. Eur J Plant Pathol 117(2):97–107

Piche LM, Singh RP, Nie X, Gudmestad NC (2004) Diversity among PVY field isolates obtained from potatoes grown in the United States. Phytopathology 94:1368–1375

Ranki H, Hepojoki J, Lankinen H, Tian YP, Rännäli M Valkonen JPT (2008) Epitopes of the *Potato virus Y* (PVY) coat protein recognized by monoclonal antibodies. 17th Triennal conference of European Association for Potato Research, Roumania, p 55

Rigotti S, Gugerli P (2007) Rapid identification of *Potato virus Y* strains by one-step triplex RT-PCR. J Virol Methods 140:90–94

Rolland M (2008) Impact de la capacité nécrogène sur la valeur adaptative du virus Y de la pomme de terre. Thèse de l'Université de Rennes I, p. 214

Rolland M, Glais L, Kerlan C, Jacquot E (2008a) A multiple single nucleotide polymorphisms interrogation assay for reliable *Potato virus Y* group and variant characterization. J Virol Methods 147(1):108–117

Rolland M, Lacroix C, Blanchard A, Baldwin T, Kerlan C, Jacquot E (2008b) *Potato virus Y* (PVY): from its discovery to the latest outbreaks. Virologie 12(4):261–273

Rose DG, Hubbard AL (1986) Production of monoclonal antibodies for the detection of *Potato virus Y*. Ann Appl Biol 109:317–321

Rose DG, McCarra S, Mitchell DH (1987) Diagnosis of Potato virus Y^{N} – a comparison between polyclonal and monoclonal antibodies and a biological assay. Plant Pathol 36(1):95–99

Rosner A, Maslenin L (1999) Differentiating PVY^{NTN} by unique single-restriction cleavage of PCR products. Potato Res 42:215–221

Rosner A, Maslenin L (2001) Differentiating PVY^{NTN} from PVY^{N} by annealing to reference RNA transcripts. J Virol Methods 97:125–131
Rowley JS, Gray SM, Karasev AV (2015) Screening potato cultivars for new sources of resistance to *Potato virus Y*. Am J Potato Res 92:38–48
Rupar M, Kogovsek P, Pompe-Novak M, Gutierrez-Aguirre I, Delaunay A, Jacquot E, Ravnikar M (2013) Assessment of SNaPshot and single step RT-qPCR methods for discriminating *Potato virus Y* (PVY) subgroups. J Virol Methods 189(1):93–100
Saiki RK, Gelfand DH, Stoffel S, Scharf SJ, Higuchi R, Horn GT, Mullis KB, Erlich HA (1988) Primer directed enzymatic amplification of DNA with a thermostable DNA-polymerase. Science 239(4839):487–491
Sanz A, Cambra M, Desanroman CP, Miguet JG, Cortes E, Gorris MT, Vela C (1990) Preparation of additional monoclonal antibodies for detection and discrimination of *Potato virus Y* isolates infecting potato. Potato Res 33(3):365–375
Schena M, Shalon D, Davis RW, Brown PO (1995) Quantitative monitoring of gene-expression patterns with a complementary-DNA microarray. Science 270:467–470
Schoen CD, Knorr D, Leone G (1996) Detection of *Potato leaf roll virus* in dormant tubers by immunocapture and a fluoregenic 5 nuclease RT-PCR assay. Phytopathology 86:993–999
Schubert J, Fomitcheva V, Sztangret-Wisniewski J (2007) Differentiation of *Potato virus Y* using improved sets of diagnostic PCR- primers. J Virol Methods 140:66–74
Singh M, Singh RP (1996a) Nucleotide sequence and genome organization of a Canadian isolate of the common strain of potato virus Y (PVY^{O}). Can J Plant Pathol 18(3):209–224
Singh M, Singh RP (1996b) Factors affecting detection of PVY in dormant tubers by reverse transcription polymerase chain reaction and nucleic acid spot hybridization. J Virol Methods 60(1):47–57
Singh RP, Kurz J, Boiteau G (1996) Detection of stylet-borne and circulative potato viruses in aphids by duplex reverse transcription polymerase chain reaction. J Virol Methods 59(1-2):189–196
Singh RP, Valkonen JPT, Gray SM, Boonham N, Jones RAC, Kerlan C, Schubert J (2008) Discussion paper: the naming of Potato virus Y strains infecting potato. Arch Virol 153(1):1–13
Sip M, Bystricka D, Kmoch S, Noskova L, Hartmannova H, Dedic P (2010) Detection of viral infections by an oligonucleotide microarray. J Virol Methods 165:64–70
Smith KM (1931) On the composite nature of certain potato virus diseases of the mosaic group as revealed by the use of plant indicators and selective methods of transmission. Proc R Soc Sery B 109:251–267
Smith KM, Dennis RWG (1940) Some notes on a suspected variant of *Solanum virus 2* (*Potato virus Y*). Ann Appl Biol 27:65–70
Studholme DJ, Glover RH, Boonham N (2011) Application of high-throughput DNA sequencing in phytopathology. Annu Rev Phytopathol 49:87–105
Szajko K, Chrzanowska M, Witek K, Strzelczyk-Zyta D, Zagorska H, Gebhardt C, Hennig J, Marczewski W (2008) The novel gene *Ny-1* on potato chromosome IX confers hypersensitive resistance to *Potato virus Y* and is an alternative to *Ry* genes in potato breeding for PVY resistance. Theor Appl Genet 116:297–303
Szajko K, Strzelczyk-Zyta D, Marczewski W (2012) The development of diagnostic markers for selection of hypersensitive response (HR) gene for resistance to PVY in the cultivated potato. SOL2012, From the Bench to Innovative Applications, Neuchâtel, Switzerland. p 156.
Szajko K, Strzelczyk-Zyta D, Marczewski W (2014) *Ny-1* and *Ny-2* genes conferring hypersensitive response to *Potato virus Y* (PVY) in cultivated potatoes: mapping and marker-assisted selection validation for PVY resistance in potato breeding. Mol Breed 34:267–271
Tanner NA, Zhang YH, Evans TC (2012) Simultaneous multiple target detection in real-time loop-mediated isothermal amplification. Biotechniques 53:81–89
Thompson GJ, Hoffman DCA, Prins PJ (1987) A deviant strain of *Potato virus Y* infecting potatoes in South Africa. Potato Res 30(2):219–228
Tian YP, Hepojoki J, Ranki H, Lankinen H, Valkonen JPT (2014) Analysis of *Potato virus Y* coat protein epitopes recognized by three commercial monoclonal antibodies. PLoS One 9(12):1–20

Tomczynska I, Jupe F, Hein I, Marczewski W, Sliwka J (2014) Hypersensitive response to *Potato virus Y* in potato cultivar Sarpo Mira is conferred by the *Ny-Smira* gene located on the long arm of chromosome IX. Mol Breed 34:471–480

Tribodet M, Glais L, Kerlan C, Jacquot E (2005) Characterization of *Potato virus Y* (PVY) molecular determinants involved in the vein necrosis symptom induced by PVY^N isolates in infected *Nicotiana tabacum* cv. Xanthi. J Gen Virol 86:2101–2105

Valkonen JPT (1997) Novel resistances to four potyviruses in tuber-bearing potato species, and temperature-sensitive expression of hypersensitive resistance to *Potato virus Y*. Ann Appl Biol 130:91–104

Wei T, Pearson MN, Blohm D, Nolte M, Armstrong K (2009) Development of a short oligonucleotide microarray for the detection and identification of multiple potyviruses. J Virol Methods 162:109–118

Weidemann HL (1988) Importance and control of Potato virus Y^N (PVY^N) in seed potato production. Potato Res 31(1):85–94

Weilguny H, Singh RP (1998) Separation of Slovenian isolates of PVY^{NTN} from the North American isolates of PVY^N by a 3-primer PCR. J Virol Methods 71:57–68

Wu Q, Ding SW, Zhang Y, Zhu S (2014) Identification of viruses and viroids by next-generation sequencing and homology-dependent and homology-independent algorithms. Annu Rev Phytopathol 53:425–444

Chapter 6
Transmission and Epidemiology of *Potato virus Y*

Christophe Lacomme, Jon Pickup, Adrian Fox, Laurent Glais, Brice Dupuis, Thomas Steinger, Jean-Louis Rolot, Jari P.T. Valkonen, Kerstin Kruger, Xianzhou Nie, Spela Modic, Natasa Mehle, Maja Ravnikar, and Maurice Hullé

Abstract As obligate parasites, plant viruses, require in order to survive, to be transmitted to another plant. Experimentally, viruses such as *Potato virus Y* (PVY) can be transmitted by mechanical means such as wounding and grafting. In its natural environment, PVY transmission is mediated by sap-feeding aphid vector, or vegetatively through propagated organs such as potato tubers. A vast number of aphid species have been reported to transmit PVY in a non-persistent manner with variable efficiency to a large number of solanaceous and non-solanaceous plant species including weeds and ornamentals. Several sensory stimuli will influence

C. Lacomme (✉) • J. Pickup
Science and Advice for Scottish Agriculture (SASA),
1 Roddinglaw Road, Edinburgh EH12 9FJ, UK

National Institute of Biology, Vecna pot 111, SI-1000 Ljubljana, Slovenia
e-mail: Christophe.Lacomme@sasa.gsi.gov.uk

A. Fox
Fera Science Ltd, Sand Hutton, York YO41 1LZ, UK

L. Glais
FN3PT/RD3PT, 43-45 rue de Naples, 75008 Paris, France

IGEPP, Agrocampus Ouest, INRA, Université de Rennes 1, F-35650 Le Rheu, France

B. Dupuis • T. Steinger
Agroscope ACW, Route de Duiller 50, Case postale 1012, CH-1260 Nyon, Switzerland

J.-L. Rolot
CRA-W, 6800 Libramont, Belgium

J.P.T. Valkonen
Department of Agricultural Sciences, University of Helsinki,
PO Box 27, FI-00014 Helsinki, Finland

K. Kruger
Department of Zoology & Entomology, University of Pretoria, 0002 Pretoria, South Africa

X. Nie
Agriculture and Agri-Food Canada, 850 Lincoln Road, PO Box 20280,
Fredericton, NB E3B 4Z7, Canada

C. Lacomme et al. (eds.), *Potato virus Y: biodiversity, pathogenicity, epidemiology and management*, DOI 10.1007/978-3-319-58860-5_6

host selection and feeding behaviour of the aphid and will strongly influence virus epidemiology. The interactions between the virus, its vector, and the environment are complex and are the focus of many studies aiming to understand the molecular basis of these interactions and their impact on disease development. This chapter will present the current knowledge of PVY transmission, epidemiology, and its management in different countries.

1 Introduction

Plant viruses are obligate parasites, and their survival depends on their transmission (often via a vector) from an infected plant to another plant. For the vast majority of plant viruses, their infectious cycle is tightly associated with that of their vector(s) and their host(s). When a plant is infected, a virus can persist in seed or in vegetatively propagated organs, such as potato tubers, bulbs, corms, at the end of the growing season. These offer ways for a virus to survive adverse conditions and initiate new infections during the following growing season. Plants acquire virus through two pathways. The first, termed vertical or secondary transmission, occurs when a virus moves from infected planting material, *e.g.*, potato tuber into the growing plant and, in the case of potato, into the daughter tubers. The second pathway termed horizontal transmission occurs when a plant is infected mechanically or through a vector, usually an invertebrate such as an insect. Both pathways of infection by *Potato virus Y* (PVY) occur with potato but foliar symptoms on infected plants are generally more severe with vertical transmission than with horizontal transmission. Nearly 765 species of aphids (Hemiptera: Aphididae) have been reported to transmit PVY with variable efficiency. PVY has a broad host range infecting solanaceous and non-solanaceous plant species including weeds and ornamentals (Kerlan 2006). Biennial and perennial arable weeds can act as reservoirs for aphid-mediated transmission (Beemster and de Bokx 1987) and can contribute to the rapid dissemination of PVY within field crops and its persistence in the environment. The interactions between the virus, its vector, and the environment are complex and are the focus of many studies aiming to understand the molecular basis of these interactions and their impact on disease development.

This chapter is intended to give an overview of the current knowledge of the transmission of PVY, and its interaction with its aphid vectors, its host (potato) and the environment.

S. Modic
Agricultural Institute of Slovenia, Hacquetova ulica 17, SI-1000 Ljubljana, Slovenia

N. Mehle • M. Ravnikar
National Institute of Biology, Vecna pot 111, SI-1000 Ljubljana, Slovenia

M. Hullé
IGEPP, Agrocampus Ouest, INRA, Université de Rennes 1, 35650 Le Rheu, France

2 Transmission of PVY

2.1 *"Natural" PVY Transmission*

2.1.1 Vertical Transmission: Seed-Borne Infection

The infectious cycle of a virus in a plant host starts from a single infected cell. After an initial stage of uncoating (*i.e.*, virion CP disassembly exposing the viral RNA genome), genome replication occurs, followed by cell-to-cell (local) movement through plasmodesmata to reach phloem vessels from where the virus will be transported in the phloem sap throughout the whole plant. As is the case for all plant viruses in infected plants, PVY movement throughout the plant will follow the source-sink partitioning of photoassimilates, resulting in systemic infection of sink tissues above and below the ground. Hence, developing leaves and other tissues, including new tubers, will gradually become infected (for a review, see Hull 2002).

Infected seed potato tubers can be an important source of PVY inoculum as they can maintain the virus within a crop and enable its spread within daughter growing crops and to other potato crops in following seasons. Analyses of seed potato certification data in Scotland revealed that PVY was four times more likely to be found in a crop derived from seed potatoes from a crop which contained plants with symptoms of PVY than a crop without any symptomatic plants (Fenton et al. 2012). PVY transmission through true seed or pollen of potatoes has not been recorded (de Bokx 1972).

2.2 *Horizontal Transmission*

2.2.1 Transmission by Aphids

As with about 380 other plant viruses, PVY is transmitted naturally by aphids (Nault 1997). Aphids are sap (phloem)-feeding hemipteran insects, and over 190 aphid species are known to transmit plant viruses (Nault 1997), with many species capable of transmitting more than one virus species (reviewed in Katis et al, 2007). Aphid populations can rapidly reach very high numbers, primarily through asexual (parthenogenetic) reproduction rather than sexual reproduction. Aphid multiplication is very variable and depends on aphid species, geographical location, temperature, and the presence of suitable hosts. The temperature threshold for aphid flights is around 15 °C for *Rhopalosiphum padi* and *Aphis fabae*; however, it does vary among and within species (Bale et al. 2007).

The interactions between aphids and the viruses they transmit have been studied at the physiological, cellular, and molecular levels and have emerged, perhaps not surprisingly, as very complex and closely related. The host selected by an aphid will be guided primarily by its perception of plant colour, volatiles, and, once having landed on a plant, through the ingestion of plant cellular contents (Moericke 1955; Kennedy et al. 1959a, b; Kring 1972; Pettersson 1970; Pettersson et al. 2007;

Giordanengo et al. 2010). Interestingly, plant viruses can promote their own transmission by manipulating host physiology to make the plant more attractive to aphids and, in some cases, increase aphid fecundity (Holmes and Bethel 1972; Ingwell et al. 2012; Casteel et al. 2014). Moreover, it has been reported that the infection of a potato plant by PVY can affect the feeding behaviour of some aphid species (Boquel et al. 2011, 2012). In the case of *Myzus persicae*, the most efficient aphid vector of PVY, the landing of an aphid on a PVY-infected plant increased its ingestion of phloem sap and reduced the time spent not probing (Boquel et al. 2011), which could increase the risk of PVY spreading. Consequently, a wide variety of tri-trophic interactions can be observed within the PVY/aphid/host system.

Process of Virus Transmission by Aphid

Virus transmission by aphids occurs by transferring virions (virus particles containing the viral genome, assembled coat protein monomers, and other associated virus-encoded proteins) from infected to healthy plants. After landing on plants, aphids probe (puncture) leaves by inserting their long, flexible stylet into epidermal cells to form a salivary and food canal in both host and non-host plants. The first probes last less than a minute and seem to provide sufficient information for either plant rejection (promoting aphid flight) or plant acceptance (secretion of gelling and watery saliva, followed by progression through different feeding stages until reaching phloem vessels and uptake of phloem). These stages of the aphid-host interaction have emerged as a very complex molecular interplay between host and aphid. As for other plant-pathogens interactions, aphids are likely to deliver effector molecules inside their host which will alter cellular processes, thus enabling successful establishment and fulfilment of their life cycle (Hogenhout et al. 2009; Giordanengo et al. 2010).

The process of virus transmission can be divided into four phases. The first phase is termed the "acquisition phase" in which an aphid acquires virions from an infected plant. The second phase is the "retention phase" in which an aphid will retain and carry the virions. The third phase is the "inoculation phase" in which an aphid will release the retained virions into another host, initiating a new infection. For some viruses, another period termed the "latent phase" occurs during which an aphid has acquired virions but is unable to transmit them to another host during this period of time (reviewed by Katis et al. 2007).

Based on the duration of these different phases, the types of virus transmission by aphids can be classified as "persistent", "semi-persistent", or "non-persistent". In persistent transmission, virus acquisition requires a relatively long time (hours or days) before an aphid becomes infective because most persistent viruses, *e.g.*, *polerovirus* such as *Potato leafroll virus*, are confined to phloem tissues and can either replicate (propagative viruses) or not (non-propagative viruses) in their aphid vector. Semi-persistent viruses, such as *Cauliflower mosaic virus*, may be acquired within minutes or hours. However, their efficiency of transmission to a new host increases with the duration of the acquisition phase (Palacios et al. 2002).

Transmission of non-persistent viruses requires only a brief period of seconds or minutes of probing of epidermal cells by an aphid's stylet before the virus becomes infective in the aphid. The retention of virions in an aphid foregut suggests that non-persistent viruses such as PVY are acquired by ingestion of cellular contents during probing of the epidermal cells (Pirone and Perry 2002). There is no latent period for semi-persistent viruses. Aphids rapidly lose their infectivity immediately after the acquisition phase and become non-infective fairly quickly when feeding on healthy plants. However, PVY particles can persist for more than 17 h in winged forms of *A. nasturtii* (Kostiw 1975). The vast majority (about 75%) of aphid-transmitted viruses are transmitted in a non-persistent manner. Non-persistent viruses include viruses of the *Alfamovirus*, *Caulimovirus*, *Cucumovirus*, *Fabavirus*, *Macluravirus,* and *Potyvirus* genera, with different virus particles shape (helical and isometric), genomes composition (mono or multipartite DNA, ssRNA+ genomes).

Identification of PVY Aphid Vectors

Since the 1980s, a total of 65 aphid species or group species have been reported as having the ability to acquire and transmit PVY in a non-persistent fashion (Table 6.1).

This list includes aphid vectors that colonise potato, such as the peach-potato aphid *M. persicae* or the potato aphid *Macrosiphum euphorbiae* (Fig. 6.1), or colonise other host plants, such as black bean aphid *A. fabae*, bird cherry-oat aphid *R. padi*, cereal aphids like the rose-grain aphid *Metopolophium dirhodum* and the grain aphid *Sitobion avenae*, but nevertheless have the ability to transmit PVY in a non-persistent manner.

All these aphid species exhibit various degrees of efficiency in transmitting PVY. This capacity is defined by a relative efficiency factor (REF). The principle of REF determination for PVY transmission to potato relies on catching live alate aphids of different species present in potato crops, starving them for a period, allowing them to probe a PVY-infected potato and finally transferring each individual aphid to a virus-free potato plant or another bait plant. The transmission rate is calculated by dividing the total number of bait plants infected for each aphid biotype by the total number of plants infected for the *M. persicae* reference biotype. REF values are then calculated for each aphid species by dividing the transmission rate value of each individual species by the value of *M. persicae* transmission rate (which will have a value of 1). Since 1980s, several researchers have applied this methodology to define the REF factor for specific aphid species and isolates of PVY^{O}, PVY^{N} strain groups, and PVY^{NTN} and $PVY^{N\text{-}Wi}$ variants (Kostiw 1979; van Hoof 1980; van Harten 1983; Sigvald 1984, 1992; Harrington and Gibson 1989; de Bokx and Piron 1990; Halbert et al. 2003: Verbeek et al. 2010) (Table 6.2).

However, differences in REF values for specific virus aphid combinations can be observed, *e.g.*, *Acyrthosiphon pisum* and PVY^{O}. Verbeek et al. (2010) concluded that the two most important factors influencing an assessment of an REF value were biotype of an aphid species and isolate of PVY, while acknowledging that a number of difficulties still occurred in standardising this type of experiment, particularly

Table 6.1 List of the 65 aphid species or group species reported in the literature as potential vectors of PVY

Aphid species / group species	Studied PVY strain group(s)	Reported by
Acyrthosiphon pisum	PVY^O, PVY^N	von Hoof (1980), Sigvald (1984), Harrington et al. (1986) and DiFonzo et al. (1997)
Acyrthosiphon primulae	PVY^N	Ragsdale et al. (2001)
Anoecia corni	PVY^O	Basky and Raccah (1990)
Aphis citricola	PVY (pepper)	Raccah et al. (1985)
Aphis craccivora	PVY^O	Basky and Raccah (1990)
Aphis fabae	PVY^O, PVY^N	van Hoof (1980), Sigvald (1984), Harrington et al. (1986), de Bokx and Piron (1990), DiFonzo et al. (1997) and Basky and Almasi (2005)
Aphis fabae cirsiacanthoides	PVY^O, PVY^N	Basky and Almasi (2005)
Aphis frangulae	PVY^O	Sigvald (1992)
Aphis glyacines	PVY^O, PVY^N, PVY^{NTN}	Davis et al. (2005)
Aphis gossypii	PVY^O	Raccah et al. (1985)
Aphis hellantti	PVY^O	DiFonzo et al. (1997)
Aphis nasturtii	PVY^O, PVY^N	Sigvald (1984), Harrington et al. (1986) and de Bokx and Piron. (1990)
Aphis pomi	PVY^O, PVY^N	van Hoof (1980), Harrington and Gibson (1989) and Basky and Almasi (2005)
Aphis rumicis	PVY^O	Basky and Raccah (1990)
Aphis sambuci	PVY^O, PVY^N	Harrington et al. (1986), de Bokx and Piron (1990)
Aphis spiraecola	PVY^N	Basky and Almasi (2005)
Aulacorthum solani	PVY^O, PVY^N	van Hoof (1980)
Brachycaudus cardui	PVY	Basky (2002)
Brachycaudus helichrysi	PVY, PVY^O, PVY^N	Edwards (1963), van Harten, (1983), Harrington et al. (1986) and de Bokx and Piron (1990)
Brevicoryne brassicae	PVY, PVY^O	Sigvald (1984), Basky and Raccah (1990)
Capitophorus elaeagni	PVY, PVY^O	DiFonzo et al. (1997), Halbert et al. (2003)
Capitophorus hippophaes	PVY^N	van Hoof (1980), de Bokx and Piron (1990)
Cavariella aegopodii	PVY^O, PVY^N	de Bokx and Piron (1990)
Cavariella pastinaca	PVY^N	Salazar (1996)
Cryptomyzus ballotae	PVY^O	Harrington et al. (1986)
Cryptomyzus galeopsidis	PVY^N	de Bokx and Piron (1990)
Cryptomyzus ribis	PVY^N	de Bokx and Piron (1990)
Diuraphis noxia	PVY, PVY^O	Halbert et al. (2003), Basky and Almasi (2005)
Drepanosiphum platanoidis	PVY^N	Powell et al. (1992)
Dysaphis plantaginea	PVY, PVY^O	Basky et Raccah (1990)
Dysaphis spp	PVY^N	de Bokx and Piron (1990)

(continued)

Table 6.1 (continued)

Aphid species / group species	Studied PVY strain group(s)	Reported by
Hayhurstia atripllicis	PVY	Basky and Raccah (1990)
Hyadaphis foeniculi	PVY^N	de Bokx and Piron (1990)
Hyalopterus pruni	PVY, PVY^O, PVY^N	de Bokx and Piron (1990), Basky and Raccah (1990)
Hyperomyzus lactucae	PVY^O, PVY^N	Harrington et al. (1986), de Bokx and Piron (1990)
Hyperomyzus pallidus	PVY	Basky and Raccah (1990)
Lipaphis erysimi	PVY^O	DiFonzo et al. (1997)
Macrosiphum euphorbiae	PVY^O, PVY^N	van Hoof (1980), van Harten (1983), Harrington et al. (1986), de Bokx and Piron (1990)
Macrosiphum rosae	PVY^N	Basky and Almasi (2005)
Metopolophium albidum	PVY^N	van Hoof (1980)
Metopolophium dirhodum	PVY, PVY^O, PVY^N	van Hoof (1980), van Harten (1983), de Bokx and Piron (1990), Sigvald (1992) and Halbert et al. (2003)
Metopolophium festucae	PVY^O	Harrington et al. (1986)
Myzaphis rosarum	PVY^O	Harrington et al. (1986)
Neomyzus circumflexus	PVY^O, PVY^N	Salazar (1996)
Myzus ascaionicus	PVYN, PVY^{NTN}, PVY^NW	Verbeek et al. (2010)
Myzus cerasi	PVY^O, PVY^N	Harrington et al. (1986), de Bokx and Piron (1990), Basky and Almasi (2005)
Myzus certus	PVY^O, PVY^N	van Hoof (1980), de Bokx and Piron (1990)
Myzus ligustri	PVY^O, PVY^N	Harrington et al. (1986), Basky and Almasi (2005)
Myzus myosotidis	PVY^O	Harrington et al. (1986)
Myzus persicae nicotianae	PVY, PVY^N	Halbert et al. (2003), Kanavaki et al. (2006)
Myzus persicae	PVY, PVY^O, PVY^N	van Hoof (1980), van Harten (1983), Harrington et al. (1986), de Bokx and Piron. (1990), Sigvald (1992), Fereres et al. (1993) and Halbert et al. (2003)
Phorodon humuli	PVY^O, PVY^N	van Hoof (1980), van Harten (1983), de Bokx and Piron. (1990) and Harrington et al. (1986)
Rhopalosiphum insertum	PVY^O, PVY^N	van Hoof (1980), van Harten (1983), Harrington et al. (1986) and de Bokx and Piron. (1990)
Rhopalosiphum maidis	PVY^O, PVY^O	DiFonzo et al. (1997), Halbert et al. (2003)
Rhopalosiphum padi	PVY, PVY^O, PVY^N	Kosow (1979), van Hoof (1980), van Harten (1983), Sigvald (1984), Harrington et al. (1986), de Bokx and Piron. (1990), DiFonzo et al. (1997), Halbert et al. (2003) and Basky et al. (2005)
Rhopalosiphum pseudobrassicae	PVY	Ragsdale et al. (2001)

(continued)

Table 6.1 (continued)

Aphid species / group species	Studied PVY strain group(s)	Reported by
Schizaphis graminum	PVY, PVY^O, PVY^N	Basky and Raccah (1990), DiFonzo et al. (1997), Halbert et al. (2003) and Basky and Almasi (2005)
Sitobion avenae	PVY^O, PVY^N	Sigvald (1984), Harrington et al. (1986), de Bokx and Piron (1990) and DiFonzo et al. (1997)
Sitobion fragariae	PVY^O, PVY^N	Harrington et al. (1986), de Bokx and Piron (1990)
Sitobion graminum	PVY^{NTN}, PVY^NW	Verbeek et al. (2010)
Staphylae tulipaellus	PVY^N	Salazar (1996)
Therioaphis trifolli / sp	PVY (pepper)	Perez et al. (1995)
Tetraneura ulmi	PVY	Basky and Raccah (1990)
Uroleucon spp	PVY^N	Harrington et al. (1986)
Uroleucon sonchi	PVY	Raccah et al. (1985)

Aphid species known to colonise potato plants are in *bold*. Adapted from Al Mrabeh et al. (2010)

Fig. 6.1 (**a**) *Myzus persicae* (peach-potato aphid); (**b**) *Metopolophium dirhodum* (rose-grain aphid); (**c**) *Rhopalosiphum padi* (bird cherry-oat aphid); (**d**) *Macrosiphum euphorbiae* (potato aphid); (**e**) *Sitobion avenae* (grain aphid); (**f**) *Brachycaudus helichrysi* (leaf curl plum aphid) (Panels **a** and **d** courtesy of SASA, Edinburgh, UK Crown copyright. Panels **b**, **c**, **e** and **f** courtesy of Rothamsted Research, Harpenden, UK)

virus load in source plants. Differences in transmission rates as measured by REFs values among biotypes of an aphid species suggested that genetic variability among biotypes may impact on the transmission of PVY (data not shown, Verbeek et al. 2010). Moreover, as observed with *A. pisum*, *A. fabae*, *Aphis nasturtii*, *Aphis spp*., *Phorodon humuli*, and *R. padi*, recombinant isolates of PVY (PVY^{NTN} and/or $PVY^{N\text{-}Wi}$) tended to be more efficiently transmitted than non-recombinant (PVY^N) isolates (Verbeek et al. 2010). These differences may occur because the transmission of PVY depends on specific interactions between receptors present in an aphid's

Table 6.2 Relative efficiency factors (REF) for the transmission of PVY^{O}, PVY^{N}, PVY^{NTN} or $PVY^{N\text{-}Wi}$ by various aphid species or group species according to the literature since 1980s.

	REF according to												
	van Hoof (1980)	van Harten (1983)	Sigvald (1984)	Harrington and Gibson (1989)	de Bokx and Pirone (1990)	Sigvald (1992)	Halbert et al. (2003)	Verbeek et al. (2010)*			Fox et al. (2016)		
Aphid species/group species	PVY^{N}	PVY^{N}	PVY^{O}	PVY^{O}	PVY^{N}	PVY^{O}	PVY^{O}	PVY^{N} (one isolate)	PVY^{NTN} (three isolates)	$PVY^{N\text{-}Wi}$ (two isolates)	PVY^{O}	PVY^{N}	PVY^{NTN}
Acyrthosiphon pisum	0.14	0.05	0.25	0.04	0.05	0.7		0.08	0.07	0.11	1.08	0.29	0.49
Aphis fabae	0.24	0.1	0.01	0.08	0.04	0.1		0.03	0.04	0.13	0.06	0	0.13
Aphis nasturtii			0.07	0.5	0.21	0.4		0.46	0.5	0.69			
Aphis pomi	0.09			0.03									
Aphis sambuci				0.04	0.03								
Aphis spp				0.06	0.03			0.01	0.03	0.08			
Aulacorthum solani	0.05							0	0.01	0			
Brachycaudus helichrysi				0.06	0.1								
Brachycaudus spp					0.2								
Brevicoryne brassicae	0		0			0.01		0	0	0	0	0	0
Capitophorus elaeagni							0.02						
Capitophorus hippophaes	0.03				0.03								
Cavariella aegopodii	0			0.002	0.002			0	0	0	0.77	0.49	0.96
Cryptomyzus ballotae				1									

(continued)

Table 6.2 (continued)

Aphid species/group species	REF according to van Hoof (1980)	van Harten (1983)	Sigvald (1984)	Harrington and Gibson (1989)	de Bokx and Pirone (1990)	Sigvald (1992)	Halbert et al. (2003)	Verbeek et al. (2010)*			Fox et al. (2016)		
	PVY^N	PVY^N	PVY^O	PVY^O	PVY^N	PVY^O	PVY^O	PVY^N (one isolate)	PVY^{NTN} (three isolates)	$PVY^{N\text{-}Wi}$ (two isolates)	PVY^O	PVY^N	PVY^{NTN}
Cryptomyzus galeopsidis					0.1								
Cryptomyzus ribis				0.14	0.12								
Diuraphis noxia							0.06						
Drepanosiphum platanoidis											0	0	0
Dysaphis spp				0.02	0.02								
Hyadaphis foeniculi					0.09								
Hyalopterus pruni	0				0.04								
Hyperomyzus lactucae	0			0.004	0.08						0	0	0
Lipaphis erysimi	0												
Macrosiphum euphorbiae	0.29	0.1		0.07	0.03			0	0	0.01	0.3	0.66	0.38
Metopolophium albidum	0.11												
Metopolophium dirhodum	0.03	0.01		0.005	0.05	0.3	0	0.02	0	0.01	0	0.87	0.07
Metopolophium festucae				0.004									

Microlophium carnosum											0	0	0
Myzaphis rosarum				0.1									
Myzus ascalonicus	0							0.01	0	0.01			
Myzus cerasi				0.03	0.02								
Myzus certus	0.71				0.22								
Myzus ligustri				0.5									
Myzus myosotidis				1									
Myzus persicae	**0.5**	**1**	**0.26**	**0.08**	**0.5**	**1**	**0.77**	**1**	**1**	**1**	**1**	**1**	**1**
Phorodon humuli	0.35	0.15		0.05	0.07			0.22	0.23	0.3			
Rhopalosiphum insertum	0.5	0.05		0.008	0.07								
Rhopalosiphum maidis							0.01						
Rhopalosiphum padi	0.02	0.02	0.01	0.02	0.07	0.4	0.03	0	0.01	0.04	0.77	0.57	0.95
Schizaphis graminum							0.05	0	0.05	0			
Sitobion avenae	0		0	0.001	0.01	0.01	0	0	0	0	0.33	1.1	0.49
Sitobion fragariae				0.005	0.07								
Tetraneura ulmi					0.06								
Uroleucon spp				0.005									

The asterisk denotes an average REF value depending on the number of isolates tested for each PVY group stylet and the PVY helper component proteinase (HC-Pro) and virion coat protein (CP). Consequently, mutations within specific sites in at least one of these two viral proteins could impair the capacity of aphids to transmit a virus (Blanc et al. 1998). Fox et al (2016) used the Verbeek method as a standardised approach to investigate PVY virus strains and aphid biotypes in the UK. This study found that two species, *Cavariella aegopodii* and *R. padi*, were able to transmit PVY^{NTN} isolates almost as efficiently as *M. persicae*. This study also highlighted that the largest source of variability in transmission efficiency within a given aphid species was the virus source plant, confirming the findings from Verbeek et al. (2010). Although differences in REFs were recorded among the various laboratories, *M. persicae* remained the most efficient aphid vector of PVY with the greatest REF value, irrespective of the isolate tested.

In addition to these studies, PCR-based methods have been used to detect PVY in individual aphids (He et al. 2006), even in their stylets (Zhang et al. 2013), enabling species that may act as potential vectors to be identified. This approach was applied to preserved viral RNA from aphids caught in pan traps in potato crops (Nie et al. 2011; Pelletier et al. 2012). Aphid catches from the various crops were amalgamated and sorted by species. The stylets of a maximum of 12 aphids from each sample were detached from their body parts and stored for RT-PCR detection of PVY. The remaining corresponding aphid body parts were kept for taxonomic identification using a DNA barcoding approach. The sequence of the *cytochrome c oxidase subunit 1* (COI) barcode region was compared with a set of ~6600 aphid sequences representing 730 species in the Barcode of Life Data Systems project (Foottit et al. 2008). Using this approach, 65 aphid taxa were identified from the samples that tested positive for PVY. Among these, 45 taxa had never been previously recorded as being able to transmit PVY and 7 had previously been labelled as non-vectors (Pelletier et al. 2012). These results indicate that many aphid species that feed on PVY-infected plants can carry PVY particles. However, detecting PVY in aphids by RT-PCR or other methods does not mean that a specific aphid is able to transmit a virus; further experiments are required to establish whether or not probing by these PVY-containing aphids can result in infection of a healthy potato plant.

Knowledge of the REF values of aphid species and the number of aphids caught in traps is of practical importance for virus management because it allows the cumulative vector pressure to be calculated and the risk of PVY transmission to potato crops to be forecast (van Harten 1983, Sigvald 1986). Indeed, while some non-potato colonising aphid species have relatively low REF values, their importance as PVY vectors is directly dependent on their phenology, because large numbers of aphids with an attributed small REF value may contribute significantly to PVY transmission in crops (DiFonzo et al. 1997; Boiteau et al. 1998; Pickup et al. 2009; Kirchner et al. 2011; Fenton et al. 2012, Fox et al. 2016). In addition to how efficient an aphid transmits virus (REF values), virus spread in potato crops will be influenced by the activity of vector species and their behaviour. These parameters need to be taken into account in virus forecasting models and control systems (Sigvald 1986). These elements will be discussed in this chapter.

Molecular Mechanisms Involved in PVY- Aphid Interactions

As with all non-persistent viruses, PVY has a basic simple virion structure with its RNA encapsidated in rod-shaped particles by a single type of CP that is required for PVY transmission. PVY requires a non-structural protein or "helper component", the helper component proteinase (HC-Pro), which acts in its homodimer active form (Thornbury et al. 1985) as a molecular "bridge" between the virions and an aphid's mouthparts. In relation to aphid transmission, HC-Pro N-terminal domain (KITC – Lysine/Isoleucine/Threonine/Cysteine) is involved in specific binding to an aphid's stylet (Blanc et al. 1998), more precisely with receptors located in the extreme tip of the stylet called the acrostyle (Uzest et al. 2007), while its C-terminal domain (PTK – Proline/Threonine/Lysine) has been shown to be involved either directly or indirectly in HC-Pro binding to the DAG motif (Aspartic acid/Alanine/Glycine) at the CP N-terminus (Fig. 6.2). Mutagenesis studies performed on the N-terminal region of the CP protein of other non-persistent viruses have demonstrated that amino acids in the vicinity of the DAG motif influence the transmission efficiency of aphids (Pirone and Blanc 1996).

HC-Pro is a multimodular protein with numerous functional domains such as a zinc-finger motif involved in viral synergism with co-infecting viruses (Vance et al. 1995; Pruss et al. 1997), nucleic acid/RNA binding domains (Urcuqui-Inchima et al. 2000), peptidyl domains involved in viral genomic RNA replication and systemic

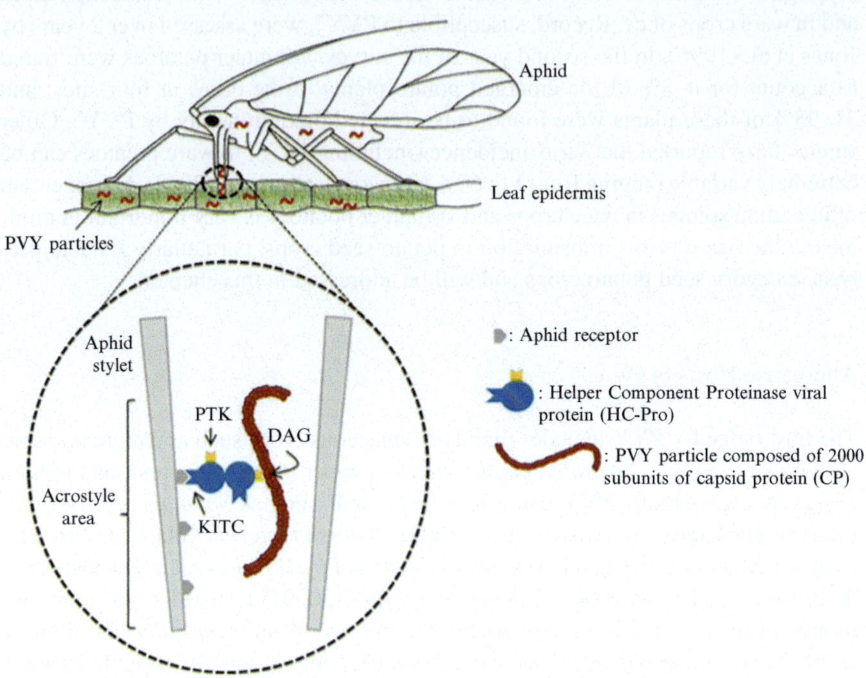

Fig. 6.2 Schematic representation of the molecular interactions between a PVY particle and an aphid stylet

movement (Cronin et al. 1995; Kasschau et al. 1997) and cysteine-protease-like domain (Oh and Carrington 1989; Guo et al. 2011). Apart from mediating aphid transmission, HC-Pro is involved in almost all processes of the viral cycle such as movement through plasmodesmata, cell-to-cell and long-distance Potyvirus movement in a plant (Kasschau and Carrington 2001; Sáenz et al. 2002), pathogenicity (Pruss et al. 1997) and suppression of antiviral RNA silencing (Llave et al. 2000).

Studies of the acquisition of purified, labelled, non-persistent viruses by aphids demonstrated that successful transmission of PVY by individual aphids of *M. persicae* to potato plants required fewer than 50 virus particles (Pirone and Thornbury 1988). In a more recent study using a competition system between infectious and non-infectious PVY isolates, it was estimated that aphid transmission of fewer than four PVY particles is sufficient to generate a successful infection (Moury et al. 2007).

2.2.2 Sources of PVY Inoculum

Volunteers and Ware Potato Crops

Volunteer plants, *i.e.*, potato plants growing from true seed or potato tubers or parts of a tuber left in the soil after the preceding crop, can be a source of PVY inoculum and, if not adequately controlled, may present a risk for PVY transmission to potato crops growing in the vicinity. In England the incidence of PVY in volunteer potatoes and in ware crops of cv. Record, susceptible to PVY^N, were assessed over 2 years by Jones et al. (1996). In the second year of the survey, volunteer potatoes were found to account for 4–8% of the emerged potato plants in the crops at four sites, and 31–93% of these plants were found to be infected predominantly by PVY^N. Other studies have reported that virus incidence (including PVY) in ware potatoes can be extremely variable ranging from 1 to 60% (Chatzivassiliou et al. 2008). Management of inoculum sources in ware crops and volunteer potatoes is very important in minimising the risk of virus transmission to potato seed crops, particularly PVY to pre-basic category seed potato crops and will be addressed in this chapter.

Alternative Hosts of PVY

The host range for PVY includes mainly solanaceous crops such as *Nicotianae* spp. (*Nicotiana tabacum, Nicotiana benthamiana*), pepper (Capsicum spp.), and tomato (*Solanum esculentum*). PVY can also infect weeds such as *Solanum sarrachoides* (hairy nightshade) (Srinivasan et al. 2008), *Solanum nigrum* (black nightshade) (Chickh-Ali et al. 2008), and ornamentals such as *Physalis floridana* (Beemster and de Bokx 1987; Kerlan 2006). Kaliciak and Syller (2009) identified a range of commonly occurring arable weeds such as *Erodium cicutarium* (redstem filaree), *Geranium pusillum* (small-flowered crane's bill), *Lactuca serriola* (prickly lettuce), and *Lamium purpureum* (purple deadnettle) that could be infected mechanically by PVY. The virus has also been detected in other weeds such as *Plantago major* (plantain), *Taraxacum* spp. (dandelion), and *Sonchus* spp. (sow thistle), *Fallopia* spp.

(knotweed), *Senecio* spp. (common groundsel), *Chenopodium album* (goosefoot) (R. van der Vlugt, personal communication). Their importance in the secondary transmission of PVY and as reservoir of the virus is likely to vary in different regions of the world where cultural practices differ. This aspect is discussed in Chap. 6.

Water as a Source of PVY Inoculum

Water sources such as canals, rivers, streams, ponds, lakes and oceans are known to be reservoirs of many viruses. Plant viruses found in natural water sources belong mainly to seven groups, but there is limited evidence about their survival in water and their potential for water-borne infection (reviewed by Mehle and Ravnikar 2012). There is minimal evidence of water transmission of potyviruses, specifically of PVY.

Recently, Mehle et al. (2014) investigated PVY^{NTN} survival in water and its water-mediated transmission. Survival of PVY^{NTN} in water was monitored under controlled conditions using macerated infected leaves in tap water. Mechanical inoculation of test plants with a PVY-infested aqueous solution confirmed that PVY^{NTN} could remain infectious in water for up to 1 week at 20 ± 4 °C. When the solution was stored at 4 °C, PVY^{NTN} remained infectious for up to 10 weeks. The role of water as a source of PVY inoculum was also assessed experimentally using a hydroponic system. Inoculated potato plants were placed in a glass tank filled with nutrient solution while healthy (bait) potato plants were placed in separate tanks. Bait plants were irrigated with nutrient solution from the tank containing inoculated plants. PVY^{NTN} was detected in the nutrient solution in the first month after starting the experiment. The virus was also found in the roots of bait plants in the first month and in the foliage 54 days after initiation of the experiment. After 4 months when the experiment was concluded, PVY^{NTN} was detected in two of the six potato bait plants. In addition, it was demonstrated that PVY^{NTN} could be released from injured tomato roots into the nutrient solution and could infect healthy tomato plants through the roots, ultimately spreading to the foliage, where it was detected more than 3 months after initiation of irrigation with the PVY-infested nutrient solution. These experiments suggest that while irrigation is not the most efficient means of transmission of PVY between plants, it can potentially represent a source of PVY inoculum at least in hydroponic systems. Hydroponic systems and/or intensive irrigation are widely used in commercial potato production; therefore, water could be an important source of PVY infection in such systems.

2.3 "Experimental" PVY Transmission

There are several "artificial" or "experimental" methods to transmit PVY. In most of the cases, these are used for research purposes, *i.e.*, studying the pathogenicity of PVY, characterising the resistance status of a given genotype of potato. Nevertheless, in some cases, this could represent a potential means of PVY dissemination in favourable conditions.

2.3.1 Mechanical Transmission

Healthy plants can be infected by PVY by manually rubbing infectious sap from a PVY-infected plant (obtained by grinding infected material with a buffered solution) onto the leaf surface together with a fine abrasive powder (such as aluminium oxide or carborundum). This method, while requiring a relatively high titre of PVY in the sap, is a convenient way of experimentally transmitting PVY to a range of host plants for studying the pathogenicity of PVY strains (see Chap. 2). For efficient transmission, it is preferable to use freshly infected biological material and young plantlets at 4–6-leaf stage. Using this method, disease symptoms usually develop 7 days after inoculation. If viral inoculum originates from freeze or dried material, the transmission rate is generally reduced and a delay in the symptom expression can be expected.

2.3.2 Transmission by Wounding

Under environmentally controlled conditions, Fageria et al. (2015) found that PVY^{O}, $PVY^{N:O}$, and PVY^{NTN} strains could be transmitted by different types of wounding (*i.e.*, squashing or squeezing leaf and stems or leaf petioles, repeated contacts of leaves through air flow between infected and bait plants). However, PVY was not transmitted by the process of cutting seed potato tubers. This study showed that wounding inflicted by the manipulation of plants in the greenhouse could result in PVY transmission and spread under particularly favourable conditions (Fageria et al. 2015; de Bokx 1972). However, de Bokx (1972) reported previously that, under field conditions, transmission of viruses through wounding is not as efficient as under greenhouse conditions, perhaps because plants are less brittle and PVY infectiveness is significantly reduced outdoors.

2.3.3 Transmission by Grafting

PVY can also be transmitted from infected to healthy plants by grafting. This method consists of joining together an infected piece of potato stem with another solanaceous plant such as potato, tobacco, tomato, or pepper. An apical scion from a virus-infected plant is placed into a slit cut the stem of the stock plant so that a vascular junction is created allowing PVY to move in the phloem sap into the healthy plant (Fig. 6.3). For PVY, this approach is essentially used to study the reaction of potato genotypes to infection by PVY. If a potato cultivar carries the hypersensitive resistance *N* gene, top necrosis symptoms will be observed in the grafted stock.

Fig. 6.3 Representation of the grafting method of potato on to potato. Photographs were provided by J.P. Dantec (INRA, Ploudaniel, France). (1) a healthy potato plant is used as stock; (2) select an apical part of a virus-infected potato plant to prepare the scion, cut downward away from the apex to expose a large area of stem tissue; (3) split the stock's stem at mid-height, keep the split slightly open, insert the scion to ensure the cambia are in contact, and hold the assembly together with Parafilm; (4) wait a few days to be sure the graft has taken (Adapted from de Bokx 1972)

3 Epidemiology and Dissemination of PVY

3.1 Characteristics of the Transmission and Dissemination of PVY Strains Under Field Conditions

The prevalence of PVY in cultivated potato crops depends on many biotic and abiotic factors. As detailed below, the PVY spread depends essentially on the incidence of PVY inoculum in the seed potatoes, climatic conditions, the genetic background of the potato cultivar (*i.e.*, relative resistance to PVY strains), and, as discussed earlier, the pressure from PVY vectors between plant emergence and haulm destruction prior to harvest (Rolot 2005; Steinger et al. 2014). In the USA, Nolte et al. (2004) reported that the incidence of PVY in daughter tubers derived from virus-free seed potatoes varied significantly among cultivars ranging from 1.9 to 13.2%. While REF values for aphid species and PVY strains (Verbeek et al. 2010) have been assessed in environmentally controlled transmission experiments, transmission of PVY strains in the field will be affected by various parameters, *e.g.*, temperature, aphid phenology, and the abundance of sources of inoculum.

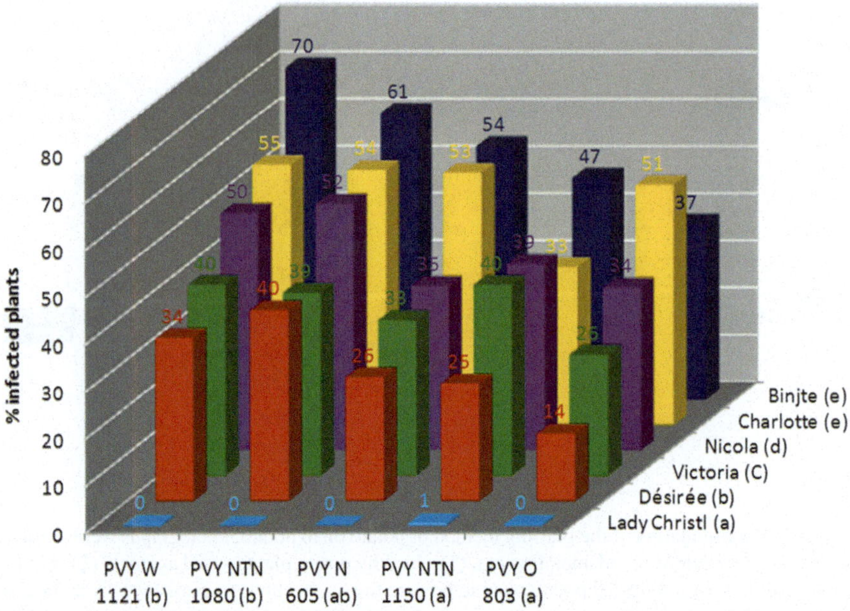

Fig. 6.4 The incidence of plant infection caused by five PVY strains ($PVY^{N\text{-}Wi}$ isolate 1121, PVY^{NTN} isolate 1080, PVY^{NTN} isolate 1150, PVY^{N} isolate 605 and PVY^{O} isolate 803) on six potato cultivars, as assessed by testing daughter tubers collected at harvest (mean of a 3 year trial). *Letters in brackets* shows the homogeneity groups of the Newman–Keuls mean comparison test

Information on the dynamics of transmission of various PVY strains in the field is limited. The transmission of five isolates of four PVY strains was studied in small plots, each consisting of 50 plants with 4 replications, over a 3-year period (Dupuis and Schwaerzel 2011). Plant infection was assessed by testing a sample of tubers collected from each plant at harvest time. The incidence of plant infection varied significantly among strains and cultivars (Fig. 6.4). The $PVY^{N\text{-}Wi}$ strain produced the highest incidence of plant infection while infection by PVY^{O} strain resulted in a significantly lower incidence. There was no significant difference in the incidence of plant infection for the PVY^{N}, PVY^{NTN}, and $PVY^{N\text{-}Wi}$ strains. However, the incidence of infection for the PVY^{NTN} strain was greater for isolate 1080 than for isolate 1150, indicating that differences in transmission characteristics may exist among isolates of the same strain. No interaction was detected between the year of experiment and the incidence of plant infection produced by the PVY strains. A second study in larger plots, each of consisting 100 plants with 6 replications, was undertaken over a 2 year period in order to compare the transmission efficiency of a $PVY^{N\text{-}Wi}$ and a PVY^{NTN} strain (Dupuis et al. 2014). The incidence of plant infection on the two cultivars was greater for $PVY^{N\text{-}Wi}$ strain than for PVY^{NTN} strain over both years of the experiment (Fig. 6.5). These results suggest that the transmission of PVY strains to potato crops may vary significantly

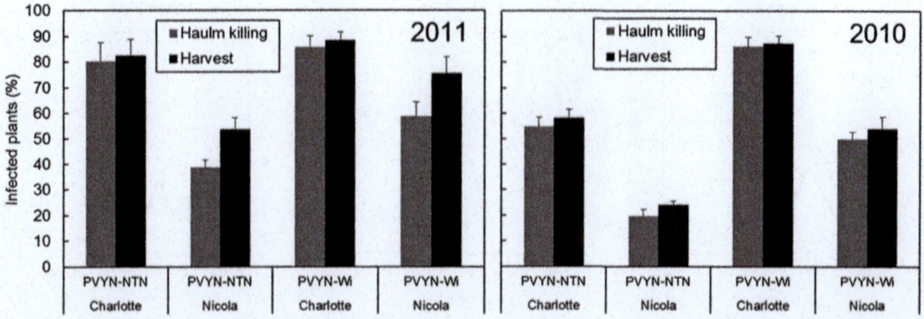

Fig. 6.5 The incidence of plant infection caused by PVY^{NTN} and $PVY^{N\text{-}Wi}$ strains in cultivars Nicola and Charlotte. The incidence was assessed by testing daughter tubers collected before haulm killing and 4 weeks later at harvest time. *Error bars* show the standard error of the mean

among strains and cultivars with an overall risk of greater transmission with PVY^{NTN} and $PVY^{N\text{-}Wi}$ strains than with PVY^{O}, potentially explaining the increased prevalence of these variants in potato crops.

3.2 *Modelling Approach to Determine the Relative Importance of Aphid Species in Transmitting PVY*

Modelling can be used to predict the relative importance of aphid species known to be PVY vectors. In the high grade (HG) seed potato production zone in Finland, over 30,000 winged aphids from 6 or 7 potato fields were caught over three growing seasons using yellow pan traps that were emptied twice a week (Kirchner et al. 2011). Of the aphids caught, 37% belonged to nine species known to transmit PVY. The seasonal increase of PVY (incidence of PVY in harvested daughter tubers compared with that in seed tubers), aphid counts in traps, aphid REF values, and PVY resistance status of potato cultivars were used as explanatory variables. Early-season flights of aphids were the most important factor accounting for the incidence of PVY. *Aphis fabae* was found to be the most important vector, because only models including this aphid species showed a strong statistical model fit with the incidence of PVY in the harvested tubers (Kirchner et al. 2011).

In a different study in Scotland, a similar approach was adopted by correlating weekly incidence of virus in experimental test plants and weekly aphid catches over several years (Pickup et al. 2009; Fenton et al. 2012). A logistic regression model based on binomial response data was used to relate PVY infection to aphid counts for individual species. The cereal aphids *M. dirhodum* and *S. avenae* were identified as the most important PVY vectors (Fig. 6.6).

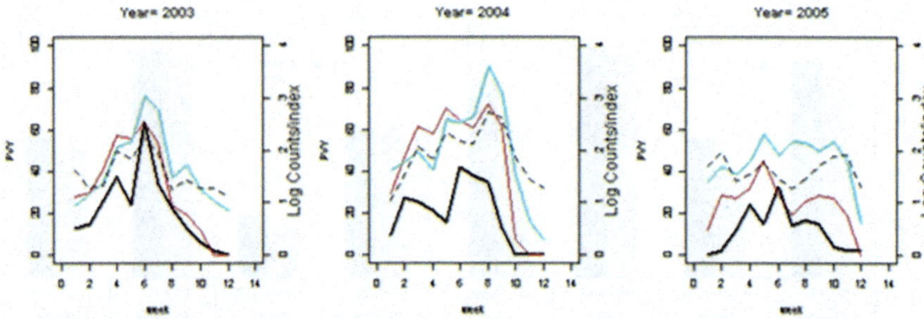

Fig. 6.6 Weekly incidence (%) of PVY^O in bait plants (*thick black line*), suction trap counts for *M. dirhodum* (*red line*) and *S. avenae* (*blue line*), and index of vector pressure (*dashed black line*) over 3 years (Fenton et al. 2012; Pickup et al. manuscript in preparation)

3.3 *Aphid Monitoring, Phenology, and Current Status of PVY Transmission Worldwide*

3.3.1 Aphid Monitoring and Aphid Trapping Techniques

The association of virus spread in potatoes with aphid flights is documented in numerous publications (Ashby 1976; Boiteau et al. 1988; DiFonzo et al. 1997; Harrington et al. 1986; Robert et al. 2000; Basky 2002; Radcliffe and Ragsdale 2002; Sigvald 1989; Steinger et al. 2015; Kirchner et al. 2011).

The activity of aphids can be monitored using several techniques:

- Sticky fishing-line traps made of a wooden frame and transparent sticky polyamide threads, the frame being fixed vertically and oriented into the wind (Labonne et al. 1983)
- Traps which collect insects flying in the air by suction at a specific flow rate (Fig. 6.7).
- Coloured pan traps (Moericke 1951) which attract and trap alate aphids (water pan traps or sticky tile traps) (Fig. 6.7).

Suction traps and sticky fishing-line traps do not rely on attracting aphids and thus sample aphids in an unbiased way. The latter can easily be placed in crops at the height of foliage providing information on aphid activity within crops. Suction traps which are around 12 m high (Johnson and Taylor 1955) might not reflect exactly the abundance and composition of aphid populations in specific crops; however, it is generally accepted that suction traps can give a good representation of aphid phenology for larger areas (up to 30–60 km wide) and are easier to manage on a daily basis. Suction traps can provide information on aphid populations more rapidly and are widely used as a warning system for the risk of virus transmission. For more localised studies, *i.e.*, the spread of a virus in a crop, the use of trap(s) positioned in a crop is recommended.

Fig. 6.7 Suction trap (Rothamsted-type 12.2 m high) (*right panel*) and water pan trap (*right panel*) currently used for aphid monitoring (Courtesy of SASA, Crown copyright)

Strong correlations between aphid catches in suction traps and yellow pan traps have been found in Belgium, even when traps were several kilometres apart (Rolot 2005). Similar studies conducted in other locations showed that the range of aphid species and numbers of each species collected differed between yellow pan traps and suction traps (Robert et al. 1987; Labonne et al. 1989, Seco et al. 1990). Numbers of *A. fabae* and *M. euphorbiae* were greater in yellow water traps than in suction traps, while *M. persicae* and *M. dirhodum* made up a greater proportion of aphids caught in suction traps.

The efficacy of green (GWT) and yellow (YWT) water pan traps and suction trap has been assessed in the Ardennes region of Belgium. Water pan traps were positioned in three different fields of potato with a suction trap (ST) situated within a few hundred metres to 3 km from the fields. Aphids were counted and identified over a 10-week period. A significant difference was found between YWT and GWT in their efficacy in catching aphids ($F_{(1,2)}$, $P = 0.0065$), with more being caught in YWT than in GWT (Fig. 6.8).

The relative abundance of each species in each type of water trap (YWT, GWT) was also analysed by testing the interaction "Species × Colour" against "Species × Colour x Sites". These results suggest that, at least in this case, trap colour might not have had a significant impact on the relative proportion of aphid species caught in each potato field ($F_{(8,16)}$, $P = 0.3849$, data not shown). However, other reports on aphid response to colour in field experiments suggest that colour, brightness, and contrast with surroundings crops might influence aphid trapping (Baldy and Rabasse 1983; Doring and Chittka 2007). The data from the water pan traps were also compared with suction trap data (Fig. 6.8 right panel). A discrepancy in aphid trapping efficacy between ST and YWT or GWT was observed for *Aphis* spp., *Brachycaudus helichrysi*, *M. persicae*, *R. padi*, and *S. avenae*. This might reflect differences in flying patterns and behaviour among various aphid species. In addition, because that *M. persicae* is the most efficient vector of PVY, the difference between traps may be important in predicting the risk of PVY transmission in potato crops.

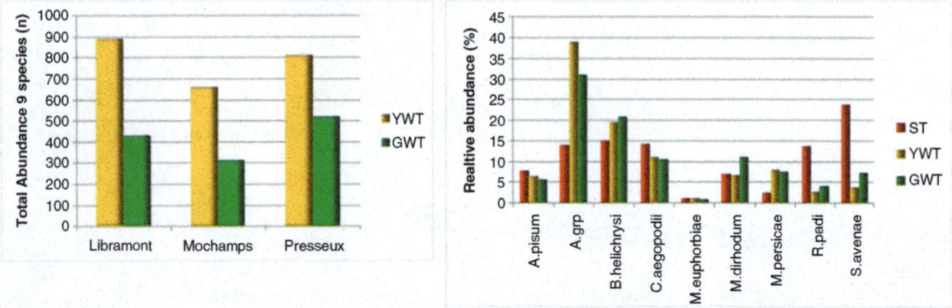

Fig. 6.8 Total number of aphids caught in *yellow water traps* (YWT) and *green water traps* (GWT) in three locations in Belgium between weeks 21 and 31 (*left panel*). Relative abundance of each of the 9 selected species expressed as a percentage of total number of aphids caught in YWT, GWT, and suction trap (ST) (*right panel*)

3.3.2 Cases Studies on the Current Status of PVY Transmission and Aphid Species Phenology Worldwide

Finland

The HG seed potato production zone in Finland (latitude 64°) is one of the northernmost intensive crop production areas of the world. PVY was not common there until 2005 when many seed potato lots were down-graded in certification due to a sudden increase in the incidence of PVY, mainly PVY^{NTN}-like strains. In a study conducted between 2007 and 2012, aphids were monitored in potato crops in the HG zone using yellow pan traps. Over 58,000 aphids covering 83 aphid taxa were caught and 34 species were further characterised by DNA barcoding. The number of aphids caught peaked every 3 years in a recurring cycle. Most of the agriculturally relevant species occurred late in the potato growing season and did not contribute significantly to the spread of PVY (Kirchner et al. 2013). However, *A. fabae* was one of the most common species caught in yellow traps being found mainly in the first 2–3 weeks after emergence of the potato plants when they are most susceptible to virus infection. *A. fabae* is known to be a vector of PVY, and modelling analysis determined that it was the main vector of PVY in the HG zone (see Chap. 4.3; Kirchner et al. 2011). The snowball tree (*Viburnum opulus*) grown in gardens is the only known winter host of *A. fabae* there. Based on this information, four approaches were designed to reduce PVY infection pressure on seed potato crops in the HG zone: (1) Straw mulch to be spread on potato crops at plant emergence using a bale chopper in order to reduce the landing of *A. fabae* on plants (Kirchner et al. 2014). This new practice was readily adopted by growers because barley is grown as an intercrop in the seed potato farms and cattle farmers use the straw and have bale choppers. (2) The seed potato growers sponsored and conducted a campaign to replace snowball trees in gardens with other ornamental species, in order to reduce the local overwintering population of *A. fabae*. (3) Production of seed potatoes of the potato cultivar most commonly grown in Finland was terminated because it did not show symptoms when infected by PVY^{NTN}-like strains and roguing of infected plants was, therefore, not possible. The cultivar was considered to be a major reservoir of PVY and a source of PVY inoculum for seed potato crops in the HG zone. 4) The recurring triennial cycle of aphid abundance in the HG zone was utilised in planning the production of seed potatoes of cultivars susceptible to PVY. Since these proposals were implemented, the incidence of PVY in seed potato crops has decreased in the HG zone and the downgrading of seed potato lots at certification is now as rare as it was prior to 2005.

Switzerland

In Switzerland, the flight activity of aphids has been monitored since 1987 using a standard Rothamsted suction trap located at the Agroscope Agricultural Research Station in Nyon. A second trap located in Reckenholz (Zürich) ceased operation in 2010. The traps are part of the Swiss seed potato certification scheme managed by Agroscope, who also contribute the aphid data to the European EXAMINE project. The following species are identified and counted on a daily basis during the main flight period of aphids (April–November): *A. pisum*, *A. fabae*, *Aphis* sp., *Aulacorthum solani*, *Brachycaudus helichrysi*, *Brevicoryne brassicae*, *Cavariella* sp., *M. euphorbiae*, *M. dirhodum*, *Metopolophium festucae*, *Myzus ascalonicus*, *M. persicae*, *Phorodon humuli*, *Rhopalosiphum insertum*, *Rhopalosiphum maidis*, *R. padi*, and *S. avenae*. In addition, the total number of aphids of the family Aphididae and non-Aphididae are counted.

Aphid data from the suction trap is used to calculate vector pressure, an indicator for the risk of PVY transmission in seed potato crops. Information is updated weekly and published on the internet platform "Agrometeo" (www.agrometeo.ch) to inform growers about virus risk. Furthermore, the aphid data is fed into a forecast model which predicts the incidence of tuber infection by PVY under a range of epidemiological conditions using factors such as cultivar, presence of virus inoculum in seed potatoes, number of mineral oil applications, altitude of the field, and date of haulm destruction. The relative contributions of these factors to PVY development in potato crops were obtained from a statistical analysis of compiled seed certification data spanning two decades (Steinger et al. 2014). The seed potato certification body at Agroscope uses model prediction along with field observations to set the optimal date of haulm destruction, which growers are obliged to follow.

The initial model employed to forecast post-harvest PVY incidence in Switzerland (named TuberPro) used vector pressure estimated from catches of 11 aphid species in suction traps as one of the input variables (Nemecek 1993). Aphid counts were weighted using REF values for each species. *Myzus persicae* had the highest weighting. Recently, a simple linear regression model, using the cumulative number of *B. helichrysi* caught as a predictor variable, was found to provide a more accurate forecast of PVY risk than *M. persicae* (Fig. 6.9) (Steinger et al. 2015). This new

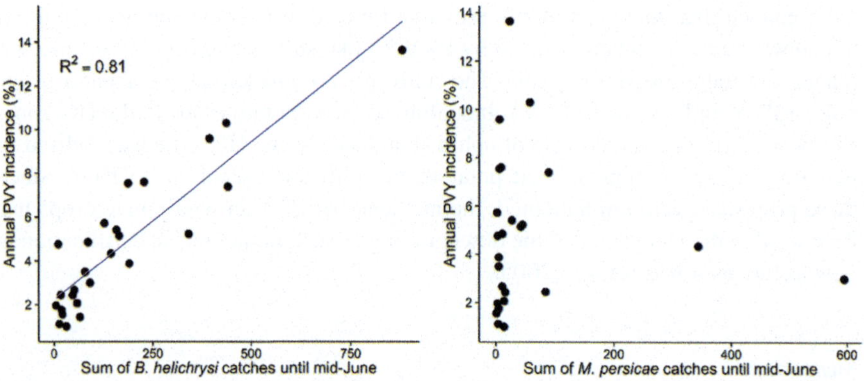

Fig. 6.9 Relationship between the incidence of PVY in tubers of susceptible potato cultivars and suction trap catches of *Brachycaudus helichrysi* and *Myzus persicae* in Switzerland. *Dots* represent the annual percentage of PVY-infected tubers averaged over all tested seed lots (N = 150–611 seed lots per year) (Adapted from Steinger et al. 2015)

model superseded TuberPro in 2013. *B. helichrysi* is an aphid species that does not colonise potato plants but flies relatively early in the growing season, which may at least partly explain its importance as a vector of PVY. Interestingly, the abundance of *M. persicae*, often considered to be the main vector of PVY, was not associated with PVY incidence in Swiss seed potatoes (Fig. 6.9).

Slovenia

While data on the aphid species responsible for PVY transmission in Slovenia is not currently available, their occurrence in potato crops was assessed between 2003 and 2006 in the four main areas of seed potato production: Komenda, Jablje, Øentvid pri Stioeni, and Libelice. Yellow water (Moericke) traps were used to determine the number and species of aphids present in potato crops during the growing season (Modic and Urek 2008). The majority of aphids of the family Aphididae that were caught belonged to the genus *Aphis*. However, some species previously unreported in Slovenia were recorded: *Aphis spiraecola* (Patch), *Amphorophora gei* (Börner), *Chaitophorus leucomelas* (Koch), *Chaitophorus populeti* (Panzer), *Drepanosiphum aceris* (Koch), *Cinara* sp., *Macrosiphum cholodkovskyi* (Mordvilko), *Macrosiphum gei* (Koch), *Myzocallis castanicola* (Baker), *Myzocallis coryli* (Goeze), *Protrama flavescens* (Koch), *Protrama ranunculi* del Guercio, *R. maidis* (Fitch), *Thelaxes dryophila* (Schrank), *Therioaphis luteola* (Börner), *Trama rara* (Mordvilko), and *Tuberolachnus salignus* (Gmelin). On the basis of previous publications and this survey, 180 species of true aphids of the family Aphididae have now been identified in Slovenia. In this study, 197 aphid species were found in seed potato crops.

France

France, as with 18 other European countries, has collaborated since 2000 in the EXAMINE programme (Exploitation of Aphid Monitoring in Europe – an EU-funded project within the Concerted Action Programme, with the remit of collecting national data on aphid distribution, phenology and abundance throughout Europe) to monitor aphid flights with a network of suction traps. In addition, the National Federation of Seed Potato Growers (FNPPPT, Fédération Nationale des Producteurs de Plants de Pomme de Terre) together with its regional producer organisations, established in 1970s an aphid trapping network of yellow water (Moericke) traps in the three main seed potato production areas. Currently, there are 18 traps: 5 in the west, 10 in the north, and 3 in the centre and the south of France. Trapping is done during the potato growing season, from April to the end of July, and the content of traps is analysed twice a week to determine the number and species of aphids. However, not all specimen are identified to the species level. Routinely, the assessment focuses on 11 species and/or group species known to transmit PVY: four that colonize potato (*M. persicae, Aulacorthum solani, M. euphorbiae, Aphis* spp.) and seven that do not colonise potato (*R. padi, S. avenae, M. dirhodum, A. pisum, Brevicoryne brassicae, Phorodon humuli, Cavariella* sp.). The data on the flights of these aphids is transmitted to potato inspectors and seed potato producers by fax with a recommendation on protection measures appropriate for the aphid pressure. In addition, this information is also integrated into plant health news bulletins which alert growers on pests present in crops. In 2006, the monitoring of nine yellow traps from west, north, and centre of France showed differences in the aphid species recorded at the various sampling sites (Fig. 6.10) (Le Hingrat 2007). *M. persicae* was the most abundant aphid species caught in the west of France, whereas in the north and the centre, the most abundant aphids were mainly the non-colonising *M. dirhodum* and *Cavariella* sp. In 2014, in west area, the situation was reversed compared with 2006, namely, a preponderance of non-colonising aphid species (*M. dirhodum, Cavariella* sp., *Aphis* spp., *S. avenae*). It was concluded that aphid species diversity depends on environment factors, notably the climate and neighbouring crops.

Although aphid flights can be monitored and the efficiency of an aphid species to transmit PVY can be evaluated, it is still a challenge to determine accurately those aphid species responsible for causing PVY infection of potato crops. Indeed, the latent period between aphid transmission of PVY to a plant and the expression of symptoms, (or when the PVY infection could be detected by ELISA), is about 10–15 days in the environmental conditions in France. To improve knowledge of the interactions between aphids, PVY, and the environment and to identify the parameters involved in PVY epidemiology, monitoring of PVY infection in seed and ware potato crops was undertaken as part of a 3-year research project (INRA-Rennes-Le Rheu and FNPPPT) (Boisgontier et al. 2013). Factors associated to landscape, meteorological data (temperature, rainfall, wind speed, and direction), aphid flights (number, species), and potato PVY infection were monitored, *e.g.*, meteorological data every 15 min each day, PVY infection survey once per week, and aphid catches

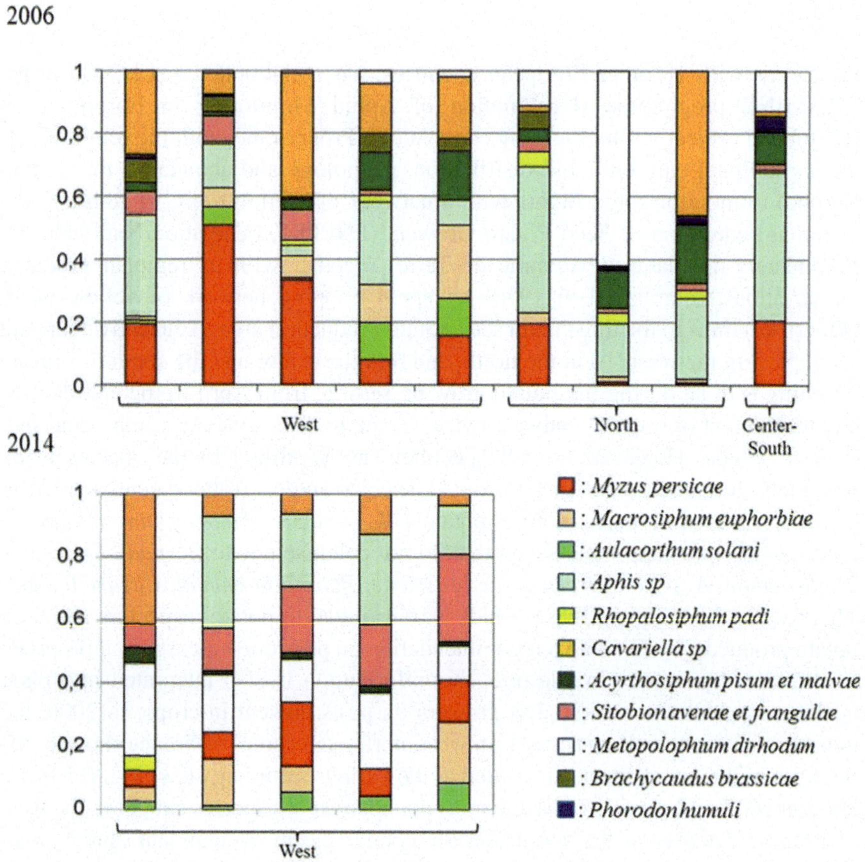

Fig. 6.10 Relative abundance of some aphid species caught in 2006 and 2014 in *yellow* water traps in the three main seed potato areas of France (West, North, Center-South)

three times per week. Depending on the season, 18–42 aphid species or group species were identified in the trap samples from potato crops, with *M. persicae*, *M. dirhodum*, *A. pisum*, and *Brevicorynae brassicae* being the most prevalent aphid species probably due to the proximity of cereal, maize, and rape crops in the vicinity of the potato crops. The analysis of quantitative and qualitative data linked to aphid flights and PVY infection in crops identified the presence of aphids throughout the growing crop season as a key factor in the incidence of PVY (Fig. 6.11). It was concluded that modelling PVY incidence at post-harvest in relation to the initial PVY incidence of the parent crop, the spread of PVY infection, and aphid phenology is very complex and additional parameters such as environmental conditions affect the model significantly. However, cultural practices (using or not insecticide and mineral oil) and initial quality of planted seed are two of the most important parameters affecting PVY incidence post-harvest.

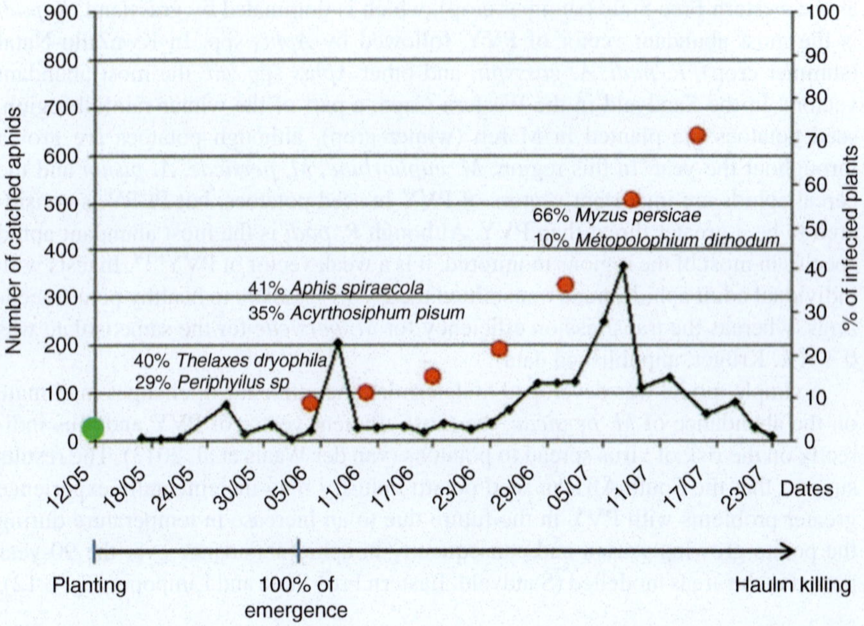

Fig. 6.11 Relationship between prevalence of aphids (*black line*) and PVY infection (*red circle*) in a ware potato crop from the planting to haulm killing. The *green circle* corresponds to the incidence of PVY in seed potatoes. Aphid species listed in the *graph* correspond to those most commonly identified in peaks of aphid numbers

South-Africa

The South African aphid monitoring network consists of 13 Rothamsted-type suction traps distributed in the Western Cape, Northern Cape, Free State, North West, and Limpopo. The national network was initiated in 2005 by Potatoes South Africa and is coordinated by the University of Pretoria. Aphid assessments are conducted on a daily or weekly basis. Bulletins are issued to growers to alert them to the risk of virus spread. These provide information on the activity of aphids and cumulative vector pressure based on the abundance of aphid species and their REF values following van Harten (1983) (mean values derived from published literature and South Africa unpublished results), together with weekly SMS notifications. The suction trap network is funded by Potatoes South Africa, the Western Cape Department of Agriculture, the KwaZulu-Natal Department of Agriculture, the Winter Cereal Trust, the Agricultural Research Council, and the Technology and Human Resources for Industry Programme (THRIP). In addition, yellow water traps are used locally to monitor the prevalence of aphids in potato crops.

The most important of the 79 species or species groups identified are *A. pisum*, *Aphis gossypii*, and other *Aphis* spp., *M. euphorbiae*, *M. persicae*, and the cereal aphids *M. dirhodum*, *R. padi*, and *S. avenae.* Their contribution to vector pressure varies with region. For example, in some areas in the summer rainfall region, such as the western Free State (summer crop), which is dominated by grassland, *R. padi* is the most abundant vector of PVY, followed by *Aphis* spp. In KwaZulu-Natal (summer crop), *R. padi*, *A. gossypii*, and other *Aphis* spp. are the most abundant vectors. In the Sandveld in the Western Cape, a part of the winter rainfall region, seed potatoes are planted in March (winter crop), although potatoes are grown throughout the year. In this region, *M. euphorbiae*, *M. persicae*, *A. pisum* and the cereal aphids are important vectors of PVY in seed potatoes, but PLRV is considered to be a greater threat than PVY. Although *R. padi* is the most abundant aphid species in most of the regions monitored, it is a weak vector of PVY^{NTN}. In tests with individual adult aphids, none transmitted the PVY^{NTN} isolate to healthy potato seedlings, whereas the transmission efficiency for *M. persicae* for the same isolate was 0.33 (K. Krüger, unpublished data).

A simple model was developed to determine the influence of changes in climate on the abundance of *M. persicae*, the most efficient vector of PVY and thus indirectly on the risk of virus spread to potatoes (van der Waals et al. 2013). The results suggest that the South African seed potato industry may incrementally experience greater problems with PVY in the future due to an increase in temperature during the potato growing season and consequently in aphid abundance over the 90-year period in the areas modelled (Sandveld, Eastern Free State and Limpopo; Fig. 6.12).

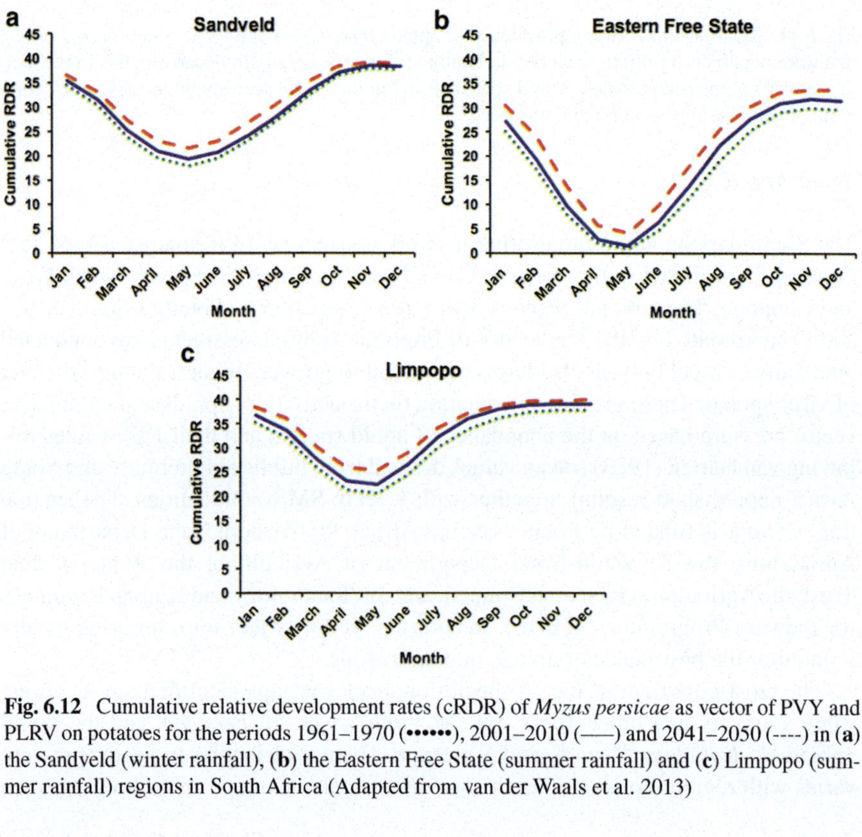

Fig. 6.12 Cumulative relative development rates (cRDR) of *Myzus persicae* as vector of PVY and PLRV on potatoes for the periods 1961–1970 (••••••), 2001–2010 (—) and 2041–2050 (----) in (**a**) the Sandveld (winter rainfall), (**b**) the Eastern Free State (summer rainfall) and (**c**) Limpopo (summer rainfall) regions in South Africa (Adapted from van der Waals et al. 2013)

Although the results suggest an increase in aphid abundance, population growth may be limited by high temperature in areas such as Limpopo and the Sandveld, where aphids may reach their thermal limits during the hottest months. Shifts in main planting times for potatoes in some regions may lessen vector pressure and consequently the risk of PVY transmission.

Scotland/UK

In Scotland, Science and Advice for Scottish Agriculture (SASA-Scottish Government) provides a diagnostic service for viruses in seed potato crops (growing crops and post-harvest) and is part of a network of suction traps operated by Rothamsted Research (England, UK) in England and Scotland. Information on the weekly catches of aphids is published in a bulletin as part of the Rothamsted Insect Survey and advice is also provided on the risk of virus transmission and the need for aphid control. The data is contributed to the EU EXAMINE program. An Aphid Monitoring programme was introduced into the Seed Potato Classification Scheme in 1992 to identify seed potato crops in which potato aphids had been poorly controlled. Subsequent classification of seed potatoes from these crops was dependent upon a satisfactory post-harvest tuber test for the presence of viruses. This programme was discontinued because it was relatively unsuccessful in identifying crops with a high risk of virus infection.

The Agriculture and Horticulture Development Board-Potatoes (AHDB-Potatoes) funds 100 water traps located in potato crops in the major seed potato growing areas of Great Britain. The contents of each trap are analysed weekly at the Food and Environment Research Agency in York (UK). Registered users receive e-mail and SMS alerts when the peach-potato aphid (*M. persicae*) is found in their region or when aphid catches in any trap in their region exceed a specified weekly threshold. Comparative information with previous seasons is also available.

The pattern of changing incidence of PVY^N in Scottish seed potato crops closely follows the incidence of three species of cereal aphid: the rose-grain aphid (*M. dirhodum*), the grain aphid (*S. avenae*), and the bird cherry-oat aphid (*R. padi*). Using a model based on these three abundant cereal aphids, the incidence of PVY^N in seed potatoes was found to be strongly correlated with the abundance of these aphids up until day 210 (29 July) (Fig. 6.13). The incidence of PVY^N in seed potato crops in Scotland can be predicted using the incidence of PVY in the previous year and geometrical mean of the catches of these three aphids prior to day 210 in three standards Rothamsted suction traps (Elgin, Dundee and Edinburgh, UK).

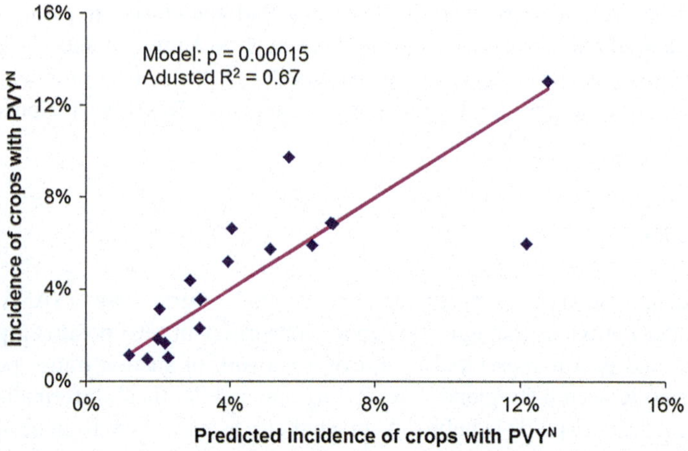

Fig. 6.13 Relationship between observed PVY^N incidence in Scottish seed potato crops and predicted PVY^N incidence based on suction trap catches of *M. dirhodum, S. avenae, and R. padi*

Belgium

In Belgium, aphid monitoring is operated using two suction traps located in Libramont (since 1983) and in Gembloux (since 1981) under the supervision of the Walloon Agriculture Research Center. Aphids are counted and identified on a daily basis each year from mid-April (week 16) until mid-October (week 41). For seed potatoes, a set of 12 species or species groups are followed during the cropping season based on their overall relative abundance in the catches and their efficiency for PVY transmission. The vegetative period for seed potatoes is generally between week 21 (first emergence of plants end of May) and week 31 (last date of destruction of foliage, end of July). A strong correlation was established between the abundance of flying aphids and the quality of the seed lots, especially regarding PVY which is the most prevalent virus in seed potato crops in Belgium. The downgrading rate of seed lots following post-harvest testing depends on: (1) the abundance of the selected species as measured by suction traps during the vegetative period, (2) the relative efficiency of each of the prescribed species in transmitting PVY (REF value), and (3) the mature plant resistance status of the plants (MPR-Table 6.3). These three factors enable a weekly index for the PVY infection pressure (IP) to be calculated using the equation:

$$\mathrm{IP_j} = \left[\sum_{i=1}^{p} \left(\mathrm{N_i} \times \mathrm{REF_i}\right)\right] \times \mathrm{MPR_j}$$

IP_j: infection pressure during week j; i….p: aphid species, N_i: aphids number of the considered species during week j, r.e.f_i: REF for species; MPRj: MPR status during week j.

Table 6.3 Mature plant resistance factor according plant age (Sigvald 1987)

	Weeks n°21 à 26	Week 27	Week 28	Week 29	Week 30	Week 31	Week 32
MPR	1	0,8	0,6	0,4	0,2	0,1	0

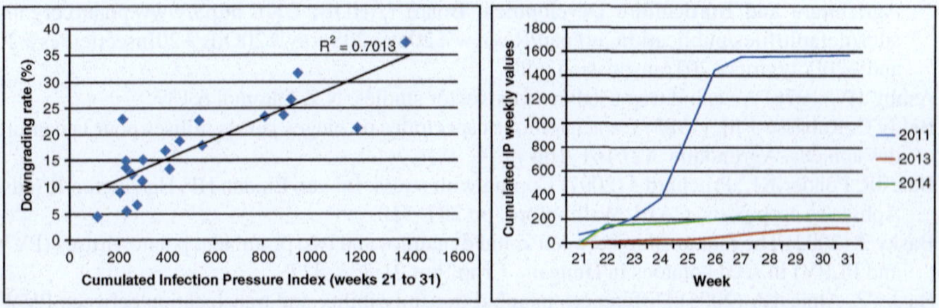

Fig. 6.14 Relation between the cumulative index of PVY infection pressure (IP) (weeks 21–31) and the downgrading rate of Elite seed potato lots after laboratory testing (*left panel*). IP index during the growing season: comparison of 2014 with two reference seasons (2013, low pressure – 2011, high pressure) (Libramont, Belgium) (*right panel*)

The weekly values can be accumulated to get a value of IP for the whole season which is strongly related to the downgrading rate of seed lots after laboratory testing. Moreover, the weekly cumulative IP can be plotted and compared in a graph with values for reference seasons (low-pressure and high-pressure seasons) so that the current IP can be compared with previous ones (Fig. 6.14) to provide advice to growers.

4 Conclusion and Future Prospects

As for many other viruses transmitted in a non-persistent fashion, the interactions between PVY and their aphid vectors and environment are very complex. As we are just beginning to uncover the molecular basis of virus-vector relationships, the ongoing and future challenge faced are to integrate the wealth of data on the molecular diversity of PVY, their interaction with their hosts-vector-environment, and how these interactions will affect transmission and spread. Ultimately, gathering knowledge from a wide variety of ecological niche should help us understand what affects PVY epidemiology at the macroscopic level and, in practice, refine PVY risk modelling.

References

Al Mrabeh A, Anderson E, Torrance L, Evans A, Fenton B (2010) A literature review of insecticide and mineral oil use in preventing the spread of non-persistent viruses in potato crops. Agriculture and Horticulture Development Board (AHDB, UK). http://www.potato.org.uk/sites/default/files/publication_upload/Review%20of%20Spray%20Oils%20Insecticides%20and%20Potyvirus%20Transmission.pdf

Ashby JW (1976) An aphid trap used in virus-vector studies. N Z Entomol 6:187

Baldy C, Rabasse J-M. (1983) Caracteristiques spectrales de pieges jaunes utilises pour la capture des aphides. Agronomie 3(2):161–166

Bale JS, Ponder KL, Pritchard J (2007) Coping with stress. In: van Emden HF, Harrington R (eds) Aphids as crop pests. CABI, Wallingford, pp 287–310

Basky Z (2002) The relationship between aphid dynamics and two prominent potato viruses (PVY and PLRV) in seed potatoes in Hungary. Crop Prot 21:823–827

Basky Z, Almasi A (2005) Differences in aphid transmissibility and translocation between PVY^N and PVY^O isolates. J Pest Sci 78:67–75

Basky Z, Raccah B (1990) Aphids colonizing peppers in Hungary and their importance as virus vectors. Acta Phytopathologica Entomol Hung 25:383–391

Beemster ABR, de Bokx JA (1987) Survey of properties and symptoms. In: de Bokx JA, van der Want JPH (eds) Viruses of potatoes and seed-potato production. Pudoc, Wageningen, pp 84–113

Blanc S, Ammar ED, Garcia-Lampasona S, Dolja VV, Llave C, Baker J, Pirone TP (1998) Mutations in the potyvirus helper component protein: effects on interactions with virions and aphid stylets. J Gen Virol 79:3119–33122

Boisgontier D, Orensanz J, Glais L, Boulard F, Laty P, Deudon O, Gaucher D, Golaz F, Molot B, Giovinazzo R, Jacquot E, Le Hingrat Y (2013) Spatialisation de l'Information Météorologique Pour l'Amélioration des Techniques Culturales (projet SIMPATIC). Innov Agronomiques 30:329–347

Boiteau G, Singh RP, Parry RH, Pelletier Y (1988) The spread of PVY^0 in New Brunswick potato fields: timing and vectors. Am Potato J 65:639–649

Boiteau G, Singh M, Singh RP, Tai GCC, Turner TR (1998) Rate of spread of PVY^N by alate *Myzus persicae* (Sulzer) from infected to healthy plants under laboratory conditions. Potato Res 41:335–344

Boquel S, Giordanengo P, Ameline A (2011) Divergent effects of PVY-infected potato plant on aphids. Eur J Plant Pathol 129(4):507–510

Boquel S, Delayen C, Couty A, Giordanengo P, Ameline A (2012) Modulation of aphid vector activity by *Potato virus Y* on *in vitro* potato plants. Plant Dis 96:82–86

Casteel CL, Yang C, Nanduri AC, De Jong HN, Whitham SA, Jander G (2014) The NIa-Pro protein of *Turnip mosaic virus* improves growth and reproduction of the aphid vector, *Myzus persicae* (green peach aphid). Plant J 77:653–663

Chatzivassiliou EK, Moschos E, Gazi S, Koutretsis P, Tsoukaki M (2008) Infection of potato crops and seeds with PVY and PLRV in Greece. J Plant Pathol 90:253–261

Chickh-Ali M, Katayama K, Maoka T, Natsuaki KT (2008) Significance of weed hosts for *Potato virus Y* protection in Syria. EPPO Bull 38:226–232

Cronin S, Verchot J, Haldeman-Cahill R, Schaad MC, Carrington JC (1995) Long-distance movement factor: a transport function of the potyvirus helper component proteinase. Plant Cell 7(5):549–559

Davis JA, Radcliffe EB, Ragsdale DW (2005) Soybean aphid, *Aphis glycines Matsumura*, a new vector of *Potato virus Y* in potato. Am J Potato Res 82(3):197–201

de Bokx JA (1972) Graft and mechanical transmission. In: C.A.P.D (ed) Viruses of potatoes and seed-potato production, pp 26–35 de Bokx JA, Piron PGM (1990) Relative efficiency of a number of aphid species in the transmission of *potato virus* Y^N in the Netherlands. Neth J Plant Pathol 96:237–246

DiFonzo CD, Ragsdale DW, Radcliffe EB, Gudmestad NC, Secor GA (1997) Seasonal abundance of aphid vectors of *Potato virus Y* in the Red River Valley of Minnesota and North-Dakota. J Econ Entomol 90:824–831

Doring TF, Chittka L (2007) Visual ecology of aphids—a critical review on the role of colours in host finding. Arthropod Plant Interact 1:3. doi:10.1007/s11829-006-9000-1

Dupuis B, Schwaerzel R (2011) *Potato virus Y* (PVY) strains spread and symptoms on plants and progeny tubers of potato cultivars. Proceedings of the The 18th Triennial Conference of the European Association for Potato Research. Oulu, Finland 115

Dupuis B, Balmelli C, Schumpp O, Steinger T (2014) Within-field spread of two strains of *potato virus Y* (PVY). In: Research EaFP (ed) Proceedings of the 19th Triennial Conference of the European Association for Potato Research. European Association for Potato Research, Brussels, p 65

Edwards AR (1963) A non-colonizing aphid vector of potato virus disease. Nature 200:1233–1234

Fageria MS, Nie X, Gallagher A, Singh M (2015) Mechanical transmission of *Potato virus Y* (PVY) through seed cutting and plant wounding. Am J Potato Res 92(1):143–147

Fenton B, Lacomme C, Collins L, Fox A, Pickup J, Evans A, Northing P, Anderson E (2012) Aphids and virus transmission in seed crops. Potato Council and Agriculture and Horticulture Development Board (AHDB, UK) project R428 final report. http://www.potato.org.uk/sites/default/files/publication_upload/R428_Final%20Report.pdf

Fereres A, Perez P, Gemeno C, Ponz F (1993) Transmission of spanish pepper- and potato-PVY isolates by aphid (Homoptera: Aphididae) vectors: epidemiological implications. Environ Entomol 22(6):1260–1265

Foottit RG, Maw HEL, Von Dohlen CD, Hebert PDN (2008) Species identification of aphids (Insecta: Hemiptera: Aphididae) through DNA barcodes. Mol Ecol Resour 8(6):1189–1201

Fox A, Collins L, Macarthur R, Blackburn LF, Northing P (2016) New aphid vectors and efficiency of transmission of *Potato virus A* and strains of *Potato virus Y* in the UK. Plant Pathol 66:325. doi:10.1111/ppa.12561

Giordanengo P, Brunissen L, Rusterucci C, Vincent C, van Bel A, Dinant S, Girousse C, Faucher M, Bonnemain JL (2010) Compatible plant-aphid interactions: how aphids manipulate plant responses. C R Biol 333(6–7):516–523

Guo B, Lin J, Ye K (2011) Structure of the autocatalytic cysteine protease domain of potyvirus helper-component proteinase. J Biol Chem 286(24):21937–21943

Halbert SE, Corsini DL, Wiebe MA (2003) *Potato virus Y* transmission efficiency for some common aphids in Idaho. Am J Potato Res 80(2):87–91

Harrington R, Gibson RW (1989) Transmission of *Potato virus Y* by aphids trapped in potato crops in southern England. Potato Res 32(2):167–174

Harrington R, Katis N, Gibson RW (1986) Field assessment of the relative importance of different aphid species in the transmission of *Potato virus Y*. Potato Res 29(1):67–76

He C, Molen TA, Xiong X, Boiteau G, Nie X (2006) Cytochrome c oxidase mRNA as an internal control for detection of *Potato virus Y* and *Potato leafroll virus* from single aphids by a co-amplification RT-PCR assay. J Virol Methods 138(1):152–159

Hogenhout SA, van der Hoorn RA, Terauchi R, Kamoun S (2009) Emerging concepts in effector biology of plant-associated organisms. Mol Plant-Microbe Interact 22:115–122

Holmes JC, Bethel WM (1972) Modification of intermediate host behaviour by parasites. In: Canning EU, Wright CA (eds) Behavioural aspects of parasite transmission. Academic Press, London, pp 123–149

Hull R (2002) Matthews' plant virology, 4th edn. Acedemic Press, London

Ingwell LL, Eigenbrode SD, Bosque-Perez NA (2012) Plant viruses alter insect behavior to enhance their spread. Sci Rep 2:578

Johnson CG, Taylor LR (1955) The development of large suction traps for airborne insects. Ann Appl Biol 43:51–61

Jones DAC, Woodford JAT, Main SC, Pallett D, Barker H (1996) The role of volunteer potatoes in the spread of PVY^N in ware crops of cv record. Ann Appl Biol 129:471–478

Kaliciak A, Syller J (2009) New hosts of *Potato virus Y* (PVY) among common wild plants in Europe. Eur J Plant Pathol 124:707–713

Kanavaki OM, Margaritopoulos JT, Katis NI, Skouras P, Tsitsipis JA (2006) Transmission of *Potato virus Y* in tobacco plants by *Myzus persicae nicotianae* and *M-persicae s.str*. Plant Dis 90(6):777–782

Kasschau KD, Carrington JC (2001) Long-distance movement and replication maintenance functions correlate with silencing suppression activity of potyviral HC-pro. Virology 285(1):71–81

Kasschau KD, Cronin S, Carrington JC (1997) Genome amplification and long-distance movement functions associated with the central domain of tobacco etch potyvirus helper component-proteinase. Virology 228(2):251–262

Katis N, Tsitsipis JA, Stevens M, Powell G (2007) Transmission of plant viruses. In: van Emden HF, Harrington R (eds) Aphids as crop pests. CABI, Wallingford, pp 353–390

Kennedy JS, Booth CO, Kershaw WJS (1959a) Host finding by aphids in the field. I. Gynoparae of *Myzus persicae* (Sulzer). Ann Appl Biol 47:410–423

Kennedy JS, Booth CO, Kershaw WJS (1959b) Host finding by aphids in the field. II. *Aphis fabae* Scop. (gynoparae) and *Brevicoryne brassicae* L.; with a re-appraisal of the role of host-finding behaviour in virus spread. Ann Appl Biol 47:424–444

Kerlan C (2006) *Potato virus Y*. Descriptions of plant viruses, no. 414. http://www.dpvweb.net/dpv/showdpv.php?dpvno=414

Kirchner SM, Döring TF, Hiltunen LH, Virtanen E, Valkonen JPT (2011) Information theory-based model selection for determining the main vector and period of transmission of *Potato virus Y* in potato crops. Ann Appl Biol 159:414–427

Kirchner SM, Hiltunen L, Döring TF, Virtanen E, Palohuhta JP, Valkonen JPT (2013) Seasonal phenology and species composition of the aphid fauna in a northern crop production area. PLoS One 8:e71030

Kirchner SM, Hiltunen LH, Santala J, Döring TF, Ketola J, Kankaala A, Virtanen E, Valkonen JPT (2014) Comparison of straw mulch, insecticides, mineral oil, and birch extract for control of transmission of *Potato virus Y* in seed potato crops. Potato Res 57:59–75

Kostiw M (1975) Investigation on the retention of *potato viruses M* and *Y* in two species of aphids (*Myzus persicae* Sulz. and *Aphis nasturtii* Kalt.) Potato Res 18:637–640

Kostiw M (1979) Transmission of *Potato virus Y* by *Rhopalosiphum padi L*. Potato Res 22:237–238

Kring JB (1972) Flight behaviour of aphids. Annu Rev Entomol 17:461–492

Labonne G, Fauvel G, Quiot JB (1983) Intérêt des pièges à fils dans l'étude des populations de pucerons ailés. Agronomie 3:315–326

Labonne G, Lauriaut F, Quiot J (1989) Comparison of 3 types of traps to sample alate aphid populations. Agronomie 9:547–557

Le Hingrat Y (2007) Identifier les pucerons pour adapter la lutte. La Pomme de Terre Française:36–37

Llave C, Kasschau KD, Carrington JC (2000) Virus-encoded suppressor of posttranscriptional gene silencing targets a maintenance step in the silencing pathway. Proc Natl Acad Sci 97:13401–13406

Mehle N, Ravnikar M (2012) Plant viruses in aqueous environment: survival, water mediated transmission and detection. Water Res 46:4902–4917

Mehle N, Gutierrez-Aguirre I, Prezelj N, Delić D, Vidic U, Ravnikar M (2014) Survival and transmission of *Potato virus Y*, *Pepino mosaic virus*, and *Potato spindle tuber viroid* in water. Appl Environ Microbiol 80(4):1455–1462

Modic S, Urek G (2008) Contribution to the knowledge of the aphid fauna (Sternorrhyncha: Aphidoidea) of Slovenia. Acta Entomol Slov 16(1):87–97

Moericke V (1951) Eine Farballe zur Kontrolle des Fluges von Blattlausen insbesondere der pfirschblattlaus, Myzodes persicae (Sulz). Nachrichtenbl Dtsch Pflanzenschutzdienst 3:23–24
Moericke V (1955) Über die Lebensgewohnheiten der geflügelten Blattläuse (Aphidina) unter besonderer Berücksichtigung des Verhaltens beim Landen. Z Angew Entomol 37:29–91
Moury B, Fabre F, Senoussi R (2007) Estimation of the number of virus particles transmitted by an insect vector. Proc Natl Acad Sci U S A 104(45):17891–17896
Nault LR (1997) Arthropod transmission of plant viruses: a new synthesis. Ann Entomol Soc Am 90:521–541
Nemecek T (1993) The role of aphid behaviour in the epidemiology of *Potato virus Y*: a simulation study. PhD thesis ETH Zürich
Nie X, Pelletier Y, Mason N, Dilworth A, Giguère MA (2011) Aphids preserved in propylene glycol can be used for reverse transcription-polymerase chain reaction detection of *Potato virus Y*. J Virol Methods 175(2):224–227
Nolte P, Whitworth JL, Thornton MK, Mcintosh CS (2004) Effect of seedborne potato virus Y on performance of russet Burbank, russet Norkotah, and Shepody potato. Plant Dis 88:248–252
Oh CS, Carrington JC (1989) Identification of essential residues in potyvirus proteinase HC-pro by site-directed mutagenesis. Virology 173(2):692–699
Palacios I, Drucker M, Blanc S, Leite S, Moreno A (2002) *Cauliflower mosaic virus* is preferentially acquired from the phloem by its aphid vectors. J Gen Virol 83:3163–3171
Pelletier Y, Nie X, Giguère MA, Nanayakkara U, Maw E, Foottit R (2012) A new approach for the identification of aphid vectors (Hemiptera: Aphididae) of *Potato virus Y*. J Econ Entomol 105(6):1909–1914
Perez P, Collar JL, Avilla C, Duque M, Fereres A (1995) Estimation of vector propensity of *Potato virus Y* in open field pepper crops of Central Spain. J Econ Entomol 88(4):986–991
Pettersson J (1970) Studies on *Rhopalosiphum padi* (L.). I. Laboratory studies on olfactometric responses to the winter host *Prunus padus L.* Lantbrukshögskolans Ann 36:381–389
Pettersson J, Tjallingii WF, Hardie J (2007) Host-plant selection and feeding. In: van Emden HF, Harrington R (eds) Aphids as crop pests. CABI, Wallingford, pp 87–114
Pickup J, Davie K, Fox A, Highet F, Holmes R (2009) Epidemiology of viruses in Scottish seed potatoes. Asp Appl Biol Potatoes 94:5–10
Pirone TP, Blanc S (1996) Helper-dependent vector transmission of plant viruses. Annu Rev Phytopathol 34:227–247
Pirone TP, Perry KL (2002) Aphids: non-persistent transmission. Adv Bot Res 36:1–20
Pirone TP, Thornbury DW (1988) Quantity of virus required for aphid transmission of a *Potyvirus*. Phytopathology 78(1):104–107
Powell G, Harrington R, Spiller NJ (1992) Stylet activities and *Potato virus Y* vector efficiencies by the aphids *Brachycaudus helichrysi* and *Drepanosiphum platanoidis*. Entomol Exp Appl 62(3):293–300
Pruss G, Ge X, Shi XM, Carrington JC, Bowman Vance V (1997) Plant viral synergism: the potyviral genome encodes a broad-range pathogenicity enhancer that transactivates replication of heterologous viruses. Plant Cell 9(6):859–868
Raccah B, Galon A, Eastop VF (1985) The role of flying aphid vectors in the transmission of *Cucumber mosaic virus* and *Potato virus Y* to peppers in Israel. Ann Appl Biol 106:451–460
Radcliffe EB, Ragsdale DW (2002) Aphid-transmitted potato viruses: the importance of understanding vector biology. Am J Potato Res 79:353–386
Ragsdale DW, Radcliffe EB, DiFonzo CD (2001) Epidemiology and field control of PVY and PLRV. In: Loebenstein G, Berger PH, Brunt AA, Lawson RH (eds) Virus and virus-like diseases of potatoes and production of seed-potatoes. Klumer Academic Publisher, Dordrecht, pp 237–270
Robert Y, Turpeau E, Tanguy S (1987) Complementation of suction and yellow traps. In: Cavalloro R (ed) Aphid migration and forecasting 'Euraphid' systems in European Community countries. Published by the Commission of the European communities, Luxembourg, pp 103–111

Robert Y, Woodford JAT, Ducray-Bourdin DG (2000) Some epidemiological approaches to the control of aphid-borne virus diseases in seed potato crops in northern Europe. Virus Res 71:33–47

Rolot JL (2005) Analyse des facteurs régulant la dissémination du virus Y de la pomme de terre (PVY) en vue de stratégies de lutte raisonnées. Dissertation. Faculté Universitaire des Sciences Agronomiques, Gembloux, Belgium

Sáenz P, Salvador B, Simón-Mateo C, Kasschau KD, Carrington JC, García JA (2002) Host-specific involvement of the HC protein in the long-distance movement of potyviruses. J Virol 76(4):1922–1931

Salazar LF (1996) Aphid transmission of potato viruses. In: C.I.P (ed) Potato viruses and their control. International Potato Center, Lima. 214 p

Seco V, Hulle M, Nafria JMN (1990) Numbers, richness and diversity of aphids trapped in Moericke and suction traps in Leon, Spain. Acta Phytopathol Entomol Hung 25:159–165

Sigvald R (1984) The relative efficiency of some aphid species as vectors of potato virus Y^{O} (PVY^{O}). Potato Res 27(3):285–290

Sigvald R (1986) Forecasting the incidence of potato virus Y^{O}. In: McLean GD, Garrett RG, Ruesink WG (eds) Plant virus epidemics, monitoring, modelling and predicting outbreaks. Academic Press, Sydney, pp 419–441

Sigvald R (1987) Forecasting *Potato virus Y*. Waxtskyddsnotiser 51:131–137

Sigvald R (1989) Relationship between aphid occurrence and spread of potato virus YO (PVYO) in field in southern Sweden. J Appl Ent 108:35–43

Sigvald R (1992) Progress in aphid forecasting systems. Neth J Plant Pathol 98(suppl. 2):55–62

Srinivasan R, Alvarez JM, Bosque-Pérez NA, Eigenbrode SD, Novy RG (2008) Effect of an alternate weed host, hairy nightshade, *Solanum sarrachoides*, on the biology of the two most important potato leafroll virus (*Luteoviridae*: *Polerovirus*) vectors, *Myzus persicae* and *Macrosiphum euphorbiae* (Aphididae: Homoptera). Environ Entomol 37(2):592–600

Steinger T, Gilliand H, Hebeisen T (2014) Epidemiological analysis of risk factors for the spread of potato viruses in Switzerland. Ann Appl Biol 164:200–207

Steinger T, Goy G, Gilliand H, Hebeisen T, Derron J (2015) Forecasting virus disease in seed potatoes using flight activity data of aphid vectors. Ann Appl Biol 166:410–419

Thornbury DW, Hellmann GM, Rhoads RE, Pirone TP (1985) Purification and characterization of potyvirus helper component. Virology 144(1):260–267

Urcuqui-Inchima S, Maia IG, Arruda P, Haenni AL, Bernardi F (2000) Deletion mapping of the potyviral helper component-proteinase reveals two regions involved in RNA binding. Virology 268(1):104–111

Uzest M, Gargani D, Drucker M, Hebrard E, Garzo E, Candresse T, Fereres A, Blanc S (2007) A protein key to plant virus transmission at the tip of the insect vector stylet. Proc Natl Acad Sci U S A 104(46):17959–17964

Vance VB, Berger PH, Carrington JC, Hunt AG, Shi XM (1995) 5' proximal potyviral sequences mediate potato virus X/potyviral synergistic disease in transgenic tobacco. Virology 10:583–590

van der Waals JE, Krüger K, Franke AC, Haverkort AJ, Steyn JM (2013) Climate change and potato production in contrasting south African agro-ecosystems 3. Effects on relative development rates of selected pathogens and pests. Potato Res 56:67–84

van Harten A (1983) The relation between aphid flights and the spread of potato virus Y^{N} (PVY^{N}) in the Netherlands. Potato Res 26(1):1–15

van Hoof HA (1980) Aphid vectors of potato virus Y^{N}. Neth J Plant Pathol 86:159–162

Verbeek M, Piron PGM, Dullemans AM, Cuperus C, van der Vlugt RAA (2010) Determination of aphid transmission efficiencies for N, NTN and Wilga strains of *Potato virus Y*. Ann Appl Biol 156(1):39–49

Zhang J, Nie X, Nanayakkara U, Boquel S, Giguère MA, Pelletier Y (2013) Detection of *Potato virus Y* from the stylets of a single aphid by one-step reverse transcription polymerase chain reaction. Entomol Exp Appl 147(1):93–97

Chapter 7
Potato virus Y: Control, Management and Seed Certification Programmes

Brice Dupuis, Claude Bragard, Stuart Carnegie, John Kerr, Laurent Glais, Mathuresh Singh, Phillip Nolte, Jean-Louis Rolot, Kürt Demeulemeester, and Christophe Lacomme

Abstract The management of *Potato virus Y* (PVY) in potato crops poses a continual challenge due to the non-persistent mode of transmission of the virus and the propagation of seed potato tubers over several generations in the field. While PVY-resistant cultivars remain the most efficient way to protect potato crops against PVY, a vast majority of cultivars grown do not display significant resistance to PVY. Due to the short time period for PVY transmission by non-colonising aphid vectors, efficient control of PVY relies on preventing aphids landing on a crop and on adopting precautionary measures by ensuring that crops are grown in areas of

B. Dupuis (✉)
Agroscope, Institute for Plant Production Sciences IPS, 1260 Nyon 1, Switzerland

Earth and Life Institute – Applied Microbiology – Phytopathology, Université catholique de Louvain UCL, B-1348 Louvain-la-Neuve, Belgium
e-mail: brice.dupuis@agroscope.admin.ch

C. Bragard
Earth and Life Institute – Applied Microbiology – Phytopathology, Université catholique de Louvain UCL, B-1348 Louvain-la-Neuve, Belgium

S. Carnegie • J. Kerr • C. Lacomme
Science and Advice for Scottish Agriculture (SASA),
1 Roddinglaw Road, Edinburgh EH12 9FJ, UK

L. Glais
FN3PT/RD3PT, 43-45 rue de Naples, F-75008 Paris, France

IGEPP, Agrocampus Ouest, INRA, Université de Rennes 1, F-35650 Le Rheu, France

M. Singh
Agricultural Certification Services, 1030 Lincoln Road, Fredericton, NB E3B 8B7, Canada

P. Nolte
University of Idaho, Idaho Falls 83402, USA

J.-L. Rolot
Walloon Agricultural Research Centre, 100, rue du Serpont, 6800 Libramont, Belgium

K. Demeulemeester
INAGRO, 87 Ieperseweg, 8800 Rumbeke-Beitem, Belgium

C. Lacomme et al. (eds.), *Potato virus Y: biodiversity, pathogenicity, epidemiology and management*, DOI 10.1007/978-3-319-58860-5_7

low aphid and low virus pressure and limiting field generation. Prophylactic measures such as roguing and early haulm destruction limit PVY spread but are not efficient alone. Among all existing control methods, spraying potato crops with mineral oils can offer significant protection against PVY spread, but their efficacy do vary in field conditions. The combination of several control methods such as mineral oil treatments, crop borders, intercropping, straw mulching or insecticide treatments can increase protection. These emphasise the importance of controlling virus through appropriate monitoring methods and crop management enforced by seed certification schemes through the use of 'clean' input seed and, when possible, the segregation of seed and ware crops to minimise the risk of virus transmission. This chapter presents and discusses the most widely used techniques of control and management of PVY, their effectiveness and their mode of action. This chapter also presents the history, objectives and principles of seed potato certification schemes and their role in minimising the spread of viruses within potato crops worldwide.

1 Introduction

The management of *Potato virus Y* (PVY) in potato crops poses a continual challenge due to the non-persistent mode of transmission of PVY and the propagation of seed potato tubers over several generations in the field, which presents a risk for primary and secondary infections of plants by PVY.

Aphids transmit the virus by flying from plant to plant within or between crops (Sigvald 1984; Boiteau 1997). An aphid will probe an infected plant, acquire the virus and then fly to a healthy plant, probe again and transmit the virus. The length of time required by an aphid for acquisition and subsequent transmission of PVY is very short, with each step generally accomplished in seconds (Fereres and Moreno 2009; Robert et al. 2000; Bragard et al. 2013) (see Chap. 6). The challenge in controlling PVY efficiently lies in either preventing aphids landing on a crop or in promoting a rapid (almost instantaneous) deleterious effect on them after landing to prevent any further transmission. These aspects will be discussed in the following sections.

Potatoes are vegetatively multiplied by seed potato growers in order to maintain the characteristics of a cultivar and to bulk up sufficient quantities of certified seed potatoes to meet the requirements of the market (Frost et al. 2013). This method of propagation enables PVY to be transmitted from one generation of a crop to the next through its translocation from an infected seed tuber into growing plant and daughter tubers (Basky and Almasi 2005). While it is still not possible to prevent translocation of the virus, it is, however, possible to minimise the multiplication of an infected seed lot by a systematic control of its quality. This can be done by the implementation of seed potato certification programmes and appropriate virus testing regimes.

A number of methods have been developed to control the spread of PVY in potato crops, and some of them, either individually or in combination, are now used by seed potato producers. This chapter will present and discuss the most widely used techniques of control and management of PVY, their effectiveness and their mode of action. This chapter will also present the history, objectives and principles of seed potato certification programmes and the role of these programmes in minimising the spread of viruses within potato crops worldwide.

2 Cultural Methods

2.1 Prophylactic Measures

2.1.1 Use of Virus-Free Seed Potatoes

The risk of PVY spread can be reduced by planting seed potatoes which are virus free or have a very low incidence of tuber infection by PVY, thus minimising the number of inoculum sources within a crop (Kerlan et al. 1987; Rolot 2005; Steinger et al. 2014). This will be discussed in the paragraph relating to seed potato certification programmes.

2.1.2 Use of Virus-Resistant Cultivars

The most effective means of controlling virus in potato crops is to grow resistant potato cultivars. This strategy allows potatoes to be produced in areas where aphid pressure is high and hence unsuitable for potato production because of disease. There are several breeding programmes worldwide whose goal is to breed new potato cultivars that are resistant to PVY (see Chap. 8). In summary, there are several different types of resistance (see Chaps. 2 and 8): (i) extreme resistance which does not allow a virus to multiply in a plant, (ii) hypersensitive response is defined as a mechanism of resistance against pathogen invasion by the rapid death of cells at the infection site preventing/reducing significantly the spread of pathogens to other parts of the plant and (iii) tolerance which allows a virus to multiply and spread within a plant without it expressing visible symptoms. However, breeders develop their selection strategy independently depending on market requirements, and the specific disease and pest pressures in the countries intended for marketing. In the USA and Canada, breeders have developed a number of widely grown cultivars which are tolerant to PVY, such as cv. Russet Norkotah, which shows mild or 'latent' symptoms of PVY (Whitworth et al. 2010), and cv. Red LaSoda, which is fully susceptible to infection by PVY but does not express symptoms (Draper et al. 2002). This could be one of the reasons for the re-emergence of PVY in these potato production areas (Gray et al. 2010; Schramm et al. 2011). In Germany and France, the selection programmes for new cultivars are different. Breeders mainly aim to develop potato cultivars resistant to the multiplication of PVY and avoid the selection of cultivars susceptible to potato tuber necrotic ringspot disease (PTNRD) (Le Romancer and Kerlan 1991). However, it is widely believed that breeding for cultivars displaying high level of PVY resistance might often not be seen as a major goal due to poor economic return on investments (Ruedi Schwaerzel, personal communication).

2.1.3 Isolation from Virus Sources

Location of crops is important in determining the risk of PVY transmission by aphids (Kerlan et al. 1987). Low-risk areas are usually those with low aphid populations, often as a consequence of low mean temperatures (Robert et al. 2000; Gabriel 1965). Klueken et al. (2009) studied the flight behaviour of alate cereal aphid *Sitobion avenae* (Fabricius) and demonstrated that, for a 16-h period, no aphid flight occurred at 10°C, while the proportion of aphids flying increased to 70% at 15°C to reach 100% at 20°C. They also showed that the temperature threshold was 3°C higher for *Metopolophium dirhodum* (Walker) (bird cherry oat aphid), another cereal aphid. This implies that, for cold temperature, these aphid species are likely to stay on their 'winter host' instead of colonising crops ('summer hosts'). Under these conditions, vector pressure and aphid transmission of PVY are minimised. In Switzerland, it was shown that an increase in altitude from 400 m to 800 m resulted in a decrease in infection of potatoes by 57%, mainly due to an average colder temperature at the higher altitude (Steinger et al. 2014). In addition, strong wind speeds (> 0.8 km/h) are not favourable for alate aphid displacement but also for aphid feeding and development, so windy areas tend to be more suited to seed potato production because there is less opportunity for aphids to transmit virus (CIP 1979; Walters and Dixon 1984). The presence of plants infected by PVY in neighbouring crops can also pose a risk with regard to inoculum sources of PVY, especially if the seed potatoes planted are not certified and the crop is being grown for consumption (ware) without management measures to control aphids. Virus is likely to be more prevalent in farm-saved seed potatoes than in certified seed potatoes (Kerlan et al. 1987).

2.1.4 Time of Planting and Haulm Killing

The physiological state of a plant may affect the transmission of PVY by aphids because older plants appear to be less susceptible to PVY infection than younger plants (Gibson 1991; Robert et al. 2000; Sigvald 1985; Dupuis 2016), a phenomenon termed mature plant resistance (Beemster 1972). Sigvald (1985) reported that potato plants are at their most susceptible state up to 25 days post-emergence, becoming more resistant thereafter at the rate of 10% every week. Managing the physiology of plants could, therefore, be integrated into a programme to control PVY in potato crops. Advancing the time of planting and haulm destruction could be important in reducing the risk of PVY spread. Pre-sprouting seed potatoes can enable plants to emerge earlier and daughter tubers to bulk earlier, thus allowing earlier haulm destruction than with unsprouted tubers. However, the effectiveness of pre-sprouting in controlling PVY spread will depend to a great extent on the timing of aphid flights. If aphid flights are late in the growing season, pre-sprouting could be an effective measure. However, if spring aphid flights are earlier than normal, this could pose a greater risk to crops which have been planted with sprouted seed tubers because the plants will have emerged when aphids are flying, whereas those from unsprouted seed potatoes will not (Saucke and Doring 2004). Early haulm destruction will reduce the time of exposure of the crop to aphid flights, thereby reducing the risk of infection (Basky 2003; Kerlan et al. 1987; Steinger et al. 2014). In some conditions, it has been reported that delaying haulm destruction can increase the incidence of PVY infection by 3.5% per day (Steinger et al. 2014). For mature crops, the choice of herbicide or the use of mechanical haulm destruction has a limited impact on reducing virus transmission (Dupuis and Schwaerzel 2011). Nevertheless, rapid and total haulm destruction is required to reduce the risk of late PVY transmission. If new foliage develops on desiccated stems, transmission of PVY by aphids could be possible as the relatively immature leaves are potentially susceptible to PVY transmission (Sigvald 1985). New growth that develops after haulm destruction can be susceptible to aphid-borne infection especially when significant aphid pressure is still observed late in the growing season. However, quantitative data are still lacking to assess accurately the impact of late infection of new growth and PVY incidence at post-harvest.

2.2 *Roguing and Weed Control*

For effective control of PVY in crops, it is essential to eliminate any sources of inoculum. These can be plants from infected seed tubers, from weeds or from volunteer plants, *i.e.* potato plants derived from tubers or parts of a tuber left in the soil after harvesting a previous crop (Jones et al. 1996). The use of certified seed potatoes according to official tolerances can provide assurance regarding the maximum amount of virus disease which could develop in a crop, but cannot rule out the absence of viruses in a crop. When a crop is being grown for marketing as seed potatoes, roguing is a key component in maintaining the health of a crop. Roguing is the removal of potato plants which are atypical of the cultivar in appearance or diseased including virus (Kerlan et al. 1987). For virus diseases, roguing is efficient in reducing PVY spread, but the effect can vary from none to a 20% reduction of PVY spread (Broadbent et al. 1950). To be effective, roguing should be conducted as soon as possible after plant emergence to minimise the opportunity for aphids to acquire virus from infected plants within a crop. However, roguing will be ineffective if a cultivar is tolerant to a virus because infected plants will be symptomless and, therefore, not recognised and removed. There is a risk that excessive amounts of virus can build up in crops of such cultivars cultivated over several generations and pose a serious threat to the health of crops of susceptible cultivars from aphid transmission (Ragsdale et al. 2011). It is also possible that roguing could enhance the transmission of PVY by aphids if too many plants are removed in a small area creating bare patches. Aphids land preferentially in areas where there is a contrast between bare soil and potato plants (reviewed in Döring (2014)). Excessive gaps might promote the landing of aphids and, consequently, increase the risk of infection of plants surrounding a gap. The risk of infection increases with the size of the gap (Davis et al. 2009). The incidence of PVY infection was 13% around gaps of $\leq$ 0.6 m^2 and 29% around gaps of $\geq$ 0.6 m^2.

Volunteer plants are a major concern in potato-growing areas. It has been estimated that in 1 hectare, 20,000–300,000 tubers could remain in a potato field following harvest because a significant proportion of tubers are too small to be collected by harvesters and will remain in the soil over the winter (Yves Le Hingrat, personal communication). Some of these tubers will survive the winter if soil temperatures are not sufficiently cold enough. The extent of tuber overwintering will vary with their depth in soil, the severity and length of cold periods (Lutman 1977; Boydston et al. 2006; Cooke et al. 2011). The growth of plants from surviving tubers (volunteer potatoes) even after several years of crop rotation could be impaired by the foliar growth of the newly planted crop, depending on its capacity to cover and shade volunteer potato plants. In France between 2007 and 2011, the density of volunteer plants in fields in the Brittany region was estimated to range from none to six stems per m^2 depending on cultivar, cropping practices and succeeding crop (Rakotonindraina et al. 2011). In addition to volunteer potatoes being a source of varietal mixtures in potato crops and posing weed control issues in other field crops, they can also act as a reservoir for potato diseases and could impact on the phytosanitary status of seed potatoes. A survey was conducted in the UK in 1996 in which volunteer plants were collected from three different sites and the percentage of volunteer plants infected by PVY was found to range from 2 to 54% (Jones et al. 1996). Inoculum from within or near a crop can also originate from weeds. In the late 1990s, 36 weed species belonging to 13 different botanical families were identified as potential host plants for PVY (in natural and/or artificial conditions) (Edwardson and Christie 1997). More recently, seven additional weed species have been identified as potential hosts (Kaliciak and Syller 2009; Kazinczi et al. 2004) (see Chap. 6). It is almost certain that other weed hosts for PVY will be reported in the future. However, the epidemiological role and impact of those weeds in the field are not fully understood, and it is unclear whether the presence of weed species will contribute significantly to the spread and prevalence of PVY in potato crops.

2.3 Crop Borders

Crop borders can be planted in order to limit the amount of virus introduced into a crop (Boiteau et al. 2009; Difonzo et al. 1996). Crop borders display two distinct modes of action, serving as a 'virus barrier' and also as a 'virus sink' (Boiteau et al. 2009). Viruliferous aphids landing on a border crop will probe the plants and shed any virus particles that they might be carrying into the barrier plant. In this case, the border crop serves the role of 'virus sink' by retaining the virus. To be effective, border crops must be planted with a plant species which is not susceptible to PVY. Border plants can also act as a physical barrier interrupting aphid flight (Simons 1957). To provide this type of protection, the border plants have to be taller than the potato plants at all stages of growth. A border must also be wide enough to maximise the probability of aphids landing on it (Boiteau et al. 2009; Difonzo et al. 1996). With a soybean border of 24 m, Difonzo et al. (1996) obtained an efficacy of PVY control of 27 and 60% over the 2 years of an experiment, while Boiteau et al. (2009) obtained 32% efficacy for an experiment with a narrower border of 4 m sown with grass. This technique has the advantage of being effective for the entire cropping season, whatever the environmental conditions. Nevertheless, the use of crop borders requires relatively large areas of a field to be removed from potato production with no commercial return so the method may not be practical, especially if fields are relatively small. To solve this issue, a potato cultivar resistant to PVY can be used as a border. A border of 4 m of cv. Kennebec (known to be relatively resistant to PVY) was used in a 3-year field trial in Canada to protect a central plot of cv. Russet Burbank, susceptible to PVY, and its effectiveness was compared with applying mineral oil (Boiteau et al. 2009). The border crop alone reduced PVY transmission by about 20% in the first 2 years of the trial and 60% in the third year, while mineral oil application alone reduced PVY transmission by 20% in the first 2 years and by 70% in the third year.

2.4 Mulching and Intercropping

Straw mulching (Fig. 7.1) is effective in controlling PVY spread (Heimbach et al. 2004; Saucke and Doring 2004; Kirchner et al. 2014; Dupuis et al. 2010). The mode of action of straw mulching is not well understood, but the main hypothesis is that straw impacts on the visual perception of a crop by an aphid (Döring 2014). The contrast between potato foliage and the yellow straw is considerably lower than potato foliage and bare soil, so that potato plants in mulched plots are less easily seen by aphids (Döring 2014; Döring and Schmidt 2007). This was confirmed when fewer aphids were captured in mulched plots compared with those in bare soil plots (Heimbach et al. 2004; Saucke and Doring 2004). In the study reported by Saucke and Doring (2004), aphids were counted on the foliage of potato plants in straw-mulched and bare soil plots. An average reduction of 54% in aphid numbers was recorded in the mulched plots. This reduction of aphid populations on foliage was 18% in an independent study (Heimbach et al. 2004), while in the same experiment, 85% fewer winged aphids were captured by sticky net traps within the mulched plots. Thus, the lower number of winged aphids landing on potato plants in mulched plots might be the main factor in controlling PVY spread by mulching. PVY control by mulching is more effective in the early stages of crop development, declining as the crop canopy develops over the mulch (Heimbach et al. 2004). This was shown

Fig. 7.1 Wheat straw mulch in potatoes, 2500 kg/ha of straw in cv. Charlotte (Photo: Maud Tallant)

Fig. 7.2 Oat intercropping (*Avena sativa*, 60kg/ha) in potatoes (cv. Charlotte), 7 days after spraying oats with tepraloxydim (200g/l) to stop their development (Photo: Maud Tallant)

clearly in a field experiment undertaken over a 3-year period (Saucke and Doring 2004) in which the efficacy of PVY control recorded in the mulched plots was about 48% in the first year, 33% in the second year and 27% in the third year. In the first year of the trial, aphids were most prevalent a few days after potato emergence in mid-May, whereas in the third year of the trial, aphids were most active 2 weeks later. In the second year of the trial, aphid activity remained low during the entire growing period. Intercropping was also tested for its effectiveness in reducing PVY spread in potato crops (Dupuis et al. (2010); Fig. 7.2). Oats sown between rows of potatoes but killed before becoming large enough to provide unwanted competition for the potatoes resulted in a significant reduction in PVY spread (Fig. 7.3). The mode of action of intercropping is comparable to the 'sink effect' of a border crop

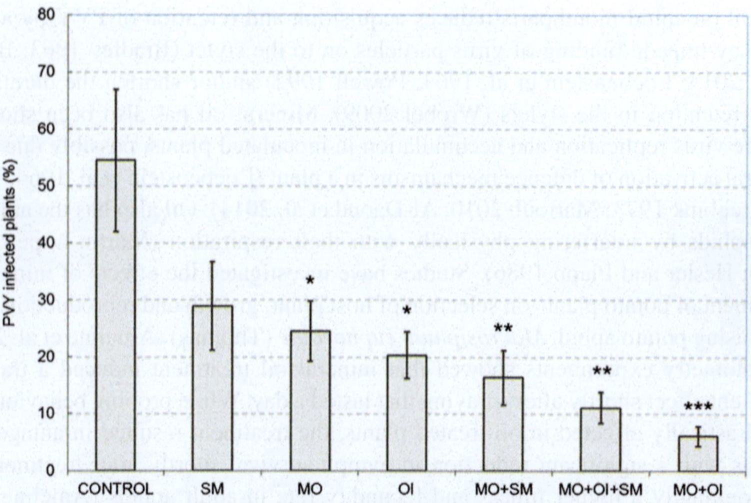

Fig. 7.3 Percentage of PVY-infected plants assessed by testing daughter tubers for six treatments and untreated (1 year; four replications of 100 plants per plot). The acronyms for the treatments are *SM* = straw mulching, *MO* = mineral oil, *OI* = oat intercropping. Error bars show the standard error, and stars show the treatments with a percentage of infection significantly lower than untreated (Dunnett's test; * for $p<0.05$ and ** for $p<0.01$) (Dupuis et al. 2010)

in which inoculum being carried by aphids is lost during probing on the associated crop (Boiteau et al. 2009). Intercropping offers an additional mode of action called the 'dilution effect' by significantly decreasing the probability of an aphid landing on potato plants. Given the same plot surface, plots containing an associated crop present many more plants to potential vectors, thereby reducing the risk of potatoes being infected.

3 Chemical Methods

3.1 Oil Treatments

Spraying oil on potato plants has been reported to provide a means of controlling the spread of PVY (Boiteau et al. 2009, Bell 1980, Bell 1989, Boiteau and Wood 1982, Bradley et al. 1962, 1966, Dewijs 1980, Dupuis et al. 2014, Hansen and Nielsen 2012, Kirchner et al. 2014, Martin-Lopez et al. 2006, Milosevic 1996, Steinger et al. 2014, Rolot 2005). Mineral oil was found to be more effective than vegetable oil (Martin-Lopez et al. 2006; Rolot 2005; Wrobel 2012), although among the latter, refined oils were more effective than raw oils (Martin-Lopez et al. 2006). The mechanism of action of the oils is not fully understood. It has been suggested that oil on aphid mouthparts reduces acquisition and retention of PVY by aphids. Oil may impede binding of virus particles on to the stylet (Bradley 1963; Boquel et al. 2013; Loebenstein et al. 1964; Powell 1992) and/or shorten the duration of virus retention in the stylets (Wróbel 2009). Mineral oil has also been shown to reduce virus replication and accumulation in inoculated plants, possibly due to the general activation of defence mechanisms in a plant (Loebenstein et al. 1964; Peters and Lebbink 1973; Martoub 2010; Al-Daoud et al. 2014). Oil also has the ability to kill aphids by interfering physically with their respiration (Martin-Lopez et al. 2006; Hesler and Plapp 1986). Studies have investigated the effects of mineral oil treatment of potato plants on selection of host plant, growth and reproduction of the colonising potato aphid, *Macrosiphum euphorbiae* (Thomas) (Ameline et al. 2009). Olfactometry experiments showed that mineral oil treatment induced a transient repellent effect shortly after spraying that lasted a day. While probing behaviour was not drastically affected in oil-treated plants, the treatment resulted in antagonistic effects with a significant reduction in nymph survival shortly after treatment and concomitantly a higher fitness and fecundity rate in adult aphids (Ameline et al. 2009). Other studies have reported antagonistic effects of mineral oil on *M. euphorbiae* that were dose dependent; topical contact at the highest concentration of oil resulted in complete mortality, while lower concentrations and exposure to oil volatiles enhanced aphid fecundity but had no effect on aphid survival (Martoub 2010). Tan et al. (2005) suggested that, after spraying, oil is 'absorbed' in various plant tissues and cells and transported throughout the plant, resulting in a decrease in local concentrations and the induction of physiological changes (Tan et al. 2005), altering photosynthesis (Helson and Minshall 1951; Wedding et al. 1952), and triggering the expression of pathogenesis-related proteins (Kachroo et al. 2001; Lin et al. 1996).

In practice, the protection achieved with mineral oil spraying varies greatly depending on the time of application (Figs. 7.3, 7.5 and 7.6). Mineral oils are more effective on older plants in reducing the speed with which PVY moves in the vascular system (Al-Daoud et al. 2014). The reduction in the transmission of PVY by oil treatments is likely to vary among different potato cultivars and with different aphid and virus pressures. Steinger et al. (2014) reported that oil treatments reduced the average incidence of PVY by 39% ($p<0.001$) over a 4-year period. The protective effect of mineral oil was slightly greater in the year in which the incidence of PVY was greatest (50% decrease in infection $N_{treated} = 432$, $N_{non\text{-}treated} = 86$) than in other years and with the susceptible cvs Bintje and Charlotte (54% decrease in infection, $N_{treated} = 819$, $N_{non\text{-}treated} = 79$) compared with more resistant cultivars. In a separate study over 3 years of field trials in the UK, mineral oil reduced PVY infection two- to threefold in cvs King Edward and Maris Piper (N = 160 plants per cultivar per treatment per year) during a year of relatively high virus pressure (overall 30% incidence of plants infected by PVY) (Dawson et al. 2015). However, there was no significant effect of mineral oil treatments on the incidence of PVY in years of low virus pressure (3–13% PVY incidence). In this study, significant variation in the effectiveness of the oil treatment in controlling PVY suggests that local differences in aphid phenology and aphid vector pressure strongly influenced the effectiveness of the treatment. These results are consistent with the expected reduction in the incidence of PVY for oil-based treatments compared with untreated plants, usually between 30 and 60% (reviewed by Al-Mrabeh et al. (2010)).

Mineral oils are usually applied weekly to a seed potato crop. The protection provided by oil is generally limited to the leaf surfaces exposed to the spray being applied (Simons et al. 1977). Effective coverage of foliage is easier to achieve on older plants than on younger plants on which new leaves are constantly developing. The rate of foliar development should be considered when determining the optimum interval between sprays for effective application of oil. An increase in the frequency of foliar applications on young plants should be considered for maximum effectiveness in order to protect recently opened leaves (Fageria et al. 2014a, b; Demeulemeester 2013). The usual practice in France to protect high-grade seed is to spray mineral oil three times a week starting at 30% plant emergence and continuing until complete emergence followed by weekly applications thereafter. In Belgium, the practice is very similar except that only two applications are recommended during the early period of crop growth followed by one each week until haulm destruction is complete (Yves Le Hingrat and Pierre Lebrun, personal communication).

Mineral and vegetable oils must be used with caution to avoid plant phytotoxicity (Kirchner et al. 2014). It is essential to use a labelled product (Dewijs 1980). Paraffinic oils are preferably used for the treatment of plants instead of naphthenic oils (phytotoxic) and aromatic oils (phytotoxic and unstable) (Rolot 2005). The linear structure of saturated paraffinic oils is relatively stable and not phytotoxic. Dewijs et al. (1979) demonstrated that the viscosity of the oil, which is related to the number of carbon atoms in the molecular chain, is an important character determining the efficacy of the oil. Nevertheless, when the chains have more than 25 carbon atoms, phytotoxicity can be observed (Walsh 2000). The phytotoxicity of the paraffinic oil also depends on its degree of refining. Less than 8% of sulphonated residues (residues reacting with sulphuric acid) are required for plant treatments (Walsh 2000). For treatments with paraffinic oils, it is also required to not exceed the maximum rate and to avoid application during hot weather because oil can become so hot that it can burn potato foliage on application after being heated in the sprayer pipes during prolonged sun exposure. Mineral oil treatment can alter plant physiology and, in some cases, be phytotoxic. This can have an adverse effect on the appearance of plants and potentially might affect crop inspection by reducing a seed potato inspector's ability to identify cultivars and virus symptoms visually in a growing crop. A recent study in the UK of the effect of oil treatments on a range of virus-infected and healthy plants of various potato cultivars concluded that the ability of inspectors to identify both cultivars and virus symptoms was not diminished by applications of mineral oil to foliage (Dawson et al. 2015), even though phytotoxic symptoms (localised necrotic spots) were occasionally observed (Fig.7.4). However, some loss of tuber yield was reported for some cultivars after mineral oil treatment, emphasising the necessity for a cautious use of mineral oil to control PVY (Kirchner et al. 2014). Delaying haulm destruction by several days could compensate for this reduction in yield (Dawson et al. 2015).

Fig. 7.4 Phytotoxic symptoms (localised necrotic spots and leaf midrib necrosis) developing on leaves of potato cv. Desiree following mineral oil spraying (Dawson et al. 2015). Note the beading of rainwater on the foliage of sprayed plants (Photo: Courtesy of SASA Crown copyright)

3.2 Insecticide Treatments

The acquisition and inoculation periods of non-persistent viruses like PVY by their aphid vectors are extremely short (seconds to minutes) (Bragard et al. 2013). To be effective, an insecticide has to kill or incapacitate an aphid very quickly to limit the infection of a plant. However, the effect of an insecticide may impair the capacity of a viruliferous aphid to fly from a treated plant to another plant. Various insecticides and their formulations have been tested for their effectiveness in controlling PVY transmission in the field (Table 7.1). Pyrethroids have a near-instantaneous 'knock-down' effect and can provide a reduction in PVY transmission in controlled conditions (Perring et al. 1999; Gibson et al. 1982; Collar et al. 1997). Unfortunately, results obtained in field-grown potato crops with the same products have proven to be variable (Table 7.1). Other group of insecticides with different modes of action can potentially interfere with the transmission of viruses either by repelling aphids (deltamethrin, Rice et al. (1983)), altering feeding behaviour (flonicamid, imidacloprid, pymetrozine and thiamethoxam, Morita et al. (2007), Cho et al. (2011), Boquel et al. (2014)) and reducing aphid's movement (aldicarb, Boiteau et al. (1985)). However, no significant reduction of PVY transmission was reported in field trials for imidacloprid, pymetrozine (Table 7.1) and flonicamid (Fig. 7.5). Pyrethroids such as deltamethrin (Gibson et al. 1982) and cypermethrin (Collar et al. 1997) can provide a significant degree of PVY protection in controlled conditions, but it is not known whether virus acquisition or transmission is affected. Lambda-cyhalothrin, dimethoate and pymetrozine reduce PVY acquisition by aphids (Boquel et al. 2014;

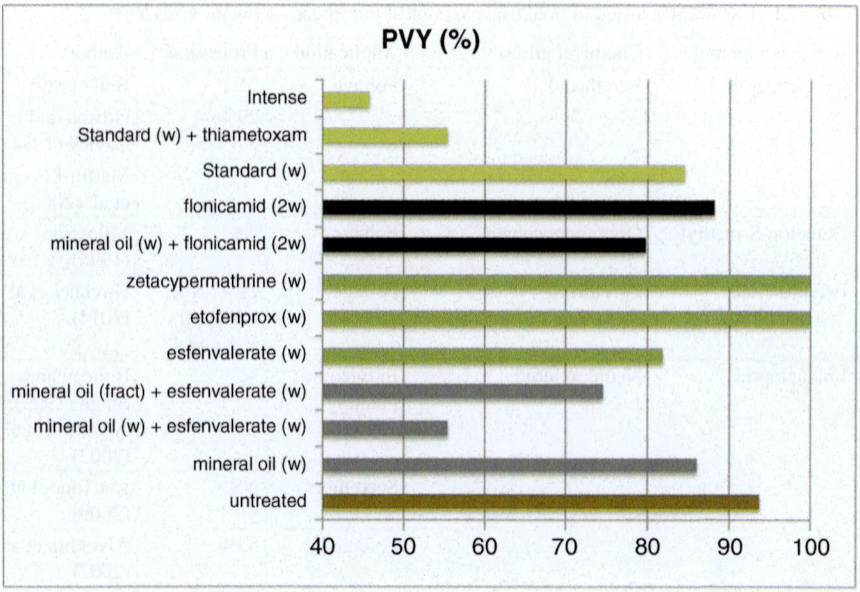

Fig. 7.5 Percentage of PVY-infected daughter tubers produced by plants treated with a range of insecticides and mineral oil. The insecticides used belong to different chemical groups such as pyrethroids (esfenvalerate, etofenprox, zeta-cypermethrin), neonicotinoids (thiacloprid, thiamethoxam) and pyridine (flonicamid, pymetrozine). The 'standard' programme refers to a weekly application of mineral oil, together with a pyrethroid in the first week after emergence (esfenvalerate) and a systemic insecticide every 2 weeks (one treatment with flonicamid followed by pymetrozine and thiacloprid). The 'intense' programme is similar to the 'standard' programme with, in addition, two applications of mineral oil in the first week after emergence. The letter (w) means weekly application, (2w) means one application every 2 weeks and (fract) means two applications per week with half dose for the first 3 weeks after emergence and thereafter one application per week (Demeulemeester 2013)

Margaritopoulos et al. 2010), but the effect of these insecticides is too slow to prevent PVY spread in the field (Table 7.1). In summary, for most of the field trial studies reported, the efficacy of an insecticide in controlling PVY spread was either not significant, of a limited impact or highly variable between independent experiments. One of the reasons that insecticides have been demonstrated to be either ineffective or only of limited effectiveness is because many aphids which transmit PVY do not colonise plants in a crop and thus only come in contact with the insecticides for a very limited period (reviewed in Perring et al. (1999)).

Table 7.1 Insecticides tested in field trials to control the spread of Potato virus Y (PVY)

Active compound	Chemical group	Application	Protection	Authors
Cypermethrin	Pyrethroid	Foliage	7%;	Bell (1989)
			29%;	Gibson and Cayley (1984)
			NS	Martin-Lopez et al. (2006)
Demeton-S-methyl	Organophosphate	Foliage	NS	Milosevic (1996)
Esfenvalerate	Pyrethroid	Foliage	29%; NS; NS (Years 1, 2 and 3)	Kirchner et al. (2014)
Imidacloprid	Neonicotinoid	In furrow at planting	NS	Boiteau and Singh (1999), Alyokhin et al. (2002)
		Seed tubers	NS	van Toor et al. (2009)
		Foliage	53%;	Alyokhin et al. (2002)
			NS	Boiteau and Singh (1999)
Lambda-cyhalothrin	Pyrethroid	Foliage	10%	Dupuis et al. (2014)
			NS	Hansen and Nielsen (2012)
			NS	van Toor et al. (2009)
Permethrin	Pyrethroid	Foliage	22%	Bell (1989)
Pymetrozine	Pyridine	Foliage	NS	van Toor et al. (2009)
				Rolot et al. (2006)
Thiomethon and tau-fluvalinate	Organophosphorus and pyrethroid	Foliage	8%	Rolot (2005)
Tau-fluvalinate and thiacloprid	Pyrethroid and neonicotinoid	Foliage	NS	Kirchner et al. (2014)
Thiamethoxam	Neonicotinoid	Foliage	NS	Kirchner et al. (2014)

The table shows the name of the active compound and its chemical group, the type of treatment applied in the trial, the authors and the corresponding protective efficiency expressed as a percentage of reduction in the incidence of tuber infection in comparison with that for untreated plants (NS = test not significant or no protective effect)

3.3 Limitation of Chemical Control: Insecticide Resistance in Aphids

The (intensive) use of insecticides has led to the development of different mechanisms of resistance in some aphid populations. The main aphid vectors displaying various levels of resistances to insecticides worldwide are *Myzus persicae* (Sulzer), *Aphis nasturtii* (Kaltenbach), *Macrosiphum euphorbiae* (Thomas) and *Sitobion avenae* (Fabricius) (Foster et al. 2014, 2002). Different types of biochemical and molecular mechanisms of insecticide resistance can be found with *Myzus persicae* (Bass et al. 2014), which includes (i) the overproduction of carboxylesterases (conferring broad-spectrum resistance to members of the organophosphate (OP), (monomethyl) carbamate and, to a much lesser extent, pyrethroid classes) that hydrolyse the active compound before it affects the nervous system of the insect; (ii) mutation of the acetylcholinesterase enzyme (MACE – modified acetylcholinesterase) resulting in the insensitivity of target site of the insecticides, the acetylcholinesterase enzyme; (iii) mutation of the voltage-gated sodium channel, also named 'knockdown resistance' (kdr or super-kdr for the enhanced allelic form), which is conferred by a mutation of a transmembrane ion channel that plays an essential role in the initiation and propagation of action potentials in neurons; and (iv) the mutation of the nicotinic acetylcholine receptor (nAChR), a neurotransmitter-gated ion channel that plays an important role in nerve signalling (Bass et al. 2014). In addition, there are new types of resistance that are currently being identified (such as resistance to pyridine compounds) (Table 7.2; Bass et al. (2014)). Table 7.2 presents the chemical groups of insecticides frequently used in potato production and the corresponding resistance mechanism displayed by *Myzus persicae*. The incidence of *Myzus persicae* susceptible to the insecticides listed in Table 7.2 is highly variable from one year to another. For example, in aphid samples collected in the UK in 2004, about 60% of the sampled aphids were genotypically associated with carboxylesterases-mediated resistance, 70% with MACE and 80% with super-kdr, while 5 years before, this proportion was about 20%–5%–80% respectively (IRAG

Table 7.2 Examples of chemical groups and active compounds available for aphid control in potatoes and types of resistance developed by *Myzus persicae* (IRAG 2014; Bass et al. 2014)

Chemical group	Active compound(s)	Resistance mechanism (period of first reported resistance)
Dimethyl carbamate	Pirimicarb	MACE (1940s) and/or carboxylesterase (1950s)
Neonicotinoid	Acetamiprid, thiacloprid, thiamethoxam	Alteration of nAChR (1990s)
Pyrethroid	Esfenvalerate, λ-cyhalothrin, etofenprox, cypermethrin	Super-kdr (1970s)
Pyridine azomethine	Pymethrozine	Action on the chordotonal organ TRPV channel modulators (1990s)
Pyridine carboxamide	Flonicamid	Action on the chordotonal organ TRPV channel modulators (2010)

2008). *Myzus persicae* individuals resistant to neonicotinoids (with an alteration of the nAChR) have not yet been found in the UK; however, a survey performed in Southern France and Northern Spain in 2010 revealed that resistance to neonicotinoids is relatively widespread in these regions, with an average of 76% of resistance to thiamethoxam (Slater et al. 2012). Combinations of the three main resistance mechanisms (MACE, carboxylesterase and kdr) can be found in individuals of *Myzus persicae*, but the most common association of resistances found in the field in the UK are MACE with super-kdr (IRAG 2008). Consequently, growers are recommended to minimise whenever possible the use of insecticide mixtures in order to prevent the outbreak of *Myzus persicae* populations displaying cross-resistances. Populations of *Aphis nasturtii* (Buckthorn-potato aphid) sampled in Belgium and northern France in the 1990s were found to be resistant to insecticide; however, resistance in *Aphis nasturtii* seems to be a relatively rare event (Duvauchelle et al. 1997). *Macrosiphum euphorbiae* has developed a resistance mechanism based on overproduction of an esterase after treatment with pirimicarb and λ-cyhalothrin. This mechanism of resistance is analogous to the one found in *M. persicae*. However, it is not known if this mode of resistance is effective when treatments are applied at the recommended rate (Foster et al. 2002). Kdr resistance to pyrethroids (λ-cyhalothrin) was found in *Sitobion avenae* in the UK (Foster et al. 2014), and the prevalence of resistant individuals in 2015 varied from 45 to 75% depending on the location of sampling (Malloch et al. 2016).

3.4 Synthetic Pheromones

Numerous aphid species secrete pheromones as a behavioural mechanism when attacked by natural enemies (Kislow and Edwards 1972; Nault et al. 1973). (E)-β-farnesene (EβF) is the main alarm pheromone (Pickett et al. 1992; Zhang et al. 1997). EβF has a repellent effect on aphid colonies and affects the development and fecundity of the aphid populations (Su et al. 2006; Gibson et al. 1984). Additionally, it was shown in the laboratory that EβF inhibits the acquisition and subsequent inoculation of PVY by *Myzus persicae* (Sulzer) (Gibson et al. 1984). Another laboratory experiment performed with tobacco plantlets using apterous aphids (*Myzus persicae* (Sulzer) and *Macrosiphum euphorbiae* (Thomas)) revealed a higher spread of PVY when the aphids are in contact with EβF (Lin et al. 2016). This result suggests an increase in the movement of apterous aphids from plant to plant in the presence of EβF. Apterous aphids are known to be able to move from plant to plant even without the overlapping of crop canopy, by walking on the soil surface (Alyokhin and Sewell 2003; Narayandas and Alyokhin 2006). Nevertheless, winged aphids are more motile in the crop by flying from plant to plant and, hence, have a higher impact on PVY spread in the field (Boiteau 1997). A 1-year experiment undertaken in field conditions showed that PVY spread was identical with or without the diffusion of EβF (Crutzen et al. 2014). While pheromone treatment might have a limited impact on the aphid transmission of PVY, the efficacy of insecticide in controlling aphid populations can be improved by using EβF treatments, as reported in Chinese cabbage fields treated simultaneously with imidacloprid and EβF (Cui et al. 2012). The strategy of combining EβF diffusion with insecticide treatments to control PVY spread has yet to be tested and might represent an interesting approach to control aphid populations and consequently virus transmission.

3.5 Elicitors

Several chemicals are able to induce systemic resistance of plants to pathogens (Kessmann et al. 1994). Among those chemicals, salicylic acid (SA) is known to induce resistance to a wide range of pathogens including viruses (Nakayama et al. 1996; Vasyukova and Ozeretskovskaya 2007). It was shown that *Potato virus X* (PVX) accumulation in tobacco leaves is reduced after SA treatment (Naylor et al. 1998). This effect was also observed in tomato leaves inoculated with PVY and treated with acibenzolar-S-methyl (Bion®), a functional analogue of SA (Petrov and Andonova 2012). Bion® was also tested in a field trial to control PVY spread in potatoes, which resulted in a relatively low (14%) but nevertheless significant reduction of PVY transmission (Dupuis et al. 2014).

4 Combined Methods

Many of the techniques cited above have often been tested in combination. Crop borders and mineral oil treatments have been tested together, and the combination of those two techniques was more effective than when they were applied individually. In the field trials of Boiteau et al. (2009), the combination of applying oil sprays to potato crops and a border crop improved the efficacy of PVY control compared with oil treatment and borders used alone: 47–59% reduction in the incidence of PVY in the first year, 57–63% in the second year and 79–97% in the third year of field trials. Insecticide treatment of a border crop did not produce a reduction in the incidence of tuber infection by PVY (Difonzo et al. 1996). A synergistic effect of applying straw mulching and mineral oil together was recorded in a 1-year trial in Switzerland (Fig. 7.3), but no synergy was found in a similar trial in Finland (Kirchner et al. 2014). In the Swiss trial, the efficacy of PVY control with straw mulching was about 47%, while spraying mineral oil produced a similar efficacy of a 55% reduction in PVY. However, an efficacy of 70% was achieved when both techniques were used together. In the Finnish trial, straw mulching produced a reduction of 26% in PVY infection, but a combined mineral oil spraying and straw mulching treatment gave only a 19% reduction. It is likely that the extent of any synergistic effect of mineral oil with straw mulch may depend upon the experimental conditions of a trial. As explained above, the efficacy of mineral oil treatment increases the older the plants are when they are sprayed, while the efficacy of straw mulching is greatest early in the season before the soil surface becomes covered by the crop canopy. Thus, the timing and location of the peak of aphid activity (data not shown) could explain the differences between those two trials. Oat intercropping was also tested with mineral oil applications, and some synergy was recorded (Fig. 7.3). Oat intercropping resulted in 63% reduction in PVY infection in the potato crop, and the efficacy of PVY reduction reached 89% when oat intercropping was used together with mineral oil spraying. The combination of insecticides and mineral oil treatments can also provide some synergy, increasing the protection of treated crops against PVY. A Canadian survey of 56 crops of seed potatoes being multiplied using distinct cropping techniques revealed that mineral oil was more effective when combined with insecticide applications, particularly when used early in the season (Mackenzie et al. 2014; 2016). This synergy was not found in all circumstances. A weekly treatment with esfenvalerate together with mineral oil gave better protection (42%) of a crop against PVY spread than a weekly spray of mineral oil alone (9%; $p<0.05$) (Fig 7.5). Contrastingly, the combination of a weekly treatment of thiomethon + fluvalinate and a twice-weekly treatment with mineral oil did not improve the protection of the crop against PVY infection ($p > 0.05$ for all years of trial) than when treatments were applied individually (Fig. 7.6). Additional studies conducted in a greenhouse have shown a significant reduction in the transmission of PVY in plants treated with mineral oil and insecticides (Martin-Lopez et al. 2006; Gibson and Rice 1986). This synergistic effect of insecticides and oil may be influenced by the particular chemistry of an insecticide, *e.g.* synthetic pyrethroids (Fig. 7.5), and by the time of application because mineral oil alone is less efficient when applied early in the season (Mackenzie et al. 2014; Gibson and Cayley 1984; Demeulemeester 2013).

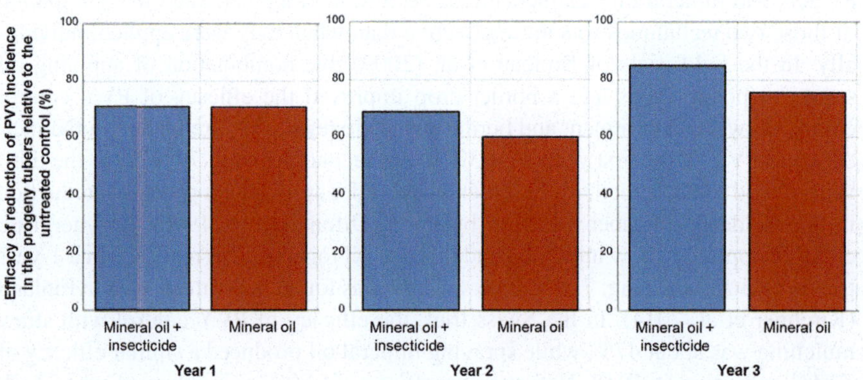

Fig. 7.6 Efficacy in the reduction of the incidence of PVY in progeny tubers in response to foliar treatments with either mineral oil (3 l twice a week) sprayed alone or in combination with a pyrethroid insecticide (0.6 l of thiomethon (200g/l) +fluvalinate (72g/l) once a week) over a 3-year period. Each year, the trials were conducted in three different locations with three replications of each treatment (Rolot 2005)

5 Seed Certification Schemes

5.1 History and Evolution

Potato is a vegetatively propagated crop, and consequently tubers can potentially be contaminated with a large range of organisms (Franc 2001). These pathogens can then be transmitted from the seed tuber to the new plant in the next generation of multiplication and also to healthy plants in the crop. The pathogens can range from fungi to viroids and can be carried within and/or on the surface of a tuber. The diseases caused by the pathogens can have a serious effect on the yield and the quality of a crop. Seed potatoes are potatoes intended for planting, unlike those intended for end uses such as consumption and processing. As commercial trade in seed potatoes developed, a perceived need arose for some form of independent verification of the quality of seed potatoes being marketed. This resulted in the establishment of seed potato certification schemes under official control with tolerances for quality aspects of seed potatoes being applied at each year of propagation (Rousselle et al. 1996; EPPO 1999). The first potato seed certification scheme was established in Europe (Germany) in the first years of the nineteenth century and about 10 years later in the USA (Shepard and Claflin 1975). Similar schemes were introduced in New Zealand in the late 1920s and in Australia in the 1930s (Maunder 2005; Crump 2008). Until early 1970s, production of the initial planting material was largely by 'clonal selection' (or 'positive selection'), which involved selecting disease-free tubers from apparently healthy mother plants (Gildemacher et al. 2011). As far as possible, testing was conducted on mother plants and tubers to ensure freedom from viruses known to be present in a country. However, appropriate testing was not available to confirm freedom from bacterial and fungal tuber pathogens. Subsequently, tissue culture techniques were developed for producing initial planting material in which greater security of plant health could be achieved through an extensive testing programme (Espinoza et al. 1984; Dodds 1988). In Scotland, the first technique used was 'stem cuttings' starting in the late 1960s with the aim of eliminating latent tuber-borne pathogens, especially those causing blackleg, from the initial planting material (Jeffries 1986; Hall 1993), but this was superseded by micropropagation to produce 'nuclear stock'. Nowadays, this method is used as the starting point in almost all certification schemes worldwide because it allows material to be tested comprehensively for quality and notifiable pathogens and pests (EPPO 1999; Donnelly et al. 2003; Frost et al. 2013). One exception is the Netherlands, where clonal selection is also used, together with micropropagation, to produce the initial planting material (NIVAP 2016). Tubers produced from the initial planting material are then multiplied for a number of generations in the field as seed potatoes under certification control until finally marketed for end use (Frost et al. 2013). In Asia, Africa and South America, 'informal' seed potato production systems account for 94% of the market (Thomas-Sharma et al. 2016). In informal seed potato production systems, seed tubers are sourced on farm, neighbouring farms, local markets and unofficial specialist producers (Hirpa et al. 2010). The health status of this type of seed potatoes can be very variable, and consequently crop yields fall far short of their potential. This is largely due to the inherently high amounts of virus in the potatoes and greater aphid pressure in more favourable environments for spread. Gildemacher et al. (2011) assessed the economic return of three different seed potato schemes: (i) the use of the farmer seed stock, (ii) the use of the 'positive selection' technique and (iii) the use of certified seed from the market. This showed that buying high-quality seed is generally more economically effective than the other schemes because the yield obtained with certified seed is usually greater. This does assume that farmers can afford buying high-quality seed potatoes and that the required cultivar is available for planting.

5.2 Objectives of Seed Potato Certification Schemes

The main goal of a potato certification scheme is to assure the quality of seed potatoes being marketed through an independent process of verification and testing conducted by a designated certifying authority. The aspects covered by certification are identity of cultivar, purity of crop, diseases and pests affecting quality or yield, external quality, physiology, size and labelling (UNECE 2015). Seed potatoes are propagated over a number of years so genealogy and traceability are key elements of a well-developed certification scheme. In order to maintain the health of seed potatoes during propagation, tolerances for disease and quality characteristics are set for each generation, being strictest for first generation and becoming slightly more relaxed for subsequent generations in the field (EPPO 1999). Verification of tolerances is largely conducted by visual inspection of plants or tubers supported by testing as appropriate (Franc 2001; Frost et al. 2013).

5.3 Principles of Certification

Although varying degrees of complexity exist, the basic principles of seed certification remain the same (EPPO 1999). Currently, the first step is the multiplication of an in vitro pathogen-free plant (nuclear stock) which can then be multiplied in an approved facility to produce large numbers of in vitro plantlets or microtubers (*i.e.* tubers produced in vitro by a micropropagated plantlet). Nuclear stock is normally subject to an extensive testing programme to ensure its freedom from a range of specified pathogens before being used to produce minitubers in insect-free greenhouses or screenhouses (Frost et al. 2013; EPPO 1999). These tubers are then planted in the field the following year to produce the first generation of seed potatoes (Franc 2001). The import of healthy plant material or clonal selection can be used as an alternative to tissue culture (Franc 2001; Schulte-Geldermann et al. 2012). Several years of field multiplication will ensue before being used by ware potato producers. In European countries, most certification schemes classify seed potatoes into three categories: pre-basic, basic and certified in accordance with United Nations Economic Commission for Europe (UNECE) Standard for Seed Potatoes (UNECE 2015). At least two classes within each category are normally set nationally although the possibility of agreed common classes is being explored by EU (2002). The number of years that seed potatoes can be multiplied is also limited in a scheme, normally less than nine (Frost et al. 2013), and with a maximum of nine generations allowed in the EU since 2016 (directives 2014/20/CE and 2014/21/CE). Some schemes, especially in North America, are based solely on generation number although the number does not always correspond to the number of generations in the field from initiation (Willem Schrage, personal communication). Each year, crops are visually examined by seed potato inspectors for varietal purity and a range of faults including virus diseases around the time when plants are flowering. Inspected crops have to comply with specific tolerances for faults for the class for which they have been entered; otherwise the crop is downgraded to the appropriate class or, in extreme circumstances, not accepted as being suitable for marketing as seed potatoes (Frost et al. 2013). The UNECE has developed an international standard for certification and marketing of seed potato that sets out a common terminology and minimum commercial quality requirements for the certification of high-quality seed intended for marketing internationally (UNECE 2015). This standard is a useful blueprint to facilitate seed potato trade between countries that already have their own certification schemes in accordance with the UNECE standard. It is also a good model for countries aiming to develop their own seed potato certification scheme.

5.4 Virus Diagnostic Methods Used in Certification Schemes

Historically, the incidence of viruses, including PVY, in seed potato crops was assessed visually at two or more inspections according to disease symptoms reported as severe mosaic or mild mosaic (see Chap. 5). Knowledge of the virus causing symptoms was not essential for certification purposes but was nevertheless collected for future analysis. If any tolerance is exceeded at inspection, a crop is downgraded as appropriate or rejected (Shepard and Claflin 1975). This method is still used in seed potato production areas where PVY pressure is low (*e.g.* Scotland). Diseased plants seen at inspection are generally the result of secondary transmission from infected tubers although symptoms of primary infection can sometimes be observed, but this can underestimate the true extent of such infection. The incidence of symptomatic plants derived from primary infection varies with environment and time of infection. Late-season infection of a susceptible potato plant will probably result in no development of foliar symptoms although virus may be translocated to daughter tubers, highlighting the necessity for post-harvest testing (Franc 2001; Frost et al. 2013). Therefore, in some environments, relying on visual inspection may be insufficient to provide assurance that seed potatoes will meet the expectations of customers (Lindner et al. 2015). Additional measures such as a targeted post-harvest assessment were introduced in some certification schemes to check for late-season virus infection or, in some cases, symptomless infection of tolerant cultivars.

There are numerous post-harvest diagnostic methods for viruses. The first method to be developed was termed 'post-harvest grow-out assay' or 'growing-on assay'. A sample of a prescribed number of tubers is collected at random from a crop before harvest or shortly after harvest. Seed pieces consisting of a single eye are removed from each tuber and planted (with or without a treatment to break dormancy) in a greenhouse or outdoors if the climate of a country is suitable. When these tests were first conducted, the plants grown from each seed piece were then examined visually for symptoms of virus infection (Franc 2001). Microscopy (Igel-Lange test) or serological (radial-immunodiffusion) techniques were used to support these assessments (Shepard and Claflin 1975; Gugerli and Fries 1983). Visual observation was later replaced by an enzyme-linked immunosorbent assay (ELISA) for the detection of virus (see Chap. 5). France and Belgium still use ELISA for this type of testing.

A variation of this method is the so-called winter test. After breaking tuber dormancy, a standard sample from a crop is shipped to a 'warmer' location for planting. After emergence, each plant is visually inspected, and/or a leaf sample is taken for further testing by serological or molecular methods. The northern states of the USA use this approach and ship seed samples to southern states for a 'winter test'. These programmes are expensive and are gradually being replaced by laboratory-based tuber tests (Shepard and Claflin 1975; Franc 2001). Laboratory-based tests of tubers were first developed in Switzerland in the early 1980s. After dormancy of the tubers is broken by fumigation with Rindite (a mixture of 2-chloroethanol, 1,2-dichloroethane and carbon tetrachloride 7:3:1), sap from the basis of the sprouts of each tested tuber is tested by ELISA (Gugerli and Gehriger 1980). In Europe, this product is banned due to its toxicity and is replaced by a solution of gibberellic acid. In this case, ELISA test is performed on recently emerged potato leaflets. ELISA is gradually being replaced by molecular techniques such as real-time RT-PCR, which has been employed in seed potato certification schemes in the Netherlands and Scotland since 2013 and in Switzerland and France since 2015 (this list of countries is not exhaustive).

5.5 The Role of Seed Certification Schemes for PVY Control and Future Challenges

Seed certification schemes are essential in managing and controlling non-persistent viruses such as PVY in seed potato production through inspection of seed potato crops and statutory virus testing. They provide a purchaser with assurance that seed potatoes have been produced according to scheme requirements and standards, particularly with regard to health and cultivar purity. A post-harvest test should provide a reasonable estimate of the incidence of PVY in the daughter crop, provided the sample collected is representative of the crop (Frost et al. 2013). Certification, therefore, offers the opportunity to discard or 'flush through' crops, which pose a risk to the health of subsequent crops. The population dynamics of PVY are changing (see references in Chap. 3). Ordinary and veinal necrotic PVY strains (*i.e.* PVY^{O} and PVY^{N}) are gradually becoming less prevalent in Europe and are being replaced by recombinant strains PVY^{NTN} and $PVY^{N\text{-}Wi}$. PVY^{O} and PVY^{N} are generally believed to be more aggressive than the now-prevalent above-mentioned recombinant strains and elicit severe foliar symptoms on potato plants, although this does vary between cultivars. On certain cultivars, the recombinant PVY^{NTN} and $PVY^{N\text{-}Wi}$ strains tend to produce milder symptoms, and in some occasions, higher copy numbers of viral RNA were found in leaves displaying milder symptoms (Lindner et al. 2015; Karasev and Gray 2013). This suggests an apparent lack of correlation between symptom severity, virus concentration and virus strains; therefore, adopting different tolerances with respect to the severity of virus symptoms might no longer be appropriate in certification schemes when it comes to control PVY (Lindner et al. 2015). Consequently, the Specialised Section on Standardisation of Seed Potatoes at UNECE has revised the UNECE Standard (S-1) Seed Potatoes and adopted new virus tolerances irrespective of the severity of viral symptoms observed in growing crops (UNECE 2014). As mild mosaic/viral symptoms may be missed during crop inspection, particularly if conditions are less favourable for visual inspection of plants (*e.g.* bright sunlight or water on foliage) or, as stated above, when plants are infected by less aggressive virus strains or if a cultivar is tolerant to some virus species, visual inspections should be complemented by post-harvest virus testing, as undertaken by most developed certification schemes worldwide.

6 Conclusion

Controlling the spread of PVY remains a challenge to the potato industry worldwide because of its non-persistent mode of transmission and the evolution of new strains and variants. Prophylactic measures such as roguing and early haulm destruction are required to limit PVY spread but are not efficient alone; the implementation of additional control strategies is needed to protect the susceptible potato cultivars. The control strategies presented in this chapter can help to reduce PVY transmission by aphids; however, each individual control strategy has its own limitations. Border crops can reduce the introduction of virus from outside the field but will not reduce the risk of transmission within a crop. While displaying variable levels of protection, mineral oil is generally used by seed potato growers to reduce aphid transmission of non-persistent viruses. Nevertheless, the effectiveness of mineral oil is dependent on the timing of aphid flights. For 'early' aphid flights that occur during the first weeks after plant emergence when plants are at their most susceptible stage, frequent oil application (two to three times per week to maximise coverage of new leaves) might provide some level of protection by reducing PVY transmission if inoculum is present. For 'late' aphid flights (3–4 weeks after emergence), most leaves on a potato plant should be more effectively covered with a film of oil, which reduces the risk of infection. The efficacy of insecticides in controlling PVY spread is highly variable and at best of a limited impact. The effectiveness of control of PVY transmission through combined insecticide and mineral oil treatment may be more effective than individual insecticide or mineral oil treatment. Due to the rapid development of leaves and the difficulty of forecasting aphid flights during early crop development, it is advisable to combine mineral oil treatments with other control techniques such as intercropping, straw mulching and insecticide treatments. These emphasise the importance of controlling virus through appropriate monitoring methods and crop management enforced by seed certification schemes through the use of 'clean' input seed (ideally virus free or seed potatoes with a low incidence of virus) and, when possible, the segregation of seed and ware crops to minimise the risk of virus transmission between them.

References

Al-Daoud F, Fageria MS, Zhang JH, Boquel S, Pelletier Y (2014) Mineral oil inhibits movement of potato virus Y in potato plants in an age-dependent manner. Am J Potato Res 91:337–345

Al-Mrabeh A, Anderson E, Torrance L, Evans A, Fenton B (2010) A literature review of insecticide and mineral oil use in preventing the spread of non-persistent viruses in potato crops In: Council P (ed) Agriculture and horticulture development board 2010. Warwickshire, UK, 65. (Council P, ed.)

Alyokhin A, Sewell G (2003) On-soil movement and plant colonization by walking wingless morphs of three aphid species (Homoptera : Aphididae) in greenhouse arenas. Environ Entomol 32:1393–1398

Alyokhin A, Sewell G, Groden E (2002) Aphid abundance and potato virus Y transmission in imidacloprid-treated potatoes. Am J Potato Res 79:255–262

Ameline A, Couty A, Martoub M, Giordanengo P (2009) Effects of mineral oil application on the orientation and feeding behaviour of Macrosiphum euphorbiae (Homoptera: Aphidae). Acta Entomol Sin 52:617–623

Basky Z (2003) Virus vector aphid activity and seed potato tuber virus infection in Hungary. Anzeiger Fur Schadlingskunde-Journal of Pest Science 76:83–88

Basky Z, Almasi A (2005) Differences in aphid transmissibility and translocation between PVYN and PVY0 isolates. J Pest Sci 78:67–75

Bass C, Puinean AM, Zimmer CT et al (2014) The evolution of insecticide resistance in the peach potato aphid, Myzus persicae. Insect Biochem Mol Biol 51:41–51

Beemster ABR (1972) Virus translocation in potato plants and mature plant resistance. In: De Bokx JA (ed) Viruses of potatoes and seed potato production. Pudoc, Wageningen, pp 144–151

Bell AC (1980) The use of mineral oil to inhibit aphid transmission of potato veinal necrosis virus: a laboratory and field experiment. Record of agricultural Research, Northern Ireland Dept. Agric. Res. 28, 13–7

Bell AC (1989) Use of oil and pyrethroid sprays to inhibit the spread of potato virus-Yn in the field. Crop Prot 8:37–39

Boiteau G (1997) Comparative propensity for dispersal of apterous and alate morphs of three potato-colonizing aphid species. Can J Zool 75:1396–1403

Boiteau G, Singh RP (1999) Field assessment of imidacloprid to reduce the spread of PVYO and PLRV in potato. Am J Potato Res 76:31–36

Boiteau G, Wood FA (1982) Persistence of mineral-oil spray deposits on potato leaves. Am Potato J 59:55–63

Boiteau G, King RR, Levesque D (1985) Lethal and sublethal effects of aldicarb on two potato aphids (Homoptera: Aphididae): *Myzus persicae* (Sulzer) and *Macrosiphum euphorbiae* (Thomas). J Econ Entomol 78:41–44

Boiteau G, Singh M, Lavoie J (2009) Crop border and mineral oil sprays used in combination as physical control methods of the aphid-transmitted potato virus Y in potato. Pest Manag Sci 65:255–259

Boquel S, Giguere MA, Clark C, Nanayakkara U, Zhang JH, Pelletier Y (2013) Effect of mineral oil on potato virus Y acquisition by Rhopalosiphum padi. Entomol Exp Appl 148:48–55

Boquel S, Zhang J, Goyer C, Giguère M-A, Clark C, Pelletier Y (2014) Effect of insecticide-treated potato plants on aphid behavior and potato virus Y acquisition. Pest Manag Sci 71:1106–1112

Boydston RA, D-Seymour M, Brown CR, Alva AK (2006) Freezing behavior of potato (Solanum tuberosum) tubers in soil. Am J Potato Res 83:305–315

Bradley RHE (1963) Some ways in which a paraffin oil impedes aphid transmission of potato virus Y. Can J Microbiol 9:369–380

Bradley RHE, Wood FA, Wade CV (1962) Aphid transmission of potato virus Y inhibitted by oils. Virology 18:327–329

Bradley RHE, Moore CA, Pond DD (1966) Spread of potato virus Y curtailed by oil. Nature 209:1370–1371

Bragard C, Caciagli P, Lemaire O, et al. (2013) Status and prospects of plant virus control through interference with vector transmission. In: Vanalfen NK (ed) Annual review of phytopathology, Vol 51. Annual Reviews, 4139 El Camino Way, Po Box 10139, Palo Alto, Ca 94303-0897 USA, 177-201. (Annual Review of Phytopathology; vol. 51.)

Broadbent L, Gregory PH, Tinsley TW (1950) Roguing potato crops for virus diseases. Ann Appl Biol 37:640–650

Cho S-R, Koo H-N, Yoon C, Kim G-H (2011) Sublethal effects of flonicamid and thiamethoxam on green peach aphid, *Myzus persicae* and feeding behavior analysis. J Korean Soc Appl Biol Chem 54:889–898

Cip IPC (1979) Insect vector transmission of potato viruses. CIP, International Potato Center

Collar JL, Avilla C, Duque M, Fereres A (1997) Behavioral response and virus vector ability of Myzus persicae (Homoptera : aphididae) probing on pepper plants treated with aphicides. J Econ Entomol 90:1628–1634

Cooke LR, HTaM S, Hermansen A et al (2011) Epidemiology and integrated control of potato late blight in Europe. Potato Res 54:183–222

Crump NS, (2008) Seed potato certification and its role in soilborne disease management. In: Vispa VCSPA (ed) ViSPA, Voctoria, 1

Crutzen F, Moreau V, Francis F, Bragard C (2014) The alarm pheromone E-β-Farnesene as biocontrol semiochemical to reduce the propagation of potato virus Y in potato fields. In: Research EaFP, ed. 19th Triennial Conference of the European Association for Potato Research. Brussels, European Association for Potato Research, 65. 1

Cui LL, Dong J, Francis F et al (2012) E-beta-farnesene synergizes the influence of an insecticide to improve control of cabbage aphids in China. Crop Prot 35:91–96

Davis JA, Radcliffe EB, Ragsdale DW (2009) Planter skips and impaired stand favors potato virus Y spread in potato. Am J Potato Res 86:203–208

Dawson G, Anderson F, Bain R, et al. (2015) Effectiveness of mineral and vegetable oils in minimising the spread of non-persistent viruses in potato seed crops in Great Britain. In Potato Council project R449 final report. Potato Council

Demeulemeester K (2013) Control of potato virus Y spread in seed potatoes. In: Virology E (ed) Proceedings of the EAPR Section Virology Meeting, Antalya, 57

Dewijs JJ (1980) The caracteristics of mineral-oils in relation to their inhibitory activity on the aphid transmission of potato virus Y. In. *Netherlands*. J Plant Pathol 86:291–300

Dewijs JJ, Sturm E, Schwinn FJ (1979) Viscosity of mineral-oils in relation to their ability to inhibit the transmission of stylet-borne viruses. In. *Netherlands*. J Plant Pathol 85:19–22
Difonzo CD, Ragsdale DW, Radcliffe EB, Gudmestad NC, Secor GA (1996) Crop borders reduce potato virus Y incidence in seed potato. Ann Appl Biol 129:289–302
Dodds JH (1988) Tissue culture technology: practical application of sophisticated methods. Am J Potato Res 65:167–180
Donnelly DJ, Coleman WK, Coleman SE (2003) Potato microtuber production and performance: a review. Am J Potato Res 80:103–115
Döring TF (2014) How aphids find their host plants, and how they don't. Ann Appl Biol 165:3–26
Döring TF, Schmidt T (2007) Response of apterous potato aphids to visual contrasts (Hemiptera : Aphididae). Entomologia Generalis 30:190–191
Draper MD, Pasche JS, Gudmestad NC (2002) Factors influencing PVY development and disease expression in three potato cultivars. Am J Potato Res 79:155–165
Dupuis B (2016) The movement of potato virus Y (PVY) in tehe vascular system of potato plants. Eur J Plant Pathol:1–9
Dupuis B, Schwaerzel R (2011) Impact of Haulm killing on potato virus Y (PVY) spread. Proceedings of the The 18th Triennial Conference of the European Association for Potato Research, . Oulu, 167
Dupuis B, Schwaerzel R, Goy G, Tallant M, Derron J (2010) Stepwise development of an efficient method to control potato virus Y spread in seed potato fields. In: Focus B (ed) Proceedings of the EAPR virology. Bioforsk Focus, Hamar, 22
Dupuis B, Schwaerzel R, Derron J (2014) Efficacy of three strategies based on insecticide, oil and elicitor treatments in controlling aphid populations and potato virus Y epidemics in potato fields. J Phytopathol 162:14–18
Duvauchelle S, Dubois L, Nguyen N (1997) Aphids and viruses on ware potatoes in northern France particularly in 1995 and 1996. Mededelingen Faculteit Landbouwkundige en Toegepaste Biologische Wetenschappen Universiteit Gent 62:545–546
Edwardson JR, Christie RG (1997) *Potyviruses*. University of Florida, Gainesville
Eppo (1999) EPPO standard PM 4/28, certification schemes – seed potatoes. Bull OEPP/EPPO 29:253–267
Espinoza N, Estrada R, Tovar P, Bryan J, Dodds JH, (1984) Tissue culture micropropagation, conservation, and export of potato germplasm. Lima, Peru: International Potato Center (CIP)
Eu (2002) Council Directive 2002/56/EC of 13 June 2002 on the marketing of seed potatoes. In: Eu (ed) Union TCOTE
Fageria M, Boquel S, Leclair G, Pelletier Y (2014a) Quantification of mineral oil accumulation and movement in potato plants and its significance in potato virus Y management. Pest Manag Sci 70:1243–1248
Fageria MS, Boquel S, Leclair G, Pelletier Y (2014b) The use of mineral oil in potato protection: dynamics in the plant and effect on potato virus Y spread. Am J Potato Res 91:476–484
Fereres A, Moreno A (2009) Behavioural aspects influencing plant virus transmission by homopteran insects. Virus Res 141:158–168
Foster SP, Hackett B, Mason N, et al (2002) Resistance to carbamate, organophosphate and pyrethroid insecticides in the potato aphid (*Macrosiphum euphorbiae*). Proceedings of the Brighton crop conference: Pests and diseases Brighton, 811–816
Foster SP, Paul VL, Slater R et al (2014) A mutation (L1014F) in the voltage-gated sodium channel of the grain aphid, Sitobion avenae, is associated with resistance to pyrethroid insecticides. Pest Manag Sci 70:1249–1253
Franc GD (2001) Seed certification as a virus management tool. Virus and virus-like diseases of potatoes and production of seed-potatoes. Kluwer, Dordrecht, pp 407–420
Frost KE, Groves RL, Charkowski AO (2013) Integrated control of potato pathogens through seed potato certification and provision of clean seed potatoes. Plant Dis 97:1268–1280
Gabriel W (1965) Influence of temperature on spread of aphid-borne potato virus diseases. Ann Appl Biol 56:461

Gibson RW (1991) The development of mature plant-resistance in 4 potato cultivars against aphid-inoculated potato-virus Yo and Yn in 4 potato cultivars. Potato Res 34:205–210

Gibson RW, Cayley GR (1984) Improved control of potato virus-Y by mineral-oil plus the pyrethroid cypermethrin applied electrostatically. Crop Prot 3:469–478

Gibson RW, Rice AD (1986) The combined use of mineral-oils and Pyrethroids to control plant-viruses transmitted nonpersistently and semi-persistently by Myzus persicae. Ann Appl Biol 109:465–472

Gibson RW, Rice AD, Sawicki RM (1982) Effect of the pyrethroid deltamethrin on the acquisition and inoculation of viruses by Myzus persicae. Ann Appl Biol 100:49–54

Gibson RW, Pickett JA, Dawson GW, Rice AD, Stribley MF (1984) Effects of aphid alarm pheromone derivatives and related compounds on non-persistent and semi-persistent plant virus transmission by *Mysus persicae*. Ann Appl Biol 104:203–209

Gildemacher PR, Schulte-Geldermann E, Borus D et al (2011) Seed potato quality improvement through positive selection by smallholder farmers in Kenya. Potato Res 54:253–266

Gray S, De Boer S, Lorenzen J et al (2010) Potato virus Y: an evolving concern for potato crops in the United States and Canada. *Plant Dis* **94**:1384–1397

Gugerli P, Fries P (1983) Characterization of monoconal antibodies to potato virus Y and their use for virus detection. J Gen Virol 64:2471–2477

Gugerli P, Gehriger W (1980) Enzyme-linked immunosorbent assay (ELISA) for the detection of potato leafroll virus and potato virus Y in potato tubers after artificial break of dormancy. Potato Res 23:353–359

Hall TD (1993) Seed potato certification in the UK. In: Council BCP (ed) Proceedings of the plant health and the Single European Market: British Crop Protection Council, pp 77–82

Hansen LM, Nielsen SL (2012) Efficacy of mineral oil combined with insecticides for the control of aphid virus vectors to reduce potato virus Y infections in seed potatoes (Solanum tuberosum). Acta Agric Scand B Soil Plant Sci 62:132–137

Heimbach U, Eggers C, Thieme T (2004) Effect of mulch on aphid populations and virus transmissions in some arable crops. In: Simon JC, Dedryver CA, Rispe C, Hulle M (eds) Proceedings of the Aphids in a new millennium. 2001 Rennes, France, pp 307–312

Helson VA, Minshall W (1951) Effects of petroleum oils on the carbon dioxide output in respiration of parsnip and mustard. Plant Physiol 31:5–11

Hesler LS, Plapp FW (1986) Uses of oils in insect control. Southwest Entomol 11:1–8

Hirpa A, Meuwissen MPM, Tesfaye A et al (2010) Analysis of seed potato systems in Ethiopia. Am J Potato Res 87:537–552

Irag (2008) Guidelines for preventing and managing insecticide resistance in aphids on potatoes. In: Council P (ed) Oxford, 4

Irag (2014) Guidelines for preventing and managing insecticide resistance in aphids on potatoes. In: Board AaHD (ed) 8

Jeffries CJ (1986) The Scottish seed potato classification scheme and the production of nucleus stocks using micropropagation. In: Council BCP (ed) Proceedings of the symposium on healthy planting material, British Crop Protection Council, 239–247

Jones DC, Woodford JT, Main SC, Pallett D, Barker H (1996) The role of volunteer potatoes in the spread of potato virus Y-N in ware crops of cv. record. Ann Appl Biol 129:471–478

Kachroo P, Shanklin J, Shah J, Whittle EJ, Klessig DF (2001) A fatty acid desaturase modulates the activation of defense signaling pathways in plants. Proc Natl Acad Sci U S A 98:9448–9453

Kaliciak A, Syller J (2009) New hosts of potato virus Y (PVY) among common wild plants in Europe. Eur J Plant Pathol 124:707–713

Karasev AV, Gray SM (2013) Continuous and emerging challenges of potato virus Y in potato. Annu Rev Phytopathol 51(51):571–586

Kazinczi G, Horvath J, Takacs AP, Gaborjanyi R, Beres I (2004) Experimental and natural weed host-virus relations. Commun Agric Appl Biol Sci 69:53–60

Kerlan C, Robert Y, Perennec P, Guillery E (1987) Survey of the level of infection by PVYo and control methods developed in France for potato seed production. Potato Res 30:651–667

Kessmann H, Staub T, Hofmann C, et al (1994) Induction of systemic acquired disease resistance in plants by chemicals. In: Cook RJ, (ed) Annual Review of Phytopathology. Annual Reviews Inc. {a}, P.O. Box 10139, 4139 El Camino Way, Palo Alto, California 94306, USA, 439–59. (Annual Review of Phytopathology; vol. 32.)

Kirchner SM, Hiltunen LH, Santala J et al (2014) Comparison of straw mulch, insecticides, mineral oil, and birch extract for control of transmission of potato virus Y in seed potato crops. Potato Res 57:59–75

Kislow CJ, Edwards LJ (1972) Repellent odours in aphids. Nature 235:108–109

Klueken AM, Hau B, Ulber B, Poehling HM (2009) Forecasting migration of cereal aphids (Hemiptera: Aphididae) in autumn and spring. J Appl Entomol 133:328–344

Le Romancer M, Kerlan C (1991) Superficial ringspot necrosis of potato tubers, a recent disease caused by potato virus Y. Agronomie 11:889–900

Lin K, Bushnell W, Szabo L, Smith A (1996) Isolation and expression of a host response gene family encoding thaumatinlike proteins in incompatible oat-stem rust fungus interactions. Mol Plant-Microbe Interact 9:511–522

Lin FJ, Bosquee E, Liu YJ, Chen JL, Liu Y, Francis F (2016) Impact of aphid alarm pheromone release on virus transmission efficiency: when pest control strategy could induce higher virus dispersion. J Virol Methods 235:34–40

Lindner K, Trautwein F, Kellermann A, Bauch G (2015) Potato virus Y (PVY) in seed potato certification. J Plant Dis Protect 122:109–119

Loebenstein G, Deutsch M, Alper M (1964) Preventing aphid-spread cucumber mosaic virus with oils. Phytopathology 54:960

Lutman PJ (1977) Investigations into some aspects of the biology of potatoes as weeds. Weed Res 17:123–132

Mackenzie TDB, Fageria MS, Nie XZ, Singh M (2014) Effects of crop management practices on current-season spread of potato virus Y. Plant Dis 98:213–222

Mackenzie TDB, Nie X, Singh M (2016) Crop management practices and reduction of on-farm spread of *potato virus Y*: a 5-Year study in commercial potato fields in New Brunswick., Canada. Am J Potato Res 93:552–563

Malloch G, Foster S, Williamson M (2016) Monitoring pyrethroid resistance (kdr) and genetic diversity in UK populations of the grain aphid, Sitobion avenae during 2015. In: Council P (ed) Kenilworth, 26

Margaritopoulos JT, Tsamandani K, Kanavaki OM, Katis NI, Tsitsipis JA (2010) Efficacy of pymetrozine against Myzus persicae and in reducing potato virus Y transmission on tobacco plants. J Appl Entomol 134:323–332

Martin-Lopez B, Varela I, Marnotes S, Cabaleiro C (2006) Use of oils combined with low doses of insecticide for the control of Myzus persicae and PVY epidemics. Pest Manag Sci 62:372–378

Martoub M (2010) Impact global de l'huile minerale blanche sur le pathosysteme plante - puceron-virus: Universite de Picardie Jule Verne PhD thesis

Maunder HB, (2005) The history of seed potato certification in New Zealand 1927–2000. Wellington, New-Zealand: Vegfed, the New Zealand Vegetable and Potato Growers' Federation

Milosevic D (1996) Efficacy of oils and insecticides in potato plant protection against infection by potato virus Y and leaf roll virus (PV and PLRV). Zaštita Bilja 47:333–342

Morita M, Ueda T, Yoneda T, Koyanagi T, Haga T (2007) Flonicamid, a novel insecticide with a rapid inhibitory effect on aphid feeding. Pest Manag Sci 63:969–973

Nakayama M, Matsuura K, Okuno T (1996) Production of salicylic acid in tobacco and cowpea plants by a systemic fungicide ferimzone and induction of resistance to virus infection. J Pestic Sci 21:69–72

Narayandas GK, Alyokhin AV (2006) Interplant movement of potato aphid (Homoptera : aphididae) in response to environmental stimuli. Environ Entomol 35:733–739

Nault LR, Edwards JL, Styer WE (1973) Aphid alarm pheromones: secretion and reception. Environ Entomol 2:101–105

Naylor M, Murphy AM, Berry JO, Carr JP (1998) Salicylic acid can induce resistance to plant virus movement. Mol Plant-Microbe Interact 11:860–868
Nivap (2016) Clonal selection – Netherlands. In: Nivap, ed. (2016.)
Perring TM, Gruenhagen NM, Farrar CA (1999) Management of plant viral diseases through chemical control of insect vectors. Annu Rev Entomol 44:457–481
Peters D, Lebbink G (1973) Effect of oil on transmission of pea enation mosaic-virus during short inoculation probes. Entomol Exp Appl 16:185–190
Petrov N, Andonova R (2012) Bion and exin as sar elicitors against potato virus Y infection in tomato. Plant Stud 2:46–49
Pickett JA, Wadhams LJ, Woodcock CM (1992) The chemical ecology of aphids. Annu Rev Entomol 37:67–90
Powell G (1992) The effect of mineral-oil on stylet activities and potato virus-Y tramsmission by aphids. Entomol Exp Appl 63:237–242
Ragsdale DW, Landis DA, Brodeur J, Heimpel GE, Desneux N, (2011) Ecology and management of the soybean aphid in North America. In: Berenbaum MR, Carde RT, Robinson GE, (eds) Annual Review of Entomology, Vol 56. Annual Reviews, 4139 El Camino Way, Po Box 10139, Palo Alto, Ca 94303-0897 USA, 375–399. (Annual Review of Entomology; vol. 56.)
Rakotonindraina T, Corbière R, Chatot C, et al. (2011) Analysis of volunteer density under the influence of cropping practices: a contribution to the modelling of primary inoculum of *Phytophthora infestans* in potato crops. Proceedings of the Thirteenth EuroBlight workshop. St. Petersburg: PPO-Special Report, 67–74
Rice AD, Gibson RW, Stirbley MF (1983) Effects of deltamethin on walking flight and potato virus Y-transmission by pyrethroid-resistant *Myzus persicae*. Ann Appl Biol 102:224–236
Robert Y, Woodford JT, Ducray-Bourdin DG (2000) Some epidemiological approaches to the control of aphid-borne virus diseases in seed potato crops in northern Europe. Virus Res 71:33–47
Rolot J-L (2005) *Analyse des facteurs régulant la dissémination du virus Y de la pomme de terre (PVY) en vue de stratégies de lutte raisonnées.* Gembloux: Faculté des Sciences agronomiques de Gembloux, PhD thesis
Rolot J-L, Seutin H, Georges G, Deveux L (2006) *Evaluation de l'efficacité du nouvel insecticide Plenum pour le contrôle de la dissémination des infections à virus Y dans une parcelle de plants*
Rousselle P, Robert Y, Crosnier JC (1996) *La pomme de terre*. INRA, Paris
Saucke H, Doring TF (2004) Potato virus Y reduction by straw mulch in organic potatoes. Ann Appl Biol 144:347–355
Schramm S, Frost K, Charkowski A, Gray S, Crockford A, Groves RL (2011) Management of potato virus Y (PVY) in Wisconsin seed potato production. University of Wisconsin. Extension (A3951),Madison 8
Schulte-Geldermann E, Gildemacher PR, Struik PC (2012) Improving seed health and seed performance by positive selection in three kenyan potato varieties. Am J Potato Res 89:429–437
Shepard JF, Claflin LE (1975) Critical analyses of the principles of seed potato certification. Annu Rev Phytopathol 13:271–293
Sigvald R (1984) The relative efficiency of some aphid species as vectors of potato virus Yo (PVYo). Potato Res 27:285–290
Sigvald R (1985) Mature-plant resistance of potato plants against potato-virus Yo (PVYo). Potato Res 28:135–143
Simons JN (1957) Effect of insecticides and physical barriers on field spread of pepper veinbanding mosaic virus. Phytopathology 47:139–145
Simons J, Mclean D, Kinsey M (1977) Effects of mineral oil on probing behavior and transmission of stylet-borne viruses by Myzus persicae. J Econ Entomol 70:6
Slater R, Paul VL, Andrews M, Garbay M, Camblin P (2012) Identifying the presence of neonicotinoidresistant peach-potato aphid (Myzus persicae) in the peach-growing regions of southern France and northern Spain. Pest Manag Sci 68:634–638
Steinger T, Gilliand H, Hebeisen T (2014) Epidemiological analysis of risk factors for the spread of potato viruses in Switzerland. Ann Appl Biol 164:200–207

Su JW, Zhu SR, Zhang ZN, Ge F (2006) Effect of synthetic aphid alarm pheromone (E)-beta-farnesene on development and reproduction of Aphis gossypii (Homoptera : Aphididae). J Econ Entomol 99:1636–1640

Tan BL, Sarafis V, Beattie GC, White R, Darley EM, Spooner-Hart R (2005) Localization and movement of mineral oil in plants by fluorescence and confocal microscopy. J Exp Bot 56:2755–2763

Thomas-Sharma S, Abdurahman A, Ali S et al (2016) Seed degeneration in potato: the need for an integrated seed health strategy to mitigate the problem in developing countries. Plant Pathol 65:3–16

Unece (2014) Virus tolerances, proposals by the United States. In: Unece (ed) Proceedings of the meeting of the extended bureau of the specialized section on standardization of seed potatoes, Melbourne, UNECE, 5

Unece (2015) United Nations international standard for certification and marketing of seed potatoes. In: Unece (ed) Geneva, Switzerland, 2

Van Toor RF, Drayton GM, Lister RA, Teulon DJ (2009) Targeted insecticide regimes perform as well as a calendar regime for control of aphids that vector viruses in seed potatoes in New Zealand. Crop Prot 28:599–607

Vasyukova NI, Ozeretskovskaya OL (2007) Induced plant resistance and salicylic acid: a review. Appl Biochem Microbiol 43:367–373

Walsh D (2000) Horticultural spray oils: useful year-round tactic for IPM systems. Agrichem Environ News 165:5–7

Walters KFA, Dixon AFG (1984) The effect of temperature and wind on the flight activity of cereal aphids. Ann Appl Biol 104:17–26

Wedding RT, Riehl LA, Rhaods WA (1952) Effect of petroleum oil spray on photosynthesis and respiration in citrus leaves. Plant Physiol 60:3–12

Whitworth JL, Hamm PB, Mcintosh CS (2010) Effect of potato virus Y on yield of a clonal selection of russet norkotah. Am J Potato Res 87:310–314

Wróbel S (2009) The retention of PVY in the stylet of *Myzus persicae* Sulz. After the application of mineral oil on potato plants. Plant Breed Seed Sci 60:3–12

Wrobel S (2012) Comparison of mineral oil and rapeseed oil used for the protection of seed potatoes against pvy and pvm infections. Potato Res 55:83–96

Zhang Z, Tu M, Du Y et al (1997) Behavioral and electrophysiological response of Myzus persicae to stimulus of (E)- beta -farnesene. Acta Entomol Sin 40:40–44

Chapter 8
Resistance to *Potato virus Y* in Potato

Jari P.T. Valkonen, Christiane Gebhardt, Ewa Zimnoch-Guzowska, and Kazuo N. Watanabe

Abstract Growing potato cultivars resistant to *Potato virus Y* (PVY) offers the easiest and the most cost-effective solution to prevent the losses caused by PVY. Genes for resistance to PVY can be found in wild and cultivated potato species. Resistance genes functioning on the "gene-for-gene" basis are commonly used in breeding for resistance to PVY, because of their dominant and simple inheritance. Ten genes for resistance to PVY have been mapped to four potato genome segments on chromosomes IV, IX, XI and XII. Genes for hypersensitive resistance (HR) are usually PVY strain-specific, whereas the genes for extreme resistance (ER) are effective against a broad spectrum of PVY strains. The need for broad-spectrum resistance to PVY has been widely realized since the turn of the millennium when new PVY strains able to overcome strain-specific resistance to the common PVY strains became prevalent in potato crops worldwide. Field resistance to PVY is widely present in potato cultivars and based almost exclusively on minor genes contributing small effects. Marker-assisted selection provides an efficient approach for the selection of traits governed by major genes or quantitative trait loci (QTLs) with large effects. Genetically engineered resistance based on RNA silencing, the basal antiviral defense system of plants, is another option to protect potatoes against PVY.

J.P.T. Valkonen (✉)
Department of Agricultural Sciences, University of Helsinki,
PO Box 27, FI-00014 Helsinki, Finland
e-mail: jari.valkonen@helsinki.fi

C. Gebhardt
Max-Planck-Institute for Plant Breeding Research,
Carl-von-Linné-Weg 10, 50829 Cologne, Germany

E. Zimnoch-Guzowska
Plant Breeding and Acclimatization Institute (IHAR), Młochów Research Center,
05-831 Młochów, Poland

K.N. Watanabe
Gene Research Center, University of Tsukuba,
1-1-1 Tennoudai, Tsukuba, Ibaraki 305-8572, Japan

C. Lacomme et al. (eds.), *Potato virus Y: biodiversity, pathogenicity, epidemiology and management*, DOI 10.1007/978-3-319-58860-5_8

1 Introduction

Potato cultivars differ in their response to *Potato virus Y* (PVY) infection and may appear as sensitive (symptoms develop), tolerant (no symptoms develop, but modest yield losses may occur) or resistant to PVY (see Chap. 2). Growing potato cultivars resistant to PVY offers the easiest and the most cost-effective solution to preventing the losses caused by PVY (Świeżyński 1994; Thiele et al. 2010). In this respect, the number of potato cultivars expressing broad-spectrum resistance to all strains of PVY is surprisingly low (Table 8.1). In developed countries, this may be because their production and certification schemes are consistently able to supply farmers with seed potatoes sufficiently healthy to limit the potential losses caused by PVY

Table 8.1 Potato cultivars expressing extreme resistance to PVY

Breeding country[a]	Resistance gene	Source of resistance[b]	Cultivar
Denmark	Ry_{sto}	*I-1039 (sto)*	Tivoli,
		sto x Frühmolle	Canasta, Thor, Wotan, Ydun
		MPI 46.152/1 (stohybrid)	
Germany	Ry_{sto}	*sto x Erika*	Alwara, Arosa, Bettina, Bison, Cordia, Esta, Fanal, Franzi, Heidrun, Jaqueline, Petra, Pirola, Solara, Wega
		MPI 47.174/3	
		sto x Frühmolle	Amado, Assia, Barbara, Forelle, Jumbo, Maxi, Tomba, Sibu, Oktan, Ute
		MPI 46.152/1 (stohybrid)	
		?	Django, Estrelle, Fioretta, Stärkeprofi, Toccata, Turus
Hungary	Ry_{sto}	*sto x Frühmolle*	Magyar Rózsa, Szignal, White Lady, Démon, Baltoni Rózsa, Katica, Lorett, Gòliàt, Rioja, Hópehely, Venusz Gold, Luca XL, Kànkàn, Arany Chipke
		MPI 46.152/1 (stohybrid) ?	
	Ry_{chc}	*chc x fmt*	Pannónia, Boró, Reka, Rebeka, Rachel, Boglarka
Poland	Ry_{sto}	*sto XIIB*	Dunajec, Fregata, Hinga, Nimfy, Ślęza,
		sto x Frühmolle	Ania, Anielka, Baszta, Beata, Bekas, Bóbr, Brda, Bzura, Kuba, Maryna, Pilica, San, Sekwana, Stobrawa
		MPI 46.152/1 (stohybrid)	
		sto x Erika	Barycz, Etiuda, Finezja, Meduza, Gabi, Gustaw, Klepa, Omulew, Owacja, Sonda, Umiak, Ursus, Vistula
		MPI 47.174/3	
		C.854	
	$Ry\text{-}f_{sto}$	*PW 243*	Ametyst, Bursztyn, Flaming, Legenda
Slovakia	Ry_{sto}	*sto x Erika*	Alva
		MPI 47.174/13	

(continued)

Table 8.1 (continued)

Breeding country[a]	Resistance gene	Source of resistance[b]	Cultivar
The Netherlands	'Ry_{sto}'	*CPC 2093*	Corine, Sante, Festien, Lady Christl,
		Y193	Cycloon,
	Ry_{sto}	*sto x Frühmolle*	Kuras
		MPI 46.152/1 (stohybrid)	
Russia	Ry_{sto}	*Stohybrid*	Bryansky Ranny, Golubizna,
		189Y-56, 189Y-95-1	Skoroplodniy
		sto x Erika	Effekt, Osen, Lira, Pransa,
		MPI 47.174/13	Zhigulevskiy
	Ry_{chc}	***chc f. garciae***	Nikulinsky, Bryanskiy nadezhiniy,
		55d	Krasnoyarskiy Ranniy,Utenok
	Ry_{adg}, Ry_{sto}	*tbr-acl-adg-dms-sto*	Resurs, Sokolskiy, Kolobok
		69.5403.259	
	?	***chc f. commersonii***	Bryanskiy delikates, Nakra, Slava,
		58d	Bryanschiny, Master,
Japan	Ry_{chc}	*chc*	Konafubuki, Sakurafubuki, Saikai 35
The USA	Ry_{adg}	*Neo-Tuberosum*	Eva, NY121, NY123, T88–19
Israel	Ry_{adg}	*Neo-Tuberosum*	Nitza
		NY66,R 112–1,	
Peru	Ry_{adg}	*Neo-Tuberosum adg*	Costanera, Muru, Yana, Chagllina-INIA, Reiche, Maria Bonita-INIA, Tacna, Primavera, Unica
		XY.7, XY.14.7, XY Bulk (XY.9, XY.13, XY.14, XY.16)	
Bangladesh	Ry_{adg}	7XY.1	Chamak from CIP material
Cameroon	Ry_{adg}	7XY.1	Ira-92 from CIP material
Rwanda	Ry_{sto}	*I-1039 (sto)*	Gikungu from CIP material

[a]Sources of information: Flis et al. (2005), Song et al. (2005), Sato et al. (2006), Linder et al. (2011), Simakov et al. (2008), Valkonen et al. (2008), Chrzanowska (2000), www.plantbreeding.wur.nl/potatopedigree/, www.europotato.org, Catalog of Varieties of the International Potato Center (2010), Catalog of Russian Cultivars; and H.G. Kirk, N. Zoteyeva, and Z. Polgar (pers. com.)
[b]Abbreviations of species names: *acl Solanum acaule, adg S. tuberosum* Group Andigenum, *chc S. chacoense, dms S.demissum, fmt S. famatinae, sto S. stoloniferum, tbr S. tuberosum*

(Bradshaw 2009). Breeding for broad-spectrum resistance to PVY has been carried out especially for the needs of developing countries where healthy seed potatoes are not available or cannot be afforded. The International Potato Center (CIP) has developed candidate resistant cultivars and potato breeding populations with enhanced resistance to PVY for use in a range of climatic conditions (Gastelo et al. 2014).

The need for broad-spectrum resistance to PVY was widely realized at the turn of the millennium when new PVY strains able to overcome strain-specific resistance to the common PVY strains became prevalent in potato crops worldwide.

The new strains can cause very severe symptoms in foliage and necrosis in tubers. However, the new strains can symptomlessly infect some cultivars, which makes roguing of infected plants impossible in seed potato crops and increases the overall PVY infection pressure on potato crops (Le Romancer and Kerlan 1991; Weidemann 1993; Karasev and Gray 2013). Therefore, breeding for more durable resistance to PVY has become an important goal.

Potato breeding is a process dynamically modified by the application of advanced molecular tools simplifying selection processes and the choice of parents. Marker-assisted selection (MAS) is an efficient approach for the selection of traits governed by major genes or quantitative trait loci (QTLs) with large effects. However, selection of many complex traits with low heritability and large genotype-environment (G × E) interactions is continuing using traditional phenotypic selection schemes.

2 Taxonomy and Genetics of Potato

The most commonly cultivated potato species, *Solanum tuberosum* L. (family Solanaceae), originates in South America, where a number of other potato species have also been domesticated. The natural habitats of cultivated and wild potato species cover South and Central America and range from the cold environments of the Andean mountains 4500 m above the sea level to the tropical climates of the Amazon jungle. The phenotypic and genetic diversity of cultivated and wild potato species is enormous (Correll 1962; Hawkes 1990; Ochoa 1991). Only a very small portion of the potato genetic resources has been utilized in breeding.

2.1 Potato Taxonomy

Classification of potato species based on morphological traits has resulted in the description of nearly 240 species [235 species according to Hawkes (1990)], of which 7 are cultivated. Most species are diploid ($2n = 2 \times = 24$), but triploid, tetraploid, pentaploid and hexaploid species have also been described (Hawkes 1990). The so-called endosperm balance number determines cross-compatibility of potato genotypes (Johnston and Hanneman 1982). Many potato species are sexually compatible, but geographical isolation has prevented the species from crossing in nature. Hence, most potato "species" are not true species as defined by mutual sexual incompatibility, but can be crossed and used in breeding programs. Recent studies based on microsatellite and plastid DNA deletion marker analyses suggest that potato taxonomy could be reduced to ca. 100 wild and 4 cultivated species (Spooner et al. 2007; Ovchinnikova et al. 2011). However, the recognized morphological and other unique traits, which distinguish the different "species" in the original taxonomic system (Correll 1962; Hawkes 1990; Ochoa 1991), indicate genetic differences that are expressed at the phenotypic level. The original species names can, therefore, be helpful from the breeding point-of-view for selecting potato germplasm for further characterization and introgression of new traits into cultivars. Crossing of incompatible potato species may also be achieved using ploidy manipulation and various special techniques, such as embryo rescue and rescue pollination (Valkonen et al. 1995; Watanabe et al. 1995).

S. tuberosum is divided into two subspecies, *S. tuberosum* ssp. *tuberosum* (modern cultivated potato, Chilotanum group) and *S. tuberosum* ssp. *andigena* (Andigenum group). The cultivated potato species have a relatively narrow ecogeographical origin and are adapted to a limited range of environmental conditions and diseases (Milbourne et al. 2007). Modern potato cultivars bred in Europe after the pandemic of the late blight disease in 1845–1850 contain more Chilean germplasm adapted to long day-length conditions than Andigena germplasm adapted to a short day-length. In general, potato cultivars bred in the nineteenth and the beginning of the twentieth century had a narrow genetic base and might have originated from only a few introductions (Ross 1986). However, recent genotyping studies based on single nucleotide polymorphism (SNP) (6373 genome-wide SNPs) of 250 advanced potato breeding lines, 25 genetic stocks and 12 wild species revealed that the average heterozygosity was substantially higher in the modern potato cultivars (53–59%) than in the wild *Solanum* species (30%) (Hirsch et al. 2013).

2.2 Genetic Inheritance in Potato: Challenges of Polysomic Polyploidy

To make further progress in potato breeding, additional potato species and genotypes need to be investigated, characterized and utilized to make better use of the large and diverse potato genetic resources. Potato breeding requires large populations and takes a long time. Not only the general pitfalls in plant breeding but also the complexity of polyploid genetics of potato challenge breeding efforts (Watanabe 2015).

The tetraploid genome combined with tetrasomic inheritance is the most prominent genetic issue associated with potato breeding (Ortiz and Watanabe 2004) (Table 8.2). Often a trait controlled by a single gene is scored as quantitative based on phenotypic segregation of the progeny, although differences may be due to variation in the condition (simplex or multiplex) of the allele in a locus. Here, the term “polysomic polyploidy” is used according to Mackey (1970), instead of “autopolyploid” as used by Stebbins (1950) and other botanists. Tetrasomic behavior of chromosomes and the corresponding genetic segregation have been reported (Matsubayashi 1991; Swaminathan and Magoon 1961), but the frequency of quadrivalent formation varies in tetraploid cultivars.

Inter-locus interaction (epistasis) and heterozygosity are important factors to be considered in potato breeding, and additive components also contribute to quantitative traits. Identification of individual chromosomes has been difficult by conventional cytogenetic methods. Also, monitoring recombination and introgression is

Table 8.2 Comparison of inheritance in tetraploid potatoes and polyploidy wild *Solanum* species

Features	Species example	Segregation	Multivalent in meiosis	Chromatid segregation	Fertility	References
Tetraploid cultivars	*S. tuberosum*	Tetrasomic and variable	Yes	Possible	Low to medium	Allard (1960), Bradshaw (1994), Ortiz and Watanabe (2004)
Polyploid species	*S. acaule* (4x), *S. stoloniferum* (4x), *S. demissum* (6x)	Disomic	No	No	Very high	Watanabe and Orrillo (1994)

difficult by phenotypic evaluation at the tetraploid level. However, theoretical prediction of segregation has been supported by cytogenetic and genetic research in many cases, and the findings have been tested and applied to potato breeding (Dong et al. 2000, 2001; Peters et al. 2012; Ross 1986; Szinay et al. 2012; Tang et al. 2014).

The four important aspects of polyploids are: (1) genotypic variation occurs at a locus, (2) multiple alleles can exist in a locus of the same genotype, (3) allelic interactions can occur between multiple alleles, and iv) both chromatid segregation and chromosome segregation can occur (Allard 1960; Bradshaw 1994). Due to the tetrasomic condition, one locus accommodates four alleles. In a diallelic locus (dominant allele *A* and recessive allele *a*), five genotypic classes are possible: *AAAA*, *AAAa*, *AAaa*, *Aaaa* and *aaaa*. When an additive effect of an allele occurs, the locus can create considerable quantitative segregation. Furthermore, diverse multiple alleles can occur in a locus and each allele can have a different function, such as a_1, a_2, a_3, a_4. The different alleles can interact, and the allelic interactions may be diallelic (a_1a_2, a_1a_3, a_1a_4, a_2a_3, a_2a_4, a_3a_4), triallelic ($a_1a_2a_3$, $a_1a_3a_4$, $a_2a_3a_4$) or tetra-allelic ($a_1a_2a_3a_4$). Thus, heterosis is not only a result of the interaction of two alleles, but there is a wider range of possibilities. A locus distal to the centromere with a chance of crossing-over can be involved in chromatid segregation. For example, a locus in a diallelic condition in genotype *Aaaa* produces gametic output in the ratio *Aa*: *aa* = 1:1 by chromosome segregation, but 15*AA*: 12*Aa*: 1*aa* by chromatid segregation, assuming random assortment of the homologous chromosomes and chromatids respectively. Thus, a rare gametic genotype and zygotic progeny phenotype are possible. Another example is the *AAaa* genotype that produces the gametes *AA*, *aa* and Aa with the ratio 2:2:8, respectively, with chromosome segregation. With fully random chromatid segregation (*AAAAaaaa*) the expected gametic ratios are *AA*:*aa*:*Aa* = 3:3:8.

Tetrasomic tetraploid genetics is far more complicated than diploid genetics. Furthermore, the outcrossing nature with high heterozygosity of tetraploid potato cultivars causes additional complications in interpreting phenotypic segregation.

Potato breeding objectives often involve quantitative traits, which are strongly influenced by environment. Considering a wide range of phenotype variation with tetrasomic inheritance and the influence of environment on phenotype expression, large-sized populations of true (botanical) potato seed (TPS) need to be produced in crosses for successful potato breeding.

3 Types and Sources of Resistance to PVY in Potato Germplasm

Enrichment of the cultivated potato gene pool has been based on only a limited portion of the available biodiversity of crop wild relatives (CWR) and cultivated species (Bradshaw et al. 2006; Jansky et al. 2013). The seven largest potato collections contain more than 7100 accessions listed in the collaborative Intergenebank Potato Database (Huamán et al. 2000). Developments in molecular techniques create possibilities for a wider use of genetic resources in breeding new cultivars. Conservation and sustainable use of CWR have become a priority followed by systematic development of utilization strategies (Maxted et al. 2012; Jansky et al. 2013). Data from CWR screening for PVY resistance are available in the homepages of the main gene banks, supplemented by published studies.

The choice of methods for the assessment of resistance to PVY in potato germplasm is important, because different ways to challenge a plant with the virus are needed to reveal the various types of resistance, which are discussed below. Observation of symptoms on the inoculated potato plant, inoculation of leaf sap from a virus-challenged plant to indicator test plants, and testing a virus-challenged plant using virus-specific antibodies or molecular diagnostic methods are commonly used to assess whether or not a potato plant resists the virus.

Accessions of cultivated potato species expressing resistance to PVY are preferred as sources of PVY resistance for obvious reasons, because they are less likely to pass on undesirable traits. *S. tuberosum* Group Andigenum Hawkes (Muňoz et al. 1975; Ross 1986; Singh et al. 1994) and *S. tuberosum* Group Phureja (Ross 1986; De Maine et al. 1993; Vallejo et al. 1994; Valkonen et al. 1995) include genotypes expressing broad-spectrum resistance to various strains of PVY. Wild potato species are rich in genes for resistance to PVY and other potato viruses (Stelzner 1950; Cockerham 1970; Ross 1986; Valkonen 1997; Zimnoch-Guzowska et al. 2013). Wild potato species known to include accessions with broad-spectrum resistance to several strains of PVY include *S. acaule* Bitt. (Singh et al. 1994), *S. chacoense* Bitt. (Cockerham 1970; Bamberg et al. 1994, Valkonen 1997; Zoteyeva et al. 2012), *S. demissum* Lindl. (Ross 1986; Zoteyeva et al. 2012), *S. dolichostigma* Buk. (Zoteyeva et al. 2012; synonymous to *S. chacoense* according to Hawkes 1990), *S. hougasii* Corr., *S. microdontum* Bitt. (Cockerham 1970), *S. megistacrolobum* Bitt. (Singh et al. 1994; Valkonen 1997), *S. michoacanum* (Bitt.) Rydb. *S. neoantipovichii* Buk. (synonymous to *S. stoloniferum* Schlechtd. et Bché. according to Hawkes 1990), *S. pinnatisectum* Dun., *S. polytrichon* Rydb. (Zoteyeva et al. 2012), *S. polyadenium*

Greenm., *S. sparsipilum* (Bitt.) Juz. et Buk., *S. sucrense* Hawkes (Valkonen 1997), *S. stoloniferum* (Ross 1958; Cockerham 1970; Bamberg et al. 1994; Valkonen 1997; Zoteyeva et al. 2012), *S. tarnii* Hawkes et Hjerting (Thieme et al. 2008) and the non tuber-bearing species *S. brevidens* Phil. and *S. etuberosum* Lindl. (Gibson et al. 1990; Valkonen et al. 1992; Novy and Helgeson 1994).

3.1 Gene-for-Gene Resistance

Resistance (R) genes functioning on the "gene-for-gene" basis (Flor 1946) are very popular in breeding for resistance to PVY, because of their dominant and simple inheritance. They elicit a quick and powerful defense response, which can take the form of a hypersensitive resistance response (HR) or extreme resistance (ER) (Cockerham 1970; Ross 1986). The signaling cascade induced by HR genes leads to activation of a wide range of defense responses. Allelic variation of genes involved in virus recognition or downstream signaling for defense may cause genotype-dependent phenotypic changes in the outcome of the resistance responses (Valkonen et al. 1998).

HR does not affect virus multiplication but prevents movement of virus from cell to cell, loading of virus into the phloem, and long-distance movement of the virus in the plant. Most of the infected cells die eventually and necrotic local lesions develop on mechanically inoculated leaves. Graft inoculation bypasses the mechanism by which HR restricts the virus to the inoculated leaves. Hence, long-distance (systemic) movement of PVY can occur in the plant and causes necrotic symptoms in the young growing leaves and may kill the apex of the plant (a symptom called top necrosis). ER, however, suppresses virus multiplication in the initially infected cells, which has been shown by inoculation of potato protoplasts (Barker and Harrison 1984). Hence, usually no visible symptoms or signs of infection can be observed on the inoculated leaves of plants expressing ER. Furthermore, PVY and symptoms cannot be detected with sensitive methods, such as enzyme-linked immunosorbent assay (ELISA), in graft-inoculated plants expressing ER, except tiny, necrotic pin-point lesions that may develop in top leaves (Cockerham 1970; Barker and Harrison 1984; Ross 1986; Valkonen et al. 1996).

The HR genes *Ny*, *Nc* and/or *Nz* (Cockerham 1970; Ross 1986; Jones 1990; Rowley et al. 2015) are found in many potato cultivars and recognize subgroups of PVY strains named PVY^{O} (ordinary strains), PVY^{C} (C strains) and PVY^{Z} (Z strains) respectively (Singh et al. 2008). Since the 1940s, breeding programs have introduced the genes *Ny* and *Nc* into many potato cultivars, which has been instrumental in controlling the respective strains of PVY. So far, PVY^{Z} strains have rarely been found (Jones 1990). The most problematic PVY strains are those which have emerged and become abundant during the last two decades, as they overcome all three HR genes. Most of them induce veinal necrosis in tobacco leaves and are placed to the tobacco veinal necrosis strain group PVY^{N} (Singh et al. 2008). They have further evolved by recombining with other PVY strains and can cause severe symptoms in potato, including tuber necrosis (TN), in contrast to the original PVY^{N} strains – hence their specification as NTN strains (Karasev and Gray 2013). Additionally, a recently described PVY strain overcomes the three HR genes and does not induce veinal necrosis in tobacco, hence defining a new strain group PVY^{E} (Singh et al. 2008; Galvino-Costa et al. 2012).

The gene-for-gene concept predicts that, for each R gene, there is a specific avirulence (*Avr*) gene in the pathogen; in the presence of the matching gene, resistance is elicited and the pathogen is rendered avirulent (Flor 1946). Recently, the gene-for-gene resistance was explained by the functions of the proteins encoded by the R and Avr genes (Jones and Dangl 2006). The R protein recognizes specifically the protein encoded by the Avr gene of the pathogen. It may also recognize a plant protein modified by activity of the Avr protein, hence representing recognition of a "pathogen-associated molecular pattern" (PAMP) (Nürnberger and Lipka 2005). A specific amino acid signature in the C-proximal part of the helper component proteinase (HC-Pro) of PVY^{O} and the C-proximal region of HCPro in PVY^{C} is required to trigger *Ny* and *Nc* respectively (Moury et al. 2011; Tian and Valkonen 2013, 2015). This information is helpful in attempts to detect and control the introduction of new PVY strains that may overcome the aforementioned resistance genes. The proteinase domain of the nuclear inclusion protein a (NIaPro) of PVY may elicit the *Ry* gene-mediated ER in potato (Mestre et al. 2003), but this hypothesis, based on the observed responses to the over-expressed NIaPro protein in leaves, awaits confirmation by studies employing infectious PVY.

Strain specificity limits the effectiveness of HR genes. Another problem is that in some cases resistance controlled by HR genes is temperature-sensitive. Higher temperatures may render resistance less efficient and, consequently, PVY spreads systemically and severe (lethal) necrotic symptoms develop in the young growing tissues of the potato plant (Adams et al. 1986; Valkonen 1997; Valkonen et al. 1998). HR to PVY^{O} in potato cv. Pito is effective at temperatures around 16–18 °C, but at higher temperatures of 19–24 °C, resistance breaks down and PVY^{O} spreads systemically, causing leaf-drop and mosaic symptoms (Valkonen 1997; Valkonen et al. 1998). The HR genes *Ny-1* and *Ny-2* protect potato cultivars at 20 °C against systemic infection by the "Wilga-type" recombinant strain of PVY ($PVY^{N\text{-}Wi}$) (Chrzanowska 1991; Glais et al. 2002), but lose their effectiveness at 28 °C (Szajko et al. 2014). The wild tuber-bearing potato species *S. sparsipilum* and *S. sucrense* express HR to PVY^{N} at a low temperature, but resistance is overcome at higher temperatures (Valkonen 1997). The mechanism by which temperature influences the outcome of HR remains to be elucidated.

The recent changes in the populations of PVY call for the use of broad-spectrum resistance to PVY in potato breeding programs. The R genes (*Ry*) conferring ER to PVY fulfil this requirement, as they are not PVY strain-specific. ER to PVY is epistatic to HR (Valkonen et al. 1994). Hence, a potato genotype carrying both *Ry* and *Ny* expresses only ER when inoculated with PVY^{O}; no necrotic symptoms are observed. These observations indicate that the genes for ER act earlier and more efficiently than the genes for HR (Bendahmane et al. 1999), which is consistent with inhibition of PVY multiplication by ER (Barker and Harrison 1984).

3.2 Tolerance, Field Resistance and Basal Defense (RNA Silencing)

Tolerance Tolerance is the capacity of an infected host plant to grow without developing severe disease symptoms and incurring significant yield losses despite allowing invasion and multiplication of the virus to high titers (Cooper and Jones 1983). Potato cultivars can be tolerant to certain strains of PVY, notably the original PVY^N strains. However, this is not useful to the potato industry because the virus can reach high concentrations and, eventually, a tolerant cultivar becomes a significant source of PVY for infection of other, more sensitive cultivars (Russel 1978; Świeżyński 1994; Niks et al. 2011).

Field Resistance Field resistance to PVY is widely present in potato cultivars. This kind of resistance to virus infection is controlled almost exclusively by minor genes contributing small effects. Field exposure trials are necessary for assessing quantitative field resistance to PVY, including a set of standard cultivars for comparison. Such trials are also useful for confirming the resistance of genotypes selected for major gene resistance (Solomon 1978). Field assessment of resistance requires repeated evaluation over many growing seasons and a range of locations to take account of variation in PVY vector populations and infection pressure. Cultivar resistance to PVY is commonly described using a 1–9 scale, in which 1 is deemed to be susceptible and 9 to be the greatest (possibly extreme) resistance [in Germany the scale is reversed (BSL 2002)]. The values are relative and should be interpreted by comparing them with values obtained for established standard cultivars (Gabriel 1995; NIAB 1985; Talbot 1987; van der Woude 1987). There are differences among countries in the official varietal assessment data for PVY resistance. In some cases, data can be very inconsistent, with differences of 5 grades on the 1–9 scale, indicating that national systems differ in the rigor of their evaluation, which may be partly related to methods of evaluation and differences in the relative resistance scores ascribed to reference cultivars (Świeżyński et al. 2001). To minimize differences among countries, it would be important to agree about the sets of standard cultivars and evaluation methods used. Older potato cultivars, with field resistance scores between 5 and 7 on the 1–9 scale, have begun to be scored as more sensitive when the structure of PVY populations in a country has changed and the new, emerging PVY^{NTN} strains have replaced PVY^O strains (Bagnall and Tai 1986; Kus 1995; Schiessendoppler 1996; Zimnoch-Guzowska et al. 2013).

Basal Defense – RNA Silencing An innate immune system called RNA silencing in potato and other plants recognizes virus infection in a universal manner and provides basal defense against viruses. It recognizes long dsRNA as "non-self" in plant cells and responds to this invariant PAMP in virus-infected cells (Nürnberger and Lipka 2005). Double-stranded viral RNA is formed during replication of RNA viruses, and the gene transcripts of DNA viruses can form secondary structures. The plant uses cellular dsRNA-specific endoribonucleases (Dicers) to cleave viral dsRNA to 21, 22 and 24 nucleotides long pieces, so-called small interfering RNA (siRNA), which are loaded to slicing complexes (RISC). The siRNAs guide RISC to cut homologous single-stranded viral RNA molecules, such as the viral RNA genomes or gene transcripts (Hamilton et al. 2002). The sliced, inactive viral RNA molecules are utilized in an amplification phase of RNA silencing. In this process, the cellular RNA-dependent RNA polymerase uses the sliced viral RNA as a template to synthesize more virus-specific dsRNA, which is processed by Dicers and used to load more RISCs (Baulcombe 2007).

RNA silencing may be sufficient to prevent virus infection, or it interferes, to an extent, with viral multiplication and systemic movement, allowing new leaves to recover from virus infection (Ratcliff et al. 1997; Gammelgård et al. 2007). It is believed that RNA silencing is a basal antiviral defense system always activated in virus-infected plants, including PVY-infected potato plants, but little is known about its overall contribution to resistance in potato cultivars.

Genetic engineering can be used to pre-condition or prime plants to respond quickly and specifically to infection by a virus. This is achieved by engineering the plant to express virus-homologous RNA. The over-expressed viral RNA is targeted by RNA silencing and activated RISC complexes are formed. They attack and inactivate homologous single-stranded viral RNA genomes or gene transcripts as soon as virus infection occurs. Virus-derived transgenic resistance to PVY was engineered in potato cultivars (Lawson et al. 1990; Pehu et al. 1995) many years before RNA silencing was known to exist and represents a natural mechanism of plant resistance to a virus (Ratcliff et al. 1997). Engineered resistance to PVY and other viruses in transgenic plants can be very efficient (Waterhouse et al. 1999). It provides an alternative to *R* gene-mediated resistance. However, the virus-derived resistance described above is target-specific. Therefore, heterologous viruses may infect the plant and, while doing so, compromise transgene-derived siRNA production and resistance (Savenkov and Valkonen 2001).

The HC-Pro protein of PVY was among the first viral proteins found to suppress RNA silencing (Brigneti et al. 1998). The P1 protein preceding HCPro in the viral polyprotein may stabilize PVY HCPro and its impact on suppression of RNA silencing (Fernandez et al. 2013). The relative efficiency of antiviral RNA silencing in a potato plant and the counter-defensive suppression of RNA silencing by the virus determine the amount of virus accumulated in infected tissues. The *R* genes for PVY resistance discussed above represent an additional level of antiviral defense, which triggers as yet poorly understood mechanisms preventing viral multiplication (ER / *Ry* genes) or viral movement (HR / *Ny, Nc, Nz*). Genes *Ny* and *Nc* recognize HC-Pro of PVY (Moury et al. 2011; Tian and Valkonen 2013). Recognition seems to rely on HC-Pro conformation, which is predicted to differ between the PVY^{O} and PVY^{C} (and PVY^{N}) and explains the strain group specificity of *Ny* and *Nc* (Tian and Valkonen 2015). Since HC-Pro is a suppressor of RNA silencing, it seems that these two *R* genes have evolved to recognize an effector, *i.e.*, a protein produced by the pathogen to suppress basic defense responses (Jones and Dangl 2006). The next move on the viral side can be mutation of the target protein to escape recognition by HR or ER gene. It seems that PVY strains overcome HR genes more readily than ER genes. Finally, the R genes may be under post-transcriptional control of micro-RNAs (miRNA) produced by the plant itself. miRNAs are produced from long precursor RNA by RNA silencing, and similar to siRNA, they direct the target gene transcripts to degradation by silencing. Viral silencing suppression proteins can interfere with miRNA production in potato, which may reactivate R-gene-mediated resistance following virus infection (Shivaprasad et al. 2012). Possible involvement of this mechanism in resistance to PVY in potato remains to be studied (Yin et al. 2016).

4 PVY Resistance Loci in the Potato Genome and Marker-Assisted Selection for PVY Resistance

Early pioneering studies showed that resistance to PVY in many cultivated and wild potato species was conferred by single dominant genes (Cockerham 1970; Ross 1986). Recessive resistance to PVY exists in other Solanaceous species such as tomato (Parrella et al. 2002), pepper (Ruffel et al. 2002) and tobacco (Acosta-Leal and Xiong 2008) and is also likely to be present in potato. However, due to the polyploid and non-inbred nature of potato, recessive resistance is difficult to uncover and maintain in breeding programs. Recessive genes have, therefore, not played a role in the quest for potato genes for resistance to PVY.

Ry_{sto} identified in *S. stoloniferum* (Cockerham 1943; Ross 1986) governs ER to all known PVY strains, but is associated with cytoplasmic male sterility (Ross 1986). However, another gene ($Ry\text{-}f_{sto}$), also derived from *S. stoloniferum*, confers ER to PVY and is associated with male fertility (Flis et al. 2005) (see Sect. 4.2), and a male fertile hybrid of *S. stoloniferum* expressing ER to PVY has been characterized at the Vavilov Research Institute of Plant Industry (VIR) in St. Petersburg by S.M. Bukasov and co-workers. The male fertility-associated Ry_{sto} genes have been utilized intensively in European breeding programs (*e.g.*, in Germany, Hungary and Poland) due to easy phenotypic selection and durability of resistance against the various PVY strains.

The gene Ry_{adg} from *S. tuberosum* Group Andigenum (Muňoz et al. 1975) also confers ER to the main strains of PVY (Ross 1986; Hämäläinen et al. 1997). The gene was derived from Neo-Tuberosum germplasm adapted to long-day growth conditions by Simmonds (1969) and has been utilized in many breeding programs, *e.g.*, Cornell University, USA (Plaisted 1987), CIP, Peru (Mendoza and Jayashinghe 1993), James Hutton Institute (formerly Scottish Crop Research Institute), UK (Glendinning 1975; Bradshaw 2009) and Appacale SA (Ortega and Lopez-Vizcon 2012).

The Ry_{chc} gene from *S. chacoense* was identified in the Japanese potato cv. Konafubuki (Asama et al. 1982; Hosaka et al. 2001). The gene confers ER to PVY and has been exploited in a Japanese potato breeding program. The important advantage of this *Ry* source is its male fertility. Therefore, the genotypes containing Ry_{chc} may be used as female or male parents (Hosaka et al. 2001). ER to PVY has also been found in *S. chacoense* f. *garciae* and used in breeding at the All-Russian

Potato Institute in Korenevo (Simakov et al. 2008). The gene Ry_{hou} was found in *S. hougasii* (Cockerham 1970) and is effective against PVY^{O} and PVY^{C}; however, it has not been utilized in breeding.

Genetic dissection of resistance to PVY became possible with the advent of DNA-based markers and detailed molecular linkage maps of potato (Gebhardt 2005). Ten genes for resistance to PVY and one gene for resistance to PVA, a virus related to PVY, have been mapped to four potato genome segments on chromosomes IV, IX, XI and XII (Fig. 8.1). Genes for ER to PVY derived from *S.*

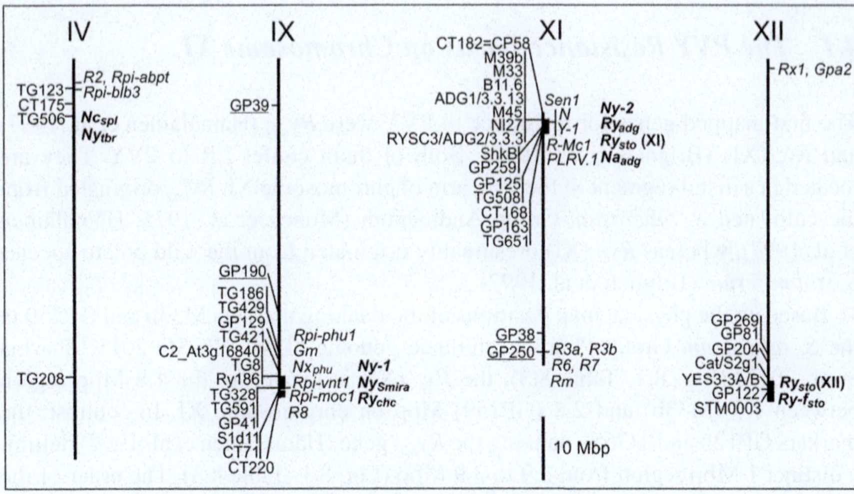

Fig. 8.1 Genomic organization of PVY resistance loci in potato. Physical maps of potato chromosomes IV, IX, XI and XII are shown as *vertical lines*. Markers genetically linked to PVY resistance genes according to the literature (Table 8.3) and cloned genes were anchored to the physical maps by BLAST searches (http://potato.plantbiology.msu.edu/integrated_searches.shtml) against the potato reference genome DM (pseudomolecules v4.03) (Sharma et al. 2013). Potato marker sequences (GP***, CP***, Nl27, Nl25, S1d11, S2 g1, 3.3.13, 3.3.3) were retrieved from the GABI Primary Database (http://www.gabipd.org/), tomato marker sequences (TG***, CT***) from the SOL Genomics Network (SGN) database (http://www.sgn.cornell.edu/), the *Arabidopsis thaliana* gene sequence At3g16840 from the TAIR database (http://www.arabidopsis.org/), and sequences for genes *R2* (FJ536332), *Rpi-abpt* (FJ536324), *Rpi-blb3* (FJ536346), *Rpi-vnt1* (FJ423044, FJ42304, FJ423046), *N* (EF091690), *Y-1* (AJ300266), *ShkB* (M95201), *R3a* (AY849382), *Rx1* (AJ011801) and *Gpa2* (AF195939) from the National Center for Biotechnology Information (NCBI) database (http://www.ncbi.nlm.nih.gov/). Markers Ry186, M39b, M33, M45, YES3-3A/B and STM0003 were anchored using published primer sequences (with expect threshold set to 1000) (references in Table 8.3). Markers linked to PVY resistance genes are shown to the *left* of the pseudomolecules. Some additional markers linked to resistance genes other than PVY resistance genes are underlined. Genes for resistance to PVY (*Ry, Ny, Na*) are shown **bold** to the *right* of the pseudomolecules. The genome segments containing the PVY resistance genes are indicated by *black bars*. Other single genes for pathogen resistance mapping to chromosomes IV, IX, XI and XII are also shown to the *right* of the pseudomolecules. Genomic positions of the genetically mapped but molecularly uncharacterized resistance genes *Rpi-phu, Gm,* Nx_{phu}*, Rpi-moc1, R8, Sen1, R-Mc1, PLRV.1, R6, R7* and *Rm* were estimated based on the physical positions of genetically linked *markers* (Brown et al. 1996; El-Kharbotly et al. 1996; Tommiska et al. 1998; Hehl et al. 1999; Marczewski et al. 2001, 2006; Smilde et al. 2005; Sliwka et al. 2006; Jo et al. 2011)

stoloniferum have been mapped to chromosomes XI and XII, as discussed below. For clarity, these genes are referred to as Ry_{sto}(XI) and Ry_{sto}(XII) respectively. The pedigrees of *Ry* genes employed in potato breeding or gene mapping are not always fully documented, and the original source of a resistance gene is sometimes disputed. Such uncertainties get, however, less important regarding the PVY resistance genes discussed here, because they are available in cultivated potato germplasm, *i.e.*, cultivars or breeding lines, and can be efficiently transferred to new cultivars using the available, tightly linked DNA markers and MAS.

4.1 The PVY Resistance Locus on Chromosome XI

The first mapped genes for resistance to PVY were Ry_{adg} (Hämäläinen et al. 1997) and Ry_{sto}(XI) (Brigneti et al. 1997). Both of them confer ER to PVY. They are located in a distal segment of the long arm of chromosome XI. Ry_{adg} originated from the cultivated *S. tuberosum* Group Andigenum (Muñoz et al. 1975; Hämäläinen et al. 1997), whereas Ry_{sto}(XI) presumably originated from the wild potato species *S. stoloniferum* (Brigneti et al. 1997).

Based on the physical map positions of the flanking markers M33b and GP259 in the *S. tuberosum* Group Phureja reference genome "DM" (PGSC 2011; Sharma et al. 2013) (Fig. 8.1, Table 8.3), the Ry_{sto}(XI) is located in the 1.8 Mbp region between 1.0 (M33b) and 2.8 (GP259) Mbp on chromosome XI. In contrast, the markers GP125 and TG651 flanking the Ry_{adg} gene (Hämäläinen et al. 1997) delimit a distinct 1 Mbp region from 2.9 to 3.9 Mbp (Fig. 8.1, Table 8.3). The order of the markers on the genetic map of Hämäläinen et al. (1997) in this region is inconsistent with the physical order of the markers in the reference DM genome, whereas the order of the markers on the genetic map of Brigneti et al. (1997) is consistent with the physical map (Fig. 8.1, Table 8.3). A rearrangement of the genome of breeding clone 2x(V-2)7, the PVY-resistant parent of the mapping population studied by Hämäläinen et al. (1997, 1998), might be one explanation for this discrepancy. Another explanation could be that the phenotype and genotype of the single recombinant plant, which was decisive for positioning marker TG651 on the genetic map of Hämäläinen et al. (1997), did not result from a true recombination event. In this case, the Ry_{adg} and Ry_{sto}(XI) genes could have quite a close chromosomal location. This presumption is supported by a study that used the sequence-characterized amplified region (SCAR) markers RYSC3 (Kasai et al. 2000) and M45 (Brigneti et al. 1997) to test 46 potato lines and cultivars for Ry_{adg} and Ry_{sto}(XI) respectively. Both markers provided identical results with only a single exception (Dalla Rizza et al. 2006). At present, it is safe to conclude that two independent studies have mapped a *Ry* gene between 1.0 and 3.9 Mbp on potato chromosome XI.

Two further genes conferring hypersensitive resistance to PVY and PVA were mapped on the same genomic region on chromosome XI in *S. tuberosum* Group Andigenum (Hämäläinen et al. 1998; Szajko et al. 2014). Na_{adg} for resistance to

Table 8.3 Markers genetically linked with PVY resistance genes and their position in the potato reference genome DM (pseudomolecules v4.03)

Gene	Marker	Locus PGSC0003DMG	Position (v4.03)	References
Ry$_{adg}$, *Ry*$_{sto}$(XI) *Na*$_{adg}$, *Ny-2*	CT182 = CP58	401013333	Chr11:00808300..814800	1, 2, 3
	M39b	404013254	Chr11:01014169..1013238	2
	M33	400013310	Chr11:01179958..1179772	3
	B11.6	400013306	Chr11:01316149..1314353	16
	ADG1/3.3.13	(a) 403016981	(a) Chr11:01425400..1430900	3
		(b) 401015682	(b) Chr11:01804770..1808180	
		(c) 400015693	(c) Chr11:01584810..1589110	
	Nl27	403016981	Chr11:01425400..1430900	16
	M45	400031476	Chr11:01494355..1494463	2, 8
	RYSC3/ ADG2/3.3.3	(a) not annotated	(a) Chr11:~01825140..1828010	3, 4, 5
		(b) 400015681	(b) Chr11:01844000..1849500	
	ShkB	402016226	Chr11:02774220..2770450	16
	GP259	400016160	Chr11:02801200..2810300	2
	GP125	400016166	Chr11:02952300..2957300	1, 3
	TG508	Intergenic	Chr11:03072200..3080400	1, 3
	CT168	400007425	Chr11:03581170..3576890	1
	GP163	400031081	Chr11:03818400..3823700	2
	TG651	Not annotated	Chr11:~03943128..3943275	3
Ry-f$_{sto}$, *Ry*$_{sto}$(XII)	GP269	Intergenic	Chr12:53279962..53280792	18
	GP81	Not annotated	Chr12:~53288206..53288754	18
	GP204	400018789	Chr12:54812070..54814050	18
	Cat/S2 g1	400029408	Chr12:57454330..57458060	11
	YES3-3AB	400016859	Chr12:59061536..59061847	12
	GP122	400004681	Chr12:59696700..59702800	9, 13, 18
	STM0003	Intergenic	Chr12:60055231..60055332	7, 13
Ry$_{chc}$, *Ny-1*, *Ny-Smira*	GP186	Not annotated	Chr09:~52791394..52791908	16
	TG429	400031817	Chr09:52885200..52899100	10
	GP129	Intergenic	Chr09:53983250..53983909	16
	TG421	400017170	Chr09:56447070..56450440	10
	C2_At3g16840	400001506	Chr09:57162900..57172500	14
	TG8	401001552	Chr09:57317400..57322500	10
	Ry186	Intergenic	Chr09:57765683..?	17
	TG328	400026431	Chr09:59361810..59365990	10
	TG591	400024365	Chr09:59513090..59514740	16
	GP41	400020592	Chr09:59604170..59607870	14
	S1d11	400021873	Chr09:59979770..59981590	16
	CT71	Not annotated	Chr09:60167714..60175726	10
	CT220	400011903	Chr09:60921980..60925590	10

(continued)

Table 8.3 (continued)

Gene	Marker	Locus PGSC0003DMG	Position (v4.03)	References
Ny_{tbr}, Nc_{spl}	TG123	400032550	Chr04:05211560..5215960	6
	CT175	400023576	Chr04:08252600..8259500	6
	TG506	400027277	Chr04:10152600..10146100	6, 15
	TG208	400002387	Chr04:57239200..57244500	6

(1) Hämäläinen et al. 1997; (2) Brigneti et al. 1997; (3) Hämäläinen et al. 1998; (4) Sorri et al. 1999; (5) Kasai et al. 2000; (6) Celebi-Toprak et al. 2002; (7) Song et al. 2005; (8) Dalla Rizza et al. 2006; (9) Witek et al. 2006; (10) Sato et al. 2006; (11) Cernak et al. 2008; (12) Song and Schwarzfischer 2008; (13) Valkonen et al. 2008; (14) Szajko et al. 2008; (15) Moury et al. 2011; (16) Szajko et al. 2014; (17) Tomczyńska et al. 2014; (18) Flis et al. 2005

PVA was closely linked (5.5 cM) to Ry_{adg} and probably lies proximal to this gene (Hämäläinen et al. 1998). Its position on the physical map is unclear due to lack of a flanking marker on the proximal site of the gene. Gene *Ny-2* for resistance to PVY co-segregated with the marker B11.6 (Szajko et al. 2014), which is at position 1.3 Mbp on the physical map and therefore in the same genome segment as Ry_{sto}(XI) (Fig. 8.1, Table 8.3).

It was recognized early on that this region on potato chromosome XI contains a number of genes with sequence similarity to the major class of dominant plant genes for pathogen resistance, namely, NB-LRR (nucleotide-binding leucine-rich repeat)-type genes, some of which share high sequence similarity with the tobacco *N* gene for resistance to tobacco mosaic virus (TMV) (Leister et al. 1996; Hehl et al. 1999). Several mapping studies also revealed that this genomic region harbors, besides the dominant genes for resistance to PVY and the nematode resistance gene *R-mc1* (Brown et al. 1996), genes conferring qualitative and quantitative resistance to the fungus *Synchytrium endobioticum* (*Sen1*) (Hehl et al. 1999; Ballvora et al. 2011; Obidiegwu et al. 2015), a major quantitative trait locus (QTL) for resistance to potato leaf roll virus (PLRV) (Marczewski et al. 2001) and QTL for resistance to the late blight (*Phytophthora infestans*) and soft rot (*Pectobacterium carotovorum*) pathogens (Gebhardt and Valkonen 2001). This region is one of the hot spots for pathogen resistance in the potato genome. Sixteen NB-LRR-type genes have been annotated in the potato reference genome between positions 1.2 and 2.9 Mbp on chromosome XI (Table 8.4), while none was found between positions 2.9 and 3.9 Mbp. Homologues of these NB-LRR-type genes are primary candidates for encoding the dominant PVY resistance genes Ry_{adg}, Ry_{sto}(XI) and *Ny-2* as well as other resistance genes. Only one of these genes has been isolated and functionally characterized in potato. The *Y-1* candidate gene for Ry_{adg} was cloned from the PVY-resistant source genotype 2x(V-2)7, sequenced and tested for complementation. However, the *Y-1* gene alone was not able to transfer ER to PVY to susceptible plants in complementation assays (Vidal et al. 2002) and probably does not encode Ry_{adg}-*1*.

Table 8.4 Candidates for PVY resistance genes based on the annotation of the *Solanum tuberosum* Group Phureja reference genome DM

Locus PGSC0003DMG	Position (v4.03)	Locus PGSC0003DMG	Position (v4.03)
400013308 (NB-LRR)	chr11:1198850..1201000	403026432 (TIR-NB-LRR)	chr09:59382000..59383800
402016979 (Bs4, N)	chr11:1333700..1339300	404026432 (TIR-NB-LRR)	chr09:59385700..59386400
403016979 (Nl27)	chr11:1340300..1347100	402026432 (NB-LRR)	chr09:59395200..59399700
401016981 (N)	chr11:1390280..1390810	400026440 (put. LB resistance)	chr09:59410040..59410990
400016982 (Bs4)	chr11:1402460..1403820	400007743 (TIR-NB-LRR)	chr09:59441500..59449200
402016981 (Bs4, N)	chr11:1407380..1411700	400007742 (NB-LRR)	chr09:59450600..59454100
400016983 (Bs4)	chr11:1412600..1417800	400024363 (Rpi-vnt1)	chr09:59505000..59508000
403016981 (N,Nl27)	chr11:1425400..1431000	400024364 (Rpi-vnt1)	chr09:59512000..59512900
400031476 (Bs4)	chr11:1492870..1496750	400024365 (Rpi-vnt1)	chr09:59513100..59514800
400031477 (N)	chr11:1506200..1508100	400024366 (Rpi-vnt1)	chr09:59518300..59519300
400015693 (Nl25)	chr11:1585000..1589000	403020585 (Rpi-vnt1)	chr09:59541600..59544600
401015682 (Nl25)	chr11:1804800..1808200	400020586 (put. LB resistance)	chr09:59547200..59548300
403015682 (Bs4)	chr11:1835200..1840600	402020585 (Rpi-vnt1)	chr09:59551700..59554400
400015681 (Y-1, ADG2/3.3.3)	chr11:1844000..1850000	400020587 (Rpi-vnt1)	chr09:59558400..59561500
400046545 (NB-LRR)	chr11:2541060..2542700	400020588 (Rpi-vnt1)	chr09:59565100..59565900
400044423 (NB-LRR)	chr11:2710610..2713370	401020585 (Rpi-vnt1)	chr09:59566000..59568000
402029345 (N isoform 2)	chr12:58535600..58538200	400020580 (TIR-NB-LRR)	chr09:59712300..59713000
401029345 (N isoform 2)	chr12:58547200..58554200	401020581 (TIR-NB-LRR)	chr09:59721300..59724300
400029344 (RPP13)	chr12:58562200..58567100	400020584 (Rpi-vnt1)	chr09:59740300..59742900
400047046 (N)	chr12:58631100..58636700	400031521 (Sw-5, B)	chr09:60087000..60091000
401004578 (put. LB resistance)	chr12:59431500..59433800	402016602 (Sw-5, C)	chr09:60608420..60612060
402004578 (put. LB resistance)	chr12:59433800..59437300	400016601 (Sw-5, A)	chr09:60622000..60632000
400004579 (NB-LRR)	chr12:59455800..59458300	400016600 (Sw-5, E)	chr09:60653000..60658000
400004722 (Cf-2.2)	chr12:60433000..60438000	400006573 (Sw-5, A)	chr09:60751000..60755000
400004725 (Hcr2-0B)	chr12:60504900..60507500	400006572 (Sw-5, B)	chr09:60780600..60783400

(continued)

Table 8.4 (continued)

Locus PGSC0003DMG	Position (v4.03)	Locus PGSC0003DMG	Position (v4.03)
402004726 (Hcr2-0A)	chr12:60508820..60509580	400006571 (Sw-5, A)	chr09:60783700..60784900
401004726 (Hcr2-0B)	chr12:60512700..60515600	400006570 (Sw-5, C)	chr09:60806200..60810800
400004727 (Hcr2-0A)	chr12:60524600..60527410	400006577 (Sw-5, C)	chr09:60841000..60843000
400005542 (Cf-2.2)	chr12:60530700..60533300	400006576 (Sw-5, A)	chr09:60863900..60866400
400005541 (Hcr2-0A)	chr12:60542200..60543900	400006581 (Sw-5, A)	chr09:60876500..60877500
400005539 (Hcr2-0A)	chr12:60559300..60561900	400006580 (Sw-5, B)	chr09:60877500..60878500
400005538 (Hcr2-0A)	chr12:60570700..60577400	400011897 (Sw-5, A)	chr09:60889900..60891100
400038990 (Hcr2-5D)	chr12:60635400..60636500	400011898 (Sw-5, B)	chr09:60891000..60896800
401023870 (Cf-2.1)	chr12:60641200..60646100	401011899 (Sw-5, B)	chr09:60897500..60898600
400023869 (Cf-2.2)	chr12:60656700..60658300	402011899 (Sw-5, B)	chr09:60899100..60900000
400023867 (Hcr2-p3)	chr12:60660400..60662500	400011900 (Sw-5, B)	chr09:60903400..60906200
401023866 (Hcr2-p4.1)	chr12:60664700..60672200	400011906 (Sw-5, D)	chr09:60965200..60967800
402023866 (Hcr2-0B)	chr12:60672200..60673300	400011907 (put. LB resistance)	chr09:60973200..60974200
400023864 (Hcr2-0B)	chr12:60694300..60697600		
400023863 (Cf-2.2)	chr12:60699700..60702000		

Sequence motifs common to the NB-LRR-type genes physically tightly linked or identical with Ry_{adg} were the basis for the first truly diagnostic DNA markers in potato (Sorri et al. 1999; Kasai et al. 2000). A diagnostic DNA marker is a DNA polymorphism (single nucleotide polymorphism [SNP] or insertion-deletion polymorphism) that is highly indicative for a particular trait, in this case ER to PVY, irrespective of the number of meiotic generations by which a given cultivar is separated from the original source of the trait, in this case a Group Andigenum accession. In other words, none or very rare recombination events have occurred between the marker and the resistance gene during introgression breeding. This was demonstrated by PCR-based screening of 50–100 PVY-resistant as well as susceptible cultivars and breeding clones for specific DNA variants in the ADG2 fragment of an NB-LRR-type gene, which distinguished the PVY-resistant genotype 2x(V-2)7 from susceptible genotypes. A 100% agreement was found between these specific DNA variants and ER to PVY originating from Group Andigenum (Sorri et al. 1999; Kasai et al. 2000). Subsequently, the diagnostic value of the SCAR marker RYSC3 developed from these specific DNA variants and its usefulness for MAS have been demonstrated in a number of breeding programs in Europe, North and South America as well as Asia (Dalla Rizza et al. 2006; Gebhardt et al. 2006; Ottoman et al. 2009; Sagredo et al. 2009; Whitworth et al. 2009; Neder et al. 2010; Ortega and Lopez-Vizcon 2012; Kaushik et al. 2013; Lopez-Pardo et al. 2013; Sharma et al. 2014; Bhardwaj et al. 2015; Fulladolsa et al. 2015; Chien et al. 2016). Additional diagnostic markers for Ry_{adg} are M45 (Dalla Rizza et al. 2006) and, to a lesser extent, B11.6 (Szajko et al. 2014).

4.2 The PVY Resistance Locus on Chromosome XII

Four independent molecular mapping experiments using different sources of ER to PVY from the wild potato species *S. stoloniferum* have placed Ry_{sto} genes on the short arm of chromosome XII (Flis et al. 2005; Song et al. 2005; Cernak et al. 2008; Valkonen et al. 2008). The source genotype PW-363 of $Ry\text{-}f_{sto}$ was male fertile (Flis et al. 2005), whereas cv. Assia, the source of Ry_{sto}(XII), was associated with cytoplasmic male sterility (Song et al. 2005; Song and Schwarzfischer 2008). The male fertility status of the third and fourth source of Ry_{sto}(XII) (breeding clone 86.61.26 and cv. White Lady) has not been reported (Cernak et al. 2008; Valkonen et al. 2008). Physical mapping of the markers linked to $Ry\text{-}f_{sto}$ and Ry_{sto}(XII) restricts the genomic region harboring this PVY resistance locus to the most distal 3 Mbp of chromosome XII (Fig. 8.1). Twenty-six candidate genes with sequence similarity to known plant resistance genes are annotated in this region of the DM reference genome between 58 Mbp and 61 Mbp (Table 8.4), none of which has been functionally characterized. A cluster of 17 genes with sequence homology to the tomato Cf genes for resistance to *Cladosporium fulvum* (Dixon et al. 1998) is located between 60.4 Mbp and 60.7 Mbp. The markers GP122, STM0003 and YES-3A/B, which are genetically most tightly linked with the Ry_{sto}(XII) genes, are all located in the 1 Mbp segment between 59 Mbp and 60 Mbp, physically very close to the candidate genes (Fig. 8.1, Table 8.3). The same markers are tightly associated with ER to PVY originating in *S. stoloniferum*, irrespective of the genetic background into which they are introgressed, and tag the same resistance locus (Flis et al. 2005; Song et al. 2005; Witek et al. 2006; Song and Schwarzfischer 2008; Valkonen et al. 2008). Whether this locus consists of a single gene or several genes organized in a tight physical cluster, as exemplified by the Cf-like gene cluster mentioned above, is unclear at present. PVY resistance from *S. stoloniferum*, together with cytoplasmic male sterility, has been incorporated mainly into cultivars bred in Europe (Ross 1986; Song and Schwarzfischer 2008; Sharma et al. 2014). Markers diagnostic for Ry_{sto}(XII) have been validated and are useful for MAS (Heldak et al. 2007; Lopez-Pardo et al. 2013; Bhardwaj et al. 2015).

4.3 The PVY Resistance Loci on Chromosome IX

A third PVY resistance locus was identified in a hot spot for pathogen resistance on the long arm of chromosome IX by mapping gene Ry_{chc} for ER to PVY that has been introgressed into the Japanese cv. Konafubuki from *S. chacoense* (Hosaka et al. 2001; Sato et al. 2006). Ry_{chc} mapped 0.9 cM distal to marker CT220, which places the gene on the physical map in an approximately 600-kbp-long genome segment distal to position 60.9 Mbp (Fig. 8.1, Table 8.3). A large cluster of 19 candidate genes homologous to the tomato tospovirus resistance gene *Sw-5* (Brommonschenkel et al. 2000) is located between 60 Mbp and 61 Mbp, only two of which (PGSC0003DMG400011906 and PGSC0003DMG400011907) are distal to position 60.9 Mbp of marker CT220 (Table 8.4). In this case, in silico mapping reduces dramatically the number of primary candidate genes for Ry_{chc}. However, resolution of the genetic map around Ry_{chc} needs to be improved before firm conclusions on the best candidate genes can be drawn. Moreover, the structure of the corresponding genome segment in *S. chacoense* accessions resistant to PVY might differ considerably from the *S. tuberosum* Group Phureja reference genome, with respect to both the physical size of the cluster and the copy number of *Sw-5*-like genes in that cluster.

Genes for HR to PVY present in simplex dosage in the cvs Rywal, Albatros, Sekwana and Sárpo Mira were also mapped to the distal region of the long arm of Chromosome IX (Szajko et al. 2008, 2014; Tomczyńska et al. 2014). Gene *Ny-1* conferring a temperature-sensitive resistance to PVY was flanked by markers GP41 and C2_At3g16840 in segregating progeny from cv. Rywal (Szajko et al. 2008) and must, therefore, be located proximal to Ry_{chc} between 57.1 Mbp and 59.6 Mbp on the physical map (Fig. 8.1, Table 8.3). This position is confirmed by mapping the same gene in progeny of cvs Albatros and Sekwana (Szajko et al. 2014). Gene *Ny-Smira* mapped distal to the marker Ry186 (Mori et al. 2011) at 57.7 Mbp and no flanking marker on the distal site was reported (Tomczyńska et al. 2014). This gene might be physically tightly linked or allelic either to *Ny-1* or to Ry_{chc}. The genomic segment between 59 and 60 Mbp harbors two further clusters of NB-LRR-type genes, which are candidates for the *Ny* resistance genes (Tomczyńska et al. 2014) (Table 8.4). One homologue of those present in *S. venturii*, *Rpi-vnt1*, encodes a functional gene for resistance to late blight (Foster et al. 2009; Pel et al. 2009). Other single genes for pathogen resistance mapping in this resistance hot spot are Nx_{phu} and *Gm* conferring resistance to potato virus X (PVX) and potato virus M (PVM) respectively (Tommiska et al. 1998; Marczewski et al. 2006), and late blight resistance genes *R8*, *Rpi-moc1* and *Rpi-phu1* = *Rpi-vnt1* (Smilde et al. 2005; Sliwka et al. 2006; Jo et al. 2011).

The markers linked to *Ny-1* seem to have diagnostic potential as they were only present in some cultivars expressing hypersensitive resistance to PVY and absent in susceptible cultivars and cultivars expressing ER to PVY (Szajko et al. 2014). The distribution of *Ny-Smira* in European cultivars was not analyzed but might be similar or identical to *Ny-1*. The markers linked to the *Ny* genes are most useful for MAS among direct descendants of the cultivars used for mapping. The same applies to Ry_{chc}, which has only been traced in some Japanese cultivars (Hosaka et al. 2001). Its distribution in other potato germplasm is currently unknown. The PCR marker Ry186 tightly linked to Ry_{chc} can be multiplexed with diagnostic markers for other dominant resistance genes and used for MAS of cultivars with multiple resistance genes (Mori et al. 2011).

4.4 The PVY Resistance Locus on Chromosome IV

Two genes for HR to PVY were located in a proximal position on the short arm of chromosome IV (Celebi-Toprak et al. 2002; Moury et al. 2011). Ny_{tbr} was found in *S. tuberosum* Group Tuberosum, whereas Ny_{spl} originated from the wild species *S. sparsipilum*. Both genes were most closely linked to the same marker TG506, which is at 10.1 Mbp on the physical map, suggesting that the genes are physically closely linked or even allelic and well separated from the late blight resistance genes more distally located on the same chromosome arm. The positions of Ny_{tbr} and Ny_{spl} on the physical map cannot be determined precisely because the flanking proximal marker TG208 is at 75.2 Mbp on the long arm, 65 Mbp away from TG506 (Fig. 8.1, Table 8.3). No reports are available on the distribution of Ny_{tbr} and Ny_{spl} in cultivars and their use in MAS.

5 Design of Potato Breeding Programs for Resistance to PVY

Potato breeding is a challenging task because of the highly heterozygous nature of *S. tuberosum* and its tetrasomic inheritance (Slater et al. 2014). Breeding programs are based mainly on a large number of pairwise crosses among cultivars and advanced breeding lines. The parents to be crossed are usually well recognized for their combining abilities and selected based on desirable, expressed complementary characteristics (Bradshaw 2007). Plants grown from true seeds obtained from crosses are subjected to preliminary selection and identification of superior individuals combining desirable features of both parents and used for further crosses. During phenotypic recurrent selection that often lasts for 10–12 years, progenies are assessed for more than 50 traits, including agronomic and quality traits as well as resistance to biotic and abiotic stresses (Milbourne et al. 2007). In the last 2 years of the process, candidate cultivars are submitted for official testing and, if successful, finally registered as cultivars (Table 8.5).

Selection efficiency can be increased by applying pre-breeding programs focused on the introgression of new sources with desired traits derived from wild species into a gene pool of breeding lines with improved agronomic characteristics (Zimnoch-Guzowska et al. 2013). Furthermore, selection efficiency is enhanced by early generation selection (Bradshaw 2007), use of parents multiplexed for desired alleles (Mackay 1987; Mendoza et al. 1996), more efficient testing methods (Bradshaw and Mackay 1994; Bradshaw et al. 2003) and MAS focused on genotypic selection (Kasai et al. 2000; Valkonen et al. 2008; Ortega and Lopez-Vizcon 2012; Zimnoch-Guzowska et al. 2013; Slater et al. 2014). For example, three to seven backcrosses are needed to transfer Ry_{sto} from an existing cultivar to a new cultivar (Ross 1986). However, a greater number of backcrosses are needed when a gene is transferred from a wild species to a potato cultivar, because the degree of domestication of the donor of the trait is low (Bradshaw et al. 2006). Neo-Tuberosum populations are derived from primitive *S. tuberosum* Group Andigenum and *S. tuberosum* Group Phureja cultivars adapted to long-day conditions (Glendinning 1979; Plaisted 1987). Cyclic recurrent mass selection and introgression of novel germplasm by hybridization and backcrossing have both been implemented in further breeding of the Neo-Tuberosum material (Ghislain et al. 2009). Yield heterosis has been observed in crosses with the long-day-adapted *S. tuberosum*; however, few new cultivars have been released from the program (Bradshaw 2007).

The portion of PVY-susceptible progenies in crosses can be reduced by producing parental lines that are multiplex heterozygote or homozygous quadruplex for the resistance locus of Ry_{adg}, as in the breeding programs in CIP, Peru (Mendoza et al. 1996), Brazil (Ribeiro et al. 2006; Andrade et al. 2009) and India (Kaushik et al. 2013). The same approach has been used to enhance breeding for resistance to potato cyst nematodes and PVY at the James Hutton Institute (formerly SCRI) in Scotland (Mackay 1987; Solomon-Blackburn and Mackay 1993). According to

Table 8.5 Outlines of conventional breeding programs at the James Hutton Institute (JHI, formerly Scottish Crop Research Institute, UK) prior to 1982 (Mackay 2005) and in Potato Breeding Zamarte – IHAR (Poland) (B. Kamiński, personal communicaton)

		JHI (UK)		HZ Zamarte (Poland)		
Year	Name	# genotypes	selection	# genotypes	selection	Phases
0	crossing program	200–300		200		*1st phase*
1	1st year seedlings in pots	100,000		80,000	Tuber size,	
2	Single hill ramsh	50,000	Visual selection	60,000–70,000	Visual selection, tuber morphology	Visual selection for highly repeatable traits, tuber morphology, short storage, resistances determined by major genes: PVY, PCN
3	Ramsh lines	4000 3–4 pl/plot	Visual selection, 1st testing for PCN	5000 7 pl/plot	Visual selection, preliminary quality tests, *in part PVY res. (mech. Inoc.)*	
4	New lines	1000	2 sites: seed and ware, yield trials, quality and diseases resistance testing, maturity	1000–1200 30 pl/plot	Visual selection, quality tests	
5	Young lines 1	500		400–500 3 × 20 pl/plot	1st tests for PCN, G. allida and wart resistances. Quality test	*2nd phase*
6	Young lines 2	200		150–200 3 × 20 pl/plot	Yield trials, quality tests	Selection for traits with lower repeatability modified by environment: yield and its structure, culinary value, chipping, French fries, storability
7	Young lines 3	60	Regional trials, 6–7 sites in UK, 2 overseas, produce approved seed stocks	30–40 3 × 30–45 pl/plot	Preliminary trials, 2 sites, 2nd testing for PCN & wart res., *meristem cultures for seed multiplication*	
8	Old lines	10		20 3 × 30–45 pl/plot	Preliminary trials, 2 sites, 3rd test for wart res.	
9	Advanced lines	5		6–7 3 × 60 pl/plot	Statutory trials, 15 sites, filed tests for resistances (*PVY, PLRV,* late blight)	*3rd phase*
10	Candidate cultivars	1–5	2 years National List Trials (UK) *Maintaining of Approved Stocks, prebasic seed testing*	4–5 3 × 60 pl /plot	Statutory trials, 15 sites, 2nd year of field tests for resistances (*PVY, PLRV,* late blight) 3rd test for PCN res.	Evaluation for specific features: suitability to mechanical harvest, tubers washing & packing, tuber greening, TGA content, semi-technological processing, field tests for resistance to PVY, PLRV, late blight
11	Candidate cultivars	1–5		1–2	*Named cultivars on National List* *Commercial seed production 3–4 ha*	
12	Named cultivar(s)	1–2	*Cultivar on National List* *Commercial seed production*	1–2	Post registration trials in 40 sites *Testing for ER res. to PVY (grafting)*	

Brown and Corsini (2001), the challenge of this approach is that the resistant parent must also have good combining ability for other agronomic traits.

In the European Union (EU), potential new cultivars must undergo 2 years of official trials before they can be registered in the national list of varieties in an EU member state and included in the Common Catalogue, which is a compilation of EU national lists. In some cases, evaluation of resistance to diseases in candidate cultivars is conducted at the end of the breeding process as part of the official trials.

5.1 Examples of Pre-breeding Programs Aiming to Enhance PVY Resistance

When the main breeding goal is resistance to PVY, selection for resistance can be done in the first year of the breeding process. Early screening of seedlings for PVY resistance can be conducted by inoculating young seedlings with leaf sap from PVY-infected tobacco plants using a spray gun, *e.g.*, as done in the Foundation for Agricultural Plant Breeding (SVP) in the Netherlands (Wiersema 1966), CIP, Peru (Mendoza and Jayashinghe 1993), James Hutton Institute, UK (Solomon 1978), Cygnet Potato Breeders Ltd., UK (formerly Plant Breeding Institute, Cambridge) (Jellis et al. 1987), and the Potato Center of Pannonia University at Keszthely, Hungary (Sarvari 1978). Seedlings with virus symptoms can be discarded 3 weeks after inoculation (Sieczka 2001), thus enabling 90% of the PVY-susceptible seedlings to be eliminated (Beekman 1987). Selected, putatively PVY-resistant progeny plants are checked for secondary infection by planting tubers, inspecting plants for symptoms and testing leaves for PVY using standard serological or molecular methods. It is possible to inoculate progeny plants with PVY and other viruses simultaneously, if the goal is to obtain advanced breeding lines or cultivars with resistance to more than one virus (Mendoza and Jayashinghe 1993; Zimnoch-Guzowska et al. 2013).

Development of multi-resistant parental lines began at Młochów Research Center of the Plant Breeding and Acclimatization Institute (IHAR), Poland, in the early 1960s (Sieczka 2001). Besides viruses, the program was aimed at pre-breeding of parental lines with resistance to late blight and bacterial soft rot, and for other agronomic traits, which would enable new resistant cultivars to be selected from the progeny of the first cross (Świeżyński 1987). The pre-breeding phase took 5 years (Table 8.6). In IHAR, putative new cultivars are subsequently tested for field resistance to PVY over two growing seasons and three locations which differ in PVY infection pressure. The field plots of putative new cultivars are surrounded by a set of infector plants and PVY-resistant standard cultivars. Those new cultivars that exhibit field resistance to PVY (resistance score 8) are graft-inoculated with tobacco scions infected with a $PVY^{N\text{-}Wi}$ isolate and tubers are tested for PVY to ensure that resistance to this strain is high (Pietrak 2001). A similar procedure is followed when selecting breeding lines for resistance to other viruses (Table 8.7).

Table 8.6 Complexity of virus resistance in parental lines offered to breeders by the Plant Breeding and Acclimatization Institute, Młochów, Poland

	Resistance to viruses						
Year of offer	PVY	PVA	PVX	PVS	PLRV	PVM	Gene combination[a]
1968	X	X	X				Ry_{sto}, Rx_{acl},
1977	X	X	X	X			Ry_{sto}, Rx_{acl}, *Ns*
1980	X	X	X	X	X		Ry_{sto}, Rx_{acl}, *Ns*, (L)
1985	X	X	X		X	X	Ry_{sto}, Rx_{acl}, *Rm*, (L)
1989	X	X	X	X	X	X	$Ry\text{-}f_{sto}$, Rx_{acl}, *Rm*, *Ns*, (L)
1993	X	X	X	X	X	X	$Ry\text{-}f_{sto}$, Rx_{acl}, *Gm*, *Ns*, *PLRV4*

[a]Genes for virus resistance:
PVY + PVA – Ry_{sto}, $Ry - f_{sto}$ (Flis et al. 2005)
PVX – Rx_{acl} (Dziewońska 1986)
PVS – *Ns* (Marczewski et al. 2002)
PVM – *Rm*, *Gm* (Marczewski et al. 2006)
PLRV – (L) field resistance from *S. tuberosum*; *PLRV4* – from the clone DW 84-1457 (Marczewski et al. 2001)

Table 8.7 The scheme of the phenotypic selection for resistance to five viruses in parental line breeding program in 5-year cycle at IHAR, Młochów, Poland

		Resistance to				
Year	Clones	PVX	PVY	PVM	PLRV	PVS
I	5000 of field grown seedlings	Spray-gun inoculation[a]				
		Visual selection (symptoms)				
II	700	Tuber indexing, ELISA				
III	100 MAS[b]		Graft inoculation	Sap inoculation	Graft inoculation	
IV	30		Tuber indexing, ELISA			
				Graft inoculation	Aphid inoculation	Graft inoculation
V	10		Reaction to PVY^{NTN}		Tuber indexing, ELISA	

[a]Spray-gun inoculation of young seedlings with $PVY^{N\text{-}Wi}$, PVX, PVM in combinations related to resistance of crossed parents
[b]MAS – marker-assisted selection applied for selection of resistance to PVY, PVS and PVM in parental lines in the 3rd year; used as an alternative to the phenotypic selection

In the last 10 years, phenotypic screening of resistant parents has been complemented with screening with molecular markers for virus resistance genes, such as Ry_{sto}, which has simplified the selection process and reduced costs. A significant output of the program has been over 60 registered cultivars, of which 27 express ER to PVY (Zimnoch-Guzowska et al. 2013).

The goal of CIP since its establishment in 1971 has been to offer new potato breeding material for the selection of new cultivars adapted to developing countries where access to healthy seed potatoes is limited or absent. High breeding efficiency is possible owing to the engagement of farmers and breeding companies in the large-scale process of selecting advanced clones in different environments. Exploitation of virus resistance is particularly important under such conditions (Thiele et al. 2010). Therefore, breeding is focused on improving potato populations and selecting cultivars that are adapted to a range of environments, produce a stable tuber quality and yield, and are resistant to the main biotic stresses caused by viruses, late blight, bacterial wilt and other pests. To achieve this goal, the potato germplasm used by CIP in the breeding programs is genetically diverse (wild species, landraces and commercial cultivars of various origins), and possesses a high frequency of desirable genes for adaptation, yield and resistance to biotic stresses. It also exhibits enhanced recombination of desirable traits (Huamán and Schmiedeche 1999). Recurrent selection and progeny testing is used for population improvement (Mendoza 1989).

CIP has developed two advanced potato breeding populations: population B, which is adapted to highland tropics and is outstanding in resistance to late blight (LB), and population LTRV, which is adapted to lowland subtropics with enhanced heat and/or drought tolerance. Furthermore, LTRV has been selected for earliness under short-day conditions. It is relatively early maturing under long-day conditions and highly resistant to PVY, PVX and PLRV (Gastelo et al. 2014). A few clones have been selected as progenitors for cultivar development and production of true potato seed (TPS). The LTRV breeding population was aimed at combining durable resistance to PLRV and ER to PVX and PVY. ER to PVY was derived from *S. tuberosum* Andigena group, Neo-Tuberosum and *S. stoloniferum* (Brown 1980). The targeted increase in the frequency of progeny expressing ER to PVY and PVX was achieved by recurrent selection of progenitors multiplex for Ry_{adg}. The selected PVY-resistant potato lines contain Ry_{adg} in a duplex condition and have been used extensively to introduce PVY resistance into the most advanced breeding populations at CIP. They have been distributed to the national breeding programs of developing countries. Crossing a parental line carrying Ry_{adg} in duplex condition with a PVY-susceptible line is expected to produce over 80% PVY-resistant progeny. Crossing genotypes in triplex condition, selected from crosses of genotypes in duplex condition, with PVY-susceptible lines increased the frequency of PVY-resistant progenies to 96% (Mendoza et al. 1996). The usefulness of the LTVR program can be assessed from the CIP catalogs of varieties and advanced clones; 14 of the 71 cultivars released express ER to PVY (Table 8.1), and 97 of the over 300 advanced clones offered by CIP for worldwide distribution are resistant to PVY and most of them originated from the LTRV population (www.cipotato.org).

References

Acosta-Leal R, Xiong Z (2008) Complementary functions of two recessive R-genes determine resistance durability of tobacco 'Virgin A Mutant' (VAM) to *Potato virus Y*. Virology 379:275–283

Adams SE, Jones RAC, Coutts RHA (1986) Effect of temperature on potato virus X infection in potato cultivars carrying different combinations of hypersensitivity genes. Plant Pathol 35:517–526

Allard RW (1960) Inheritance in autotetraploids. In: Principles of plant breeding. Wiley, New York, pp 385–399

Andrade CM, Pinto CABP, Riberio SRRP, Peixouto LS, Vilela XMS (2009) Potato clones with multiple copies of the Ry_{adg} allele conferring resistance to PVY. Crop Breed Appl Biotech 9:286–292

Asama K, Ito H, Murakami N, Itoh T (1982) New potato variety Konafubuki. Bull Hokkaido Pre Agr Exp Stn 48:75–84

Bagnall RH, Tai CCC (1986) Field resistance to potato virus Y in potato assessed by cluster analysis. Plant Dis 70:301–304

Ballvora A, Flath K, Lübeck J, Strahwald J, Tacke E et al (2011) Multiple alleles for resistance and susceptibility modulate the defense response in the interaction of tetraploid potato (*Solanum tuberosum*) with *Synchytrium endobioticum* pathotypes 1, 2, 6 and 18. Theor Appl Genet 123:1281–1292

Bamberg JBM, Martin W, Schartner JJ (1994) Elite selection of tuber- bearing Solanum species germplasm. Inter-regional Potato Introduction Station, NRSP-6, Sturgeon Bay

Barker H, Harrison BD (1984) Expression of genes for resistance to potato virus Y in potato plants and protoplasts. Ann Appl Biol 105:539–545

Baulcombe DC (2007) Amplified silencing. Science 315:199–200

Beekman AGB (1987) Breeding for resistance. In: de Box JA, van der Want JPH (eds) Viruses of potato and seed potato production. Wageningen, PUDOC, pp 162–170

Bendahmane A, Kanyuka K, Baulcombe DC (1999) The *Rx* gene from potato controls separate virus resistance and cell death responses. Plant Cell 11:781–792

Bhardwaj V, Sharma R, Dalamu SAK, Baswaraj R, Singh R, Singh BP (2015) Molecular characterization of potato virus Y resistance in potato (*Solanum tuberosum* L.) Indian J Genet Plant Breed 75:389–392

Bradshaw JE (1994) Quantitative genetics theory for tetrasomic inheritance. In: Bradshaw JE, MacKay GR (eds) Potato genetics. CAB International, Wallingford, pp 71–99

Bradshaw JE (2007) Potato-breeding strategy. Chapter 8. In: Vreugdenhil D (ed) Potato biology and biotechnology. Elsvier B.V, Amsterdam, pp 157–177

Bradshaw JE (2009) Potato breeding at the Scottish Plant Breeding Station and the Scottish Corp Research Institute: 1920–2008. Potato Res 52:141–172

Bradshaw JE, Mackay GR (1994) Breeding strategies for clonally propagated potatoes. In: Bradshaw JE, Mackay GR (eds) Potato genetics. CAB International, Cambridge, pp 467–497

Bradshaw JE, Dale MFB, Mackay GR (2003) Use of mid-parent values and progeny tests to increase the efficiency of potato breeding for combined processing quality and disease and pest resistance. Theor Appl Genet 107:36–42

Bradshaw JE, Bryan GJ, Ramsay G (2006) Genetic resources (including wild and cultivated *Solanum* species) and progress in their utilization in potato breeding. Potato Res 49:49–65

Brigneti G, Garcia-Mas J, Baulcombe DC (1997) Molecular mapping of the potato virus Y resistance gene Ry_{sto} in potato. Theor Appl Genet 94:198–203

Brigneti G, Voinnet O, Li WX, Ji LH, Ding SW, Baulcombe DC (1998) Viral pathogenicity determinants are suppressors of transgene silencing in *Nicotiana benthamiana*. EMBO J 17:6739–6746

Brommonschenkel SH, Frary A, Frary A, Tanksley SD (2000) The broad-spectrum tospovirus resistance gene *Sw-5* of tomato is a homolog of the root-knot nematode resistance gene *Mi*. Mol Plant-Microbe Interact 13:1130–1138

Brown CR (1980) Breeding for virus resistance and its integration with other traits. In: Utilization of the genetic resources of the potato III. Report of the Planning Conference 1980, International Potato Center (CIP) Lima, Peru, pp 89–109

Brown CR, Corsini D (2001) Resistance. In: Loebenstein G, Berger PH, Brunt AA, Lawson RH (eds) Virus and virus-like diseases of potatoes and production of seed potatoes. Kluwier Academic Publisher, Drodrecht/Boston/London, pp 323–340

Brown CR, Yang CP, Mojtahedi H, Santo GS, Masuelli R (1996) RFLP analysis of resistance to Columbia root-knot nematode derived from *Solanum bulbocastanum* in a BC2 population. Theor Appl Genet 92:572–576

BSL (2002) Beschreibende Sortenliste Kartoffeln 2002. Bundessortenamt, Hannover

Celebi-Toprak F, Slack SA, Jahn MM (2002) A new gene, *Ny-tbr*, for hypersensitivity to potato virus Y from *Solanum tuberosum* maps to chromosome IV. Theor Appl Genet 104:669–674

Cernak I, Decsi K, Nagy S, Wolf I, Polgar Z et al (2008) Development of a locus-specific marker and localization of the *Ry(sto)* gene based on linkage to a catalase gene on chromosome XII in the tetraploid potato genome. Breed Sci 58:309–314

Chien DH, Ganga ZN, Simongo DK, Kiswa CG, Gonzales IC, Perez J, Tung PX, Hau NT, Artola K, Valkonen JPT, Vander Zaag P (2016) The adoption of cv. Igorota in the Philippines and Vietnam. Am J Potato Res 93:296–301

Chrzanowska M (1991) New isolates of the necrotic strain of potato virus Y (PVY^N) found recently in Poland. Potato Res 34:179–182

Chrzanowska M (2000) Krańcowa odporność na wirusy Y i X ziemniaka oraz polowa odporność na wirus S ziemniaka w polskich odmianach ziemniaka. Biul Instytutu Hodowli i Aklimatyzacji Roślin 214:231–238

Cockerham G (1943) Potato breeding for virus resistance. Ann Appl Biol 30:105–108

Cockerham G (1970) Genetical studies on resistance to potato viruses X and Y. Heredity 25:309–348

Cooper JI, Jones TA (1983) Responses of plants to viruses: proposals for the use of terms. Phytopathology 73:127–128

Correll DS (1962) The potato and its wild relatives. Texas Research Foundation

Dalla Rizza M, Vilaró F, Torres D, Maeso D (2006) Detection of PVY extreme resistance genes in potato germplasm from the uruguayan breeding program. Am J Potato Res 83:297–304

De Maine MJ, Carroll CP, Stewart HE, Solomon RM, Wastie RL (1993) Disease resistance in *Solanum phureja* and diploid and tetraploid *S. tuberosum* x *S. phureja* hybrids. Potato Res 36:21–28

Dixon MS, Hatzixanthis K, Jones DA, Harrison K, Jones JDG (1998) The tomato *Cf-5* disease resistance gene and six homologs show pronounced allelic variation in leucine-rich repeat copy number. Plant Cell 10:1915–1925

Dong F, Song J, Naess SK, Helgeson JP, Gebhardt C, Jiang J (2000) Development and applications of a set of chromosome-specific cytogenetic DNA markers in potato. Theor Appl Genet 101:1001–1007

Dong FG, McGrath JM, Helgeson JP, Jiang JM (2001) The genetic identity of alien chromosomes in potato breeding lines revealed by sequential GISH and FISH analyses using chromosome-specific cytogenetic DNA markers. Genome 44:729–734

Dziewońska M (1986) Development of parental lines for breeding of potatoes resistant to viruses and associated research in Poland. In: Potato research of tomorrow. PUDOC, Wageningen, pp 96–100

El-Kharbotly A, Palomino-Sánchez C, Salamini F, Jacobsen E, Gebhardt C (1996) *R6* and *R7* alleles of potato conferring race-specific resistance to *Phytophthora infestans* (Mont.) de Bary identified genetic loci clustering with the R3 locus on chromosome XI. Theor Appl Genet 92:880–884

Fernandez FT, Gonzalez I, Doblas P, Rodriguez C, Sahana N, Kaur H, Tenllado F, Praveen S, Canto T (2013) The influence of cis-acting P1 protein and translational elements on the expression of *Potato virus Y* helper-component proteinase (HCPro) in heterologous systems and its suppression of silencing activity. Mol Plant Pathol 14:530–541

Flis B, Hennig J, Strzelczyk-Żyta D, Gebhardt C, Marczewski W (2005) The *Ry-f_{sto}* gene from *Solanum stoloniferum* for extreme resistant to *Potato virus Y* maps to potato chromosome XII and is diagnosed by PCR marker $GP122_{718}$ in PVY resistant cultivars. Mol Breed 15:95–101
Flor HH (1946) Genetics of pathogenicity in *Melampsora lini*. J Agric Res 73:335–357
Foster SJ, Park T-H, Pel M, Brigneti G, Sliwka J et al (2009) *Rpi-vnt1.1*, a *Tm-2*^{2} homolog from *Solanum venturii*, confers resistance to potato late blight. Mol Plant-Microbe Interact 22:589–600
Fulladolsa AC, Navarro FM, Kota R, Severson K, Palta JP, Charkowski AO (2015) Application of marker assisted selection for *Potato virus Y* resistance in the University of Wisconsin potato breeding program. Am J Potato Res 92:444–450
Gabriel W (1995) Une method d'utilisation possible de l'échelle de 1 à 9 pour l'évaluation de la résistance aux virus chez les cultivars de pomme de terre. Potato Res 38:297–305
Galvino-Costa SBF, Figueira AD, Camargos VV, Geraldino PS, Hu XJ, Nikolaeva OV, Kerlan C, Karasev AV (2012) A novel type of *Potato virus Y* recombinant genome, determined for the genetic strain PVY^E. Plant Pathol 61:388–398
Gammelgård E, Mohan M, Valkonen JPT (2007) Potyvirus-induced gene silencing: the dynamic process of systemic silencing and silencing suppression. J Gen Virol 88:2337–2346
Gastelo M, Kleinwechter U, Bonierbale M (2014) Global potato research for a changing world, Working paper 2014-1. International Potato Center (CIP), Lima, 43 p
Gebhardt C (2005) Potato genetics: molecular maps and more. In: Lörz H, Wenzel G (eds) Molecular marker systems in plant breeding and crop improvement. Springer, Berlin/Heidelberg, pp 215–227
Gebhardt C, Valkonen JP (2001) Organization of genes controlling disease resistance in the potato genome. Annu Rev Phytopathol 39:79–102
Gebhardt C, Bellin D, Henselewski H, Lehmann W, Schwarzfischer J et al (2006) Marker-assisted combination of major genes for pathogen resistance in potato. Theor Appl Genet 112:1458–1464
Ghislain M, Núñez J, Rosario Herrera M, Spooner DM (2009) The single Andigenum origin of neo-Tuberosum potato materials is not supported by microsatellite and plastid marker analyses. Theor Appl Genet 118:963–969
Gibson RW, Pehu E, Woods RD, Jones MGK (1990) Resistance to potato virus Y and potato virus X in *Solanum brevidens*. Ann Appl Biol 116:151–156
Glais L, Tribodet M, Kerlan C (2002) Genomic variability in potato potyvirus Y (PVY): evidence that PVY^NW and PVY^{NTN} variants are single to multiple recombinants between PVY^O and PVY^N isolates. Arch Virol 147:363–378
Glendinning DR (1975) Neo-Tuberosum: new potato breeding material. 3. Characteristics and variability of neo-tuberosum and its potential value in breeding. Potato Res 18:343–362
Glendinning DR (1979) Enriching the potato gene-pool using primitive cultivars. Proc Conf Broadening Gent Base Crops. Pudoc, Wageningen, pp 39–45
Hämäläinen JH, Watanabe KN, Valkonen JPT, Arihara A, Plaisted RL et al (1997) Mapping and marker-assisted selection for a gene for extreme resistance to potato virus Y. Theor Appl Genet 94:192–197
Hämäläinen JH, Sorri VA, Watanabe KN, Gebhardt C, Valkonen JPT (1998) Molecular examination of a chromosome region that controls resistance to potato Y and A potyviruses in potato. Theor Appl Genet 96:1036–1043
Hamilton A, Voinnet O, Chappell L, Baulcombe DC (2002) Two classes of short interfering RNA in RNA silencing. EMBO J 21:4671–4679
Hawkes JG (1990) The potato: evolution, biodiversity and genetic resources. The Smithsonian Institute Press, Washington, DC
Hehl R, Faurie E, Hesselbach J, Salamini F, Whitham S et al (1999) TMV resistance gene *N* homologues are linked to *Synchytrium endobioticum* resistance in potato. Theor Appl Genet 98:379–386

Heldak J, Bezo M, Stefunova V, Gallikova A (2007) Selection of DNA markers for detection of extreme resistance to potato virus *Y* in tetraploid potato (*Solanum tuberosum* L.) F_1 progenies. Czech J Genet Plant Breed 43:125–134

Hirsch CN, Hirsch CD, Felcher K, Coombs J, Zarka D et al (2013) Retrospective view of north American potato (*Solanum tuberosum* L) breeding in the 20th and 21st centuries. G3 Genes, Genomes, Genetics 3:1003–1013

Hosaka K, Hosaka Y, Mori M, Maida T, Matsunaga H (2001) Detection of a simplex RAPD marker linked to resistance to potato virus Y in a tetraploid potato. Am J Potato Res 78:191–196

Huamán Z, Schmiedeche P (1999) The potato genetic resources held in trust by the international potato center (CIP) in Peru. Potato Res 42:413–426

Huamán Z, Hoekstra R, Bamberg JB (2000) The inter-genebank potato database and the dimensions of available wild potato germplasm. Am J Potato Res 77:353–362

Jansky SH, Dempewolf H, Camadro EL, Simon R, Zimnoch-Guzowska E et al (2013) A case study for crop wild relative preservation and use in potato. 2013. Crop Sci 53:746–754

Jellis GJ, Boulton RE, Starling NC, Squire AM (1987) Screening for resistance to diseases in a potato breeding programme. In: Jellis DE, Richardson DE (eds) The production of new potato varieties: technological advances. Cambridge University Press, Cambridge, pp 84–85

Jo KR, Arens M, Kim T-Y, Jongsma M, Visser R et al (2011) Mapping of the *S. demissum* late blight resistance gene *R8* to a new locus on chromosome IX. Theor Appl Genet 123:1331–1340

Johnston SA, Hanneman R (1982) Manipulations of endosperm balance number overcome crossing barriers between diploid *solanum* species. Science 217:446–448

Jones RAC (1990) Strain group specific and virus specific hypersensitive reactions to infection with potyviruses in potato cultivars. Ann Appl Biol 117:93–105

Jones JDG, Dangl JL (2006) Plant immune system. Nature 444:323–329

Karasev AV, Gray SM (2013) Continuous and emerging challenges of *Potato virus Y* in potato. Annu Rev Phytopathol 51:571–586

Kasai K, Morikawa Y, Sorri VA, Valkonen JPT, Gebhardt C et al (2000) Development of SCAR markers to the PVY resistance gene Ry_{adg} based on a common feature of plant disease resistance genes. Genome 43:1–8

Kaushik SK, Sharma R, Garg ID, Singh BP, Chakrabarti SK et al (2013) Development of a triplex (YYYy) parental potato line with extreme resistance to potato virus Y using marker assisted selection. J Hortic Sci Biotechnol 88:580–584

Kus M (1995) The epidemic of the tuber necrotic ringspot strain of potato virus Y (PVYNTN) and its effect on potato crops in Slovenia. Proceedings of the 9th EAPR Virology Section Meeting, June 18–22, 1995 Bled, Slovenia, pp 159–160

Lawson C, Kaniewski W, Haley L, Rozman R, Newell C, Sanders P, Tumer NE (1990) Engineering resistance to mixed virus-infection in a commercial potato cultivar – resistance to potato virus X and potato virus Y in transgenic Russet Burbank. Bio/Technology 8:127–143

Le Romancer M, Kerlan C (1991) La maladie des necroses annulaires des tubercules: une affection de la pomme de terre due au virus Y. Agronomie 11:889–900

Leister D, Ballvora A, Salamini F, Gebhardt C (1996) A PCR-based approach for isolating pathogen resistance genes from potato with potential for wide application in plants. Nat Genet 14:421–429

Linder K, Behn A, Schwarzfischer A, Song YS (2011) Extreme Y-Resistenz im aktuellen deutschen Kartoffelsortiment. J Kult 63:97–103

Lopez-Pardo R, Barandalla L, Ritter E, de Galarreta JIR (2013) Validation of molecular markers for pathogen resistance in potato. Plant Breed 132:246–251

Mackay GR (1987) Screening for resistance to diseases and pests. In: Jellis GJ, Richardson DE (eds) The production of new potato varieties. Cambridge University Press, Cambridge, pp 88–90

Mackey J (1970) Significance of mating systems for chromosome and gametes in polyploids. Hereditas 66:165–176

Marczewski W, Flis B, Syller J, Schäfer-Pregl R, Gebhardt C (2001) A major quantitative trait locus for resistance to *Potato leaf roll virus* is located in a resistance hotspot on potato chromosome XI and is tightly linked to N-gene-like markers. Mol Plant-Microbe Interact 14:1420–1425
Marczewski W, Hennig J, Gebhardt C (2002) The *Potato virus S* resistance gene *Ns* maps to potato chromosome VIII. Theor Appl Genet 10:564–567
Marczewski W, Strzelczyk-Zyta D, Hennig J, Witek K, Gebhardt C (2006) Potato chromosomes IX and XI carry genes for resistance to potato virus M. Theor Appl Genet 112:1232–1238
Matsubayashi M (1991) Phylogenetic relationships in the potato and its relatives. In: Gupta PK, Tsuchiya T (eds) Chromosome engineering in plants: genetics, breeding, evolution, part B. Elsevier, Amsterdam, pp 93–118
Maxted N, Kell S, Ford-Lloyd B, Dulloo E, Toledo A (2012) Toward the systematic conservation of global crop wild relative diversity. Crop Sci 52:774–785
Mendoza HA (1989) Population breeding as a tool for germplasm enhancement. Am Potato J 66:639–653
Mendoza HA, Jayashinghe U (1993) Progress in breeding for combined resistance to potato viruses Y,X and leafroll. Proc. 12th Trienn Conf EAPR, Paris, pp 457–458
Mendoza HA, Mihovilovich EJ, Saguma F (1996) Identification of the triplex (*YYYy*) potato virus Y (PVY) immune progenitors derived from *Solanum tuberosum ssp. andigena*. Am Potato J 73:13–19
Mestre P, Brigneti G, Durrant MC, Baulcombe DC (2003) Potato virus Y NIa protease activity is not sufficient for elicitation of *Ry*-mediated disease resistance in potato. Plant J 36:755–761
Milbourne D, Pande B, Bryan G (2007) Chapter 12 potato. In: Kole C (ed) Genome mapping and molecular breeding in plants, vol 3. Springer Publishers, Berlin, pp 205–236
Mori K, Sakamoto Y, Mukojima N, Tamiya S, Nakao T et al (2011) Development of a multiplex PCR method for simultaneous detection of diagnostic DNA markers of five disease and pest resistance genes in potato. Euphytica 180:347–355
Moury B, Caromel B, Johansen E, Simon V, Chauvin L et al (2011) The helper component proteinase cistron of potato virus Y induces hypersensitivity and resistance in potato genotypes carrying dominant resistance genes on chromosome IV. Mol Plant-Microbe Interact 24:787–797
Muňoz FJ, Plaisted RL, Thurston HD (1975) Resistance to potato virus Y in *Solanum tuberosum ssp. andigena*. Am Potato J 52:107–115
Neder DG, Pinto C, Melo DS, Lepre AL, Peixouto LD (2010) Selection of potato clones for multiple resistances to early blight, PVY and PVX. Cienc Rural 40:1702–1708
NIAB (1985) Disease assessment manual of crop variety trials. National Institute of Agricultural Botany, Cambridge
Niks RE, Parlevliet JE, Lindhout P, Bai Y (2011) Breeding crops with resistance to diseases and pest. Wageningen Academic Publishers, Wageningen, p 198
Novy RG, Helgeson JP (1994) Resistance to potato virus Y in somatic hybrids between *Solanum etuberosum* and *S. tuberosum* x *S. berthaultii* hybrid. Theor Appl Genet 89:783–786
Nürnberger T, Lipka V (2005) Non-host resistance in plants: new insights into an old phenomenon. Mol Plant Pathol 6:335–345
Obidiegwu JE, Sanetomo R, Flath K, Tacke E, Hofferbert H-R, Hofmann A, Walkemeier B, Gebhardt C (2015) Genomic architecture of potato resistance to *Synchytrium endobioticum* disentangled using SSR markers and the 8.3k SolCAP SNP genotyping array. BMC Genet:16, 38. doi:10.1186/s12863-015-0195-y
Ochoa CM (1991) The potatoes of South America. Cambridge Press, Bolivia
Ortega F, Lopez-Vizcon L (2012) Application of molecular marker-assisted selection (MAS) for disease resistance in a practical potato breeding programme. Potato Res 55:1–13
Ortiz R, Watanabe KN (2004) Genetic contributions to breeding polyploid crops. Recent Res Dev Genet Breed 1:269–286
Ottoman RJ, Hane DC, Brown CR, Yilma S, James SR et al (2009) Validation and implementation of marker-assisted selection (MAS) for PVY resistance (Ry_{adg} gene) in a tetraploid potato breeding program. Am J Potato Res 86:304–314

Ovchinnikova A, Krylova E, Gavrilenko T, Smekalova T, Zhuk M, Knapp S, Spooner DM (2011) Taxonomy of cultivated potato (*Solanum* section *Petota:* Solanaceae). Bot J Linn Soc 165:107–155

Parrella G, Ruffel S, Moretti A, Morel C, Palloix A et al (2002) Recessive resistance genes against potyviruses are localized in colinear genomic regions of the tomato (*Lycopersicon* spp.) and pepper (*Capsicum* spp.) genomes. Theor Appl Genet 105:855–861

Pehu T, Mäki-Valkama T, Valkonen JPT, Koivu K, Lehto K, Pehu E (1995) Potato plants transformed with the potato virus Y P1 coding sequence are resistant to potato virus Y^{O}. Am Potato J 72:523–532

Pel MA, Foster SJ, Park T-H, Rietman H, van Arkel G et al (2009) Mapping and cloning of late blight resistance genes from *Solanum venturii* using an interspecific candidate gene approach. Mol Plant-Microbe Interact 22:601–615

Peters SA, Bargsten JW, Szinay D, van de Belt J, Visser RGF, Bai YL, de Jong H (2012) Structural homology in the Solanaceae: analysis of genomic regions in support of synteny studies in tomato, potato and pepper. Plant J 71:602–614

PGSC (2011) Genome sequence and analysis of the tuber crop potato. Nature 475:189–195

Pietrak J (2001) Evaluation of resistance to viruses in polish breeders selections and foreign cultivars in statutory trails. In: The methods of evaluation and selection applied in potato research and breeding, Monografie i Rozprawy Naukowe 10a/2001. IHAR, Radzików, pp 30–32

Plaisted RL (1987) Advances and limitations in the utilization of Neotuberosum in potato breeding. In: Jellis GJ, Richardson DE (eds) The production of new potato varieties. Technological advances. Cambridge University Press, Cambridge, pp 186–196

Ratcliff F, Harrison B, Baulcombe DC (1997) A similarity between viral defense and gene silencing in plants. Science 276:1558–1560

Ribeiro AM, Pinto CABP, Andrade CM, Santos JB et al (2006) SCAR marker for the selection of *Ry*-duplex potato clones immune to *Potato virus Y*. Crop Breed Appl Biotechnol 6:1–8

Ross H (1958) Inheritance of extreme resistance to virus Y in *Solanum stoloniferum* and its hybrids with *Solanum tuberosum*. In: Proceeding of 3rd conference on potato virus diseases. Veenman, Lisse-Wageningen, pp 204–211

Ross H (1986) Potato breeding – problems and perspectives, J Plant Breed Suppl 13. Paul Parey, Hamburg, 132 p

Rowley JS, Gray SM, Karasev AV (2015) Screening potato cultivars for new sources of resistance to *Potato virus Y*. Am J Potato Res 92:38–48

Ruffel S, Dussault M-H, Palloix A, Moury B, Bendahmane A et al (2002) A natural recessive resistance gene against potato virus Y in pepper corresponds to the eukaryotic initiation factor 4E (eIF4E). Plant J 32:1067–1075

Russel GE (1978) Plant breeding for pest and disease resistance. Butterworths, London/Boston

Sagredo DB, Mathias RM, Barrientos PC, Acuña BI, Kalazich BJ et al (2009) Evaluation of a SCAR RYSC3 marker of the Ry_{adg} gene to select resistant genotypes to potato virus Y (PVY) in the INIA potato breeding program. Chilean J Agric Res 69:305–315

Sarvari I (1978) Möglichkeit zur Herstellung von Pflanzgut ausgezeichneter Qualität unter Bedingungen extremer Virusinfektion und Auswirkungen auf den Ertrag. Proc.7th Trienn Conf EAPR, Warsaw, pp 275–276

Sato M, Nishikawa K, Komura K, Hosaka K (2006) *Potato virus Y* resistance gene, Ry_{chc}, mapped to the distal end of potato chromosome 9. Euphytica 149:367–372

Savenkov EI, Valkonen JPT (2001) Coat protein gene-mediated resistance to *Potato virus A* in transgenic plants is suppressed following infection with another potyvirus. J Gen Virol 82:2275–2278

Schiessendoppler E (1996) Change of susceptibility of potato cultivars to PVY-strain group as response of PVY-NTN introduction. 13th Trienn Conf of the EAPR, July 14–19, 1996, Veldhoven, Netherlands, pp 310–311

Sharma SK, Bolser D, de Boer J, Sønderkær M, Amoros W et al (2013) Construction of reference chromosome-scale pseudomolecules for potato: integrating the potato genome with genetic and physical maps. G3: Genes, Genomes, Genetics 3:2031–2047

Sharma R, Bhardwaj V, Dalamu D, Kaushik SK, Singh BP et al (2014) Identification of elite potato genotypes possessing multiple disease resistance genes through molecular approaches. Sci Hortic 179:204–211

Shivaprasad PV, Chen HM, Patel K, Bond DB, Santos BACM et al (2012) A microRNA superfamily regulates nucleotide binding site–leucine-rich repeats and other mRNAs. Plant Cell 24:859–874

Sieczka M (2001) Selection of the first year seedlings resistant to PVY,PVX and PVM using gun-spray inoculation. In: Zimnoch-Guzowska E, Syller J, Sieczka M (eds) The methods of evaluation and selection applied in potato research and breeding, Monografie i Rozprawy Naukowe IHAR, 10a. PB&AI, Radzików, pp 50–52

Simakov EA, Anisimov BV, Yashina IM, Uskov AI et al (2008) Potato breeding and seed production system development in Russia. Potato Res 51:313–326

Simmonds NW (1969) Prospects of potato improvement. Scott Plant Breed Stn Ann Rep 48:18–38

Singh M, Singh RP, Somerville TH (1994) Evaluation of tuber-bearing *Solanum* spp. for symptomology, as diagnostic host and sources of immunity to potato virus Y necrotic strain (PVY^N). Am Potato J 71:749–452

Singh RP, Valkonen JPT, Gray SM, Boonham N, Jones RAC, Kerlan K, Schubert J (2008) Discussion paper: the naming of *Potato virus Y* strains infecting potato. Arch Virol 153:1–13

Slater AT, Cogan NO, Haynes BJ, Schultz L, Dale MFB, Bryan GJ, Foster JW (2014) Improving breeding efficiency in potato molecular and quantitative genetics. Theor Appl Genet 127:2279–2292

Sliwka J, Jakuczun H, Lebecka R, Marczewski W, Gebhardt C et al (2006) The novel, major locus *Rpi-phu1* for late blight resistance maps to potato chromosome IX and is not correlated with long vegetation period. Theor Appl Genet 113:685–695

Smilde WD, Brigneti G, Jagger L, Perkins S, Jones JDG (2005) *Solanum mochiquense* Chromosome IX carries a novel late blight resistance gene *Rpi-moc1*. Theor Appl Genet 110:252–258

Solomon RM (1978) Methods of screening for resistance to viruses. In: Bradshaw JE, Mackay GR (eds) Potato genetics. CAB International, Wallingford, pp 339–363

Solomon-Blackburn RM, Mackay GR (1993) Progeny testing for resistance to virus Y: a comparison of susceptible potato cultivars for use in test-crosses with resistant parents. Potato Res 36:327–333

Song Y-S, Schwarzfischer A (2008) Development of STS markers for selection of extreme resistance (*Ry_{sto}*) to PVY and maternal pedigree analysis of extremely resistant cultivars. Am J Potato Res 85:159–170

Song Y-S, Hepting L, Schweizer G, Hartl L, Wenzel G et al (2005) Mapping of extreme resistance to PVY (Ry sto) on chromosome XII using anther-culture-derived primary dihaploid potato lines. Theor Appl Genet 111:879–887

Sorri VA, Watanabe KN, Valkonen JPT (1999) Predicted kinase-3a motif of a resistance gene analogue as a unique marker for virus resistance. Theor Appl Genet 99:164–170

Spooner DM, Nuňez J, Trujiilo G, Herrera MR, Guzmàn F, Ghislain M (2007) Extensive simple sequence repeat genotyping of potato landraces supports a major reevaluation of their gene pool structure and classification. Proc Natl Acad Sci U S A 104:19398–19403

Stebbins GL Jr (1950) Variation and evolution in plants. Columbia University Press, New York, pp 298–379

Stelzner G (1950) Virusresistenz der Wildkartoffeln. Z Pflanzenzüchtung 29:135–158

Swaminathan MS, Magoon ML (1961) Origins and cytogenetics of the commercial potato. Adv Genet 10:217–256

Świeżyński KM (1987) Potato breeding strategy in Poland. In: Jellis GJ, Richardson DE (eds) The production on new potato varieties. Cambridge University Press, Cambridge, pp 55–59

Świeżyński KM (1994) Inheritance of resistance to viruses. In: Bradshaw JE, Mackay GR (eds) Potato genetics. CAB International, Wallingford, pp 339–363

Świeżyński KM, Chrzanowska M, Domański L, Zimnoch-Guzowska E (2001) Comparison of resistance evaluation in potato variety assessment. Potato Res 44:25–31

Szajko K, Chrzanowska M, Witek K, Strzelczyk-Zyta D, Zagorska H et al (2008) The novel gene *Ny-1* on potato chromosome IX confers hypersensitive resistance to *Potato virus Y* and is an alternative to *Ry* genes in potato breeding for PVY resistance. Theor Appl Genet 116:297–303

Szajko K, Strzelczyk-Zyta D, Marczewski W (2014) *Ny-1* and *Ny-2* genes conferring hypersensitive response to *Potato virus Y* (PVY) in cultivated potatoes: mapping and marker-assisted selection validation for PVY resistance in potato breeding. Mol Breed 34:267–271

Szinay D, Wijnker E, van den Berg R, Visser RGF, de Jong H, Bai YL (2012) Chromosome evolution in *Solanum* traced by cross-species BAC-FISH. New Phytol 195:688–698

Talbot M (1987) Establishing standards in variety assessment. In: Jellis GJ, Richardson DE (eds) The production of new varieties: technological advances. Cambridge University Press, Cambridge, pp 157–162

Tang X, Datema E, Guzman MO, de Boer JM, van Eck HJ, Bachem CWB, Visser RGF, de Jong H (2014) Chromosomal organizations of major repeat families on potato (*Solanum tuberosum*) and further exploring in its sequenced genome. Mol Gen Genomics 289:1307–1319

Thiele G, Theisen K, Bonierbale M, Walker T (2010) Targeting the poor and hungry with potato science. Potato J 37:75–86

Thieme R, Rakosy-Tican E, Gavrilenko T, Antonova O et al (2008) Novel somatic hybrids (*Solanum tuberosum* L. + *Solanum tarnii*) and their fertile BC1 progenies express extreme resistance to potato virus Y and late blight. Theor Appl Genet 116:691–700

Tian YP, Valkonen JPT (2013) Genetic determinants of *Potato virus Y* required to overcome or trigger hypersensitive resistance to PVY strain group O controlled by the gene *Ny* in potato. Mol Plant-Microbe Interact 26:297–305

Tian YP, Valkonen JPT (2015) Recombination of strain O segments to HCpro-encoding sequence of strain N of *Potato virus Y* modulates necrosis induced in tobacco and in potatoes carrying resistance genes *Ny* or *Nc*. Mol Plant Pathol 16:735–747

Tomczyńska I, Jupe F, Hein I, Marczewski W, Śliwka J (2014) Hypersensitive response to *Potato virus Y* in potato cultivar is conferred by the *Ny-Smira* gene located on long arm of chromosome IX. Mol Breed 34:471–480

Tommiska TJ, Hämäläinen JH, Watanabe KN, Valkonen JPT (1998) Mapping of the gene Nx_{phu} that controls hypersensitive resistance to potato virus X in *Solanum phureja* IvP35. Theor Appl Genet 96:840–843

Valkonen JPT (1997) Novel resistance to four potyviruses in tuber bearing potato species and temperature – sensitive expression of hypersensitive resistance to potato virus Y. Ann Appl Biol 130:91–104

Valkonen JPT, Brigneti G, Salazar LF, Pehu E, Gibson RW (1992) Interactions of the *Solanum* spp. of the Etuberosa group and nine potato- infection viruses and a viroid. Ann Appl Biol 120:301–313

Valkonen JPT, Slack SA, Plaisted RL, Watanabe K (1994) Extreme resistance is epistatic to hypersensitive resistance to potato virus Y^o in a *Solanum tuberosum ssp. andigena* -derived potato genotype. Plant Dis 78:1177–1180

Valkonen JPT, Orillo M, Slack SA, Plaisted RL, Watanabe KN (1995) Resistance to viruses in F1 hybrids produced by direct crossing between diploid *Solanum* series Tuberosa and diploid *S. brevidens* (series Etuberosa) using *S. phureja* for rescue pollination. Plant Breed 114:421–426

Valkonen JPT, Jones RAC, Slack SA, Watanabe KN (1996) Resistance specificities to viruses in potato: standardization of nomenclature. Plant Breed 115:433–438

Valkonen JPT, Rokka V-M, Watanabe KN (1998) Examination of the leaf-drop symptom of virus-infected potato using anther culture-derived haploids. Phytopathology 88:1073–1077

Valkonen JPT, Wiegmann K, Hämäläinen JH, Marczewski W, Watanabe KN (2008) Evidence for utility of the same PCR-based markers for selection of extreme resistance to *Potato virus Y* controlled by Ry_{sto} of *Solanum stoloniferum* derived from different sources. Ann Appl Biol 152:121–130

Vallejo RL, Collins WW, Schiavone RD, Young JB (1994) Extreme resistance to infection by potato virus Y and potato virus X in an advanced hybrid *Solanum phureja- S. stenotomum* diploid population. Am Potato J 71:617–628

van der Woude K (1987) Variety assessment in the Netherlands. In: Jellis GJ, Richardson DE (eds) The production of new varieties: technological advances. Cambridge University Press, Cambridge, pp 119–130

Vidal S, Cabrera H, Andersson RA, Fredriksson A, Valkonen JPT (2002) Potato gene *Y-1* is an *N* gene homolog that confers cell death upon infection with potato virus Y. Mol Plant-Microbe Interact 15:717–727

Watanabe K, Orrillo M (1994) Disomic behavior of polyploid tuber-bearing Solanum species. Jpn J Genet 69:718–725

Watanabe K (2015) Potato genetics, genomics and applications. Breed Sci 65:53–68

Watanabe KN, Orillo M, Vega S, Valkonen JPT, Pehu E, Hurtado A, Tanksley SD (1995) Overcoming crossing barriers between non tuber-bearing and tuber-bearing *Solanum* species: towards potato germplasm enhancement with a broad spectrum of Solanaceous genetic resources. Genome 38:27–35

Waterhouse PM, Smith NA, Wang MB (1999) Virus resistance and gene silencing: killing the messenger. Trends Plant Sci 4:452–457

Weidemann H-L (1993) Nekrotische Ringsymptome an Kartoffelknollen. Ein neuer Stamm des Kartoffelvirus Y als Ursache. Kartoffelbau 44:308–309

Whitworth J, Novy R, Hall D, Crosslin J, Brown C (2009) Characterization of broad spectrum potato virus Y resistance in a *Solanum tuberosum ssp. andigena-derived* population and select breeding clones using molecular markers, grafting, and field inoculations. Am J Potato Res 86:286–296

Wiersema HT (1966) Early evaluation of potato seedlings in breeding for extreme resistance to virus. Qualitas Plant Mater Veg 13:349–351

Witek K, Strzelczyk-Żyta D, Hennig J, Marczewski W (2006) A multiplex PCR approach to simultaneously genotype potato towards the resistance alleles *Ry-fsto* and *Ns*. Mol Breed 18:273–275

Yin Y, Xie F, Michalaka K, Pawełkowiczc M, Zhang B, Murawska Z, Lebecka R, Zimnoch-Guzowska E (2016) Potato cultivar Etola exhibits hypersensitive resistance to PVY^{NTN} and partial resistance to PVYZ-NTN and PVYN-Wi strains and strain-specific alterations of certain host miRNAs might correlate with symptom severity. Plant Pathol 66:539–550

Zimnoch-Guzowska E, Yin Z, Chrzanowska M, Flis B (2013) Sources and effectiveness of potato resistance in IHAR's breeding research. Am J Potato Res 90:21–27

Zoteyeva N, Chrzanowska M, Flis B, Zimnoch-Guzowska E (2012) Resistance to pathogens of the potato accessions from the collection of N.I. Vavilov Institute of Plant Industry (VIR). Am J Potato Res 89:277–293

Chapter 9
Host Groups of *Potato virus Y*: Vanishing Barriers

Benoît Moury, Vincent Simon, Chantal Faure, Laurence Svanella-Dumas, Armelle Marais, and Thierry Candresse

Abstract *Potato virus Y* (PVY) is one of the rare members of the genus *Potyvirus* for which the genetic diversity is structured by the plant species of origin. This structure has been explained by the occurrence in some crop species of barriers to infection by PVY isolates originating from other plant species. A particularly strong barrier to infection has been shown between PVY isolates from pepper (*Capsicum* spp.) and potato (*Solanum tuberosum*). PVY isolates from potato belong mainly to clades N and O and are poorly or not infectious at all in pepper. By contrast, PVY isolates from pepper belong mainly to clade C1 and are poorly or not infectious at all in potato. Recent surveys and genetic analyses, however, have revealed that PVY isolates of clades N and/or O can be quite frequent in pepper crops. Conversely, PVY isolates of clade C1 are almost absent in potato crops. However, we showed that PVY isolates of clade C1 from pepper are infectious in a potato cultivar devoid of the resistance gene Nc_{tbr}. It has also been shown that, among pepper, tomato and tobacco genotypes carrying eIF4E-mediated recessive resistances, PVY evolution in one crop species could favour its future adaptation to other species. Altogether, these results suggest a lack of strong barriers to PVY host change between solanaceous crops. The contrast with older studies could be due to a range of factors, including an insufficient sampling effort in previous studies (too few isolates collected and characterised, too few plant genotypes assayed), to recent changes among PVY populations and/or to the contrasted behaviour of different genotypes within a plant species upon PVY inoculation.

B. Moury (✉) • V. Simon
Pathologie Végétale, INRA, 84140 Montfavet, France
e-mail: benoit.moury@inra.fr

C. Faure • L. Svanella-Dumas • A. Marais • T. Candresse
INRA, UMR1332 Biologie du Fruit et Pathologie, F-33140 Villenave d'Ornon, France

University Bordeaux, UMR1332 Biologie du Fruit et Pathologie,
F-33140 Villenave d'Ornon, France

C. Lacomme et al. (eds.), *Potato virus Y: biodiversity, pathogenicity, epidemiology and management*, DOI 10.1007/978-3-319-58860-5_9

1 Introduction

While PVY remains today a major viral constraint to potato cultivation worldwide, it is also a major threat to tobacco and pepper crops and, to a lesser extent, to tomato and eggplant production (Bhat et al. 1999; Aramburu et al. 2006; Sadeghi et al. 2008; Mascia et al. 2010). Because of its important economic impact, extensive efforts have been made to control PVY epidemics, including prophylaxis, the control of aphid vectors and the breeding of resistant cultivars. The PVY group includes very diverse variants that have been characterised genetically and/or pathogenically, and classification of these variants into groups is required to understand how PVY has evolved and how to control its diseases. Differences in the pathogenicity of PVY in reference cultivars carrying particular resistance genes or expressing particular symptoms when infected have led to the recognition of multiple groups of PVY isolates (see Singh et al. 2008 for a recent review). However, the large host range for PVY (Edwardson and Christie 1997), its continuously changing populations (Rolland et al. 2008) and international trade all contribute to many potential host plants being exposed to a range of very diverse isolates of PVY. This prompted us to re-examine the links between PVY diversity and host range, in terms of both plant species and plant genotypes within species.

2 Importance of Host Range in the Definition of PVY Groups

The host range of PVY is quite large, probably reflecting the diversity of variants. The natural hosts include plants from nine botanical families, whereas its potential host range, as assessed experimentally, includes 405 species belonging to 31 families (Edwardson and Christie 1997). Most of these hosts belong to *Solanaceae* (287 species), *Amaranthaceae* (including former *Chenopodiaceae*; 48 species), *Fabaceae* (25 species) and *Asteraceae* (11 species) (Kerlan 2006). Within the genus *Potyvirus*, PVY belongs to a clade of 19 virus species, referred to as the 'PVY group' (Gibbs and Ohshima 2010; Quenouille et al. 2013). Most of these viruses were isolated from *Solanaceae* (10 species), but some others were isolated from *Amaranthaceae* (3 species), *Asteraceae* (3 species), *Liliaceae* (1 species), *Amaryllidaceae* (1 species) or *Verbenaceae* (1 species). The most recent common ancestor of the genus *Potyvirus* has been tentatively dated to approximately 6560–7250 years ago (Gibbs et al. 2008; Gibbs and Ohshima 2010). The genus probably diverged from a virus of monocotyledonous plants around the Fertile Crescent of the Middle East, more or less contemporaneously with the emergence of cereal-based agriculture about 9500 years ago (Gibbs and Ohshima 2010). Given the recent diversification of the 'PVY group', the diversity of hosts from which PVY has been isolated suggests that changes in the host range must have been frequent and rapid and must have played an important role in its diversification, suggesting that similar modifications can be expected for PVY itself. Different PVY groupings have been proposed over time on the basis of the interaction between PVY isolates and a very limited proportion of its hosts, mainly a few important crop plants in the *Solanaceae*, including potato, tobacco and pepper (Singh et al. 2008).

Since PVY is a major constraint to potato production, breeders have been searching for PVY resistance sources among potato cultivars for a long time. Most of the resistance sources introgressed into various cultivars have monogenic, dominant inheritance. The differential reaction of reference potato cultivars carrying these resistance genes to infection by PVY isolates was used to classify these isolates into groups. The gene Ny_{tbr} from *S. tuberosum* triggers necrotic hypersensitive reactions (HRs) on inoculated leaves and confers resistance to PVY isolates belonging to group O. Gene Ny_{tbr} is located on potato chromosome IV (Celebi-Toprak et al. 2002). The gene Nc_{tbr}, also from *S. tuberosum*, confers resistance and HR to PVY isolates belonging to group C. The location of Nc_{tbr} in the potato genome is unknown, but a resistance gene (Nc_{spl}) with the same specificity, resistance phenotype and PVY elicitor has been described in *S. sparsipilum* and mapped to potato chromosome IV in the same genome region as Ny_{tbr} (Moury et al. 2011). Hence, Ny_{tbr}, Nc_{tbr} and Nc_{spl} could be alleles of the same gene or members of the same resistance gene cluster. The necrotic local lesions observed on inoculated leaves of potato cultivars carrying Ny_{tbr} or Nc_{tbr} are sometimes associated with systemic necrotic reactions, indicating that resistance is sometimes insufficient to limit PVY spread away from the inoculated area. PVY isolates that do not induce HR and overcome resistance in cultivars with Ny_{tbr} or Nc_{tbr} have been classified in the N group. Additional PVY groups, usually containing only a few members, have been defined based on the symptomatic and infective reactions expressed by additional potato cultivars carrying other undetermined resistance genes (Singh et al. 2008).

*2.1 PVY/***Nicotiana tabacum** *(Tobacco) Interactions*

PVY is also a major pathogen of tobacco (*Nicotiana tabacum*). Tobacco isolates of PVY have been classified according to two different criteria. Firstly, the type of symptoms (systemic necrosis or mosaic) expressed by tobacco cultivars carrying or not carrying the *Rk* gene conferring resistance to *Meloidogyne incognita* root-knot nematode allowed PVY isolates to be separated into three groups: M_SM_R (mosaic on both kinds of cultivars), M_SN_R (mosaic and necrosis on non-*Rk*- and *Rk*-carrying cultivars, respectively) and N_SN_R (necrosis on both kinds of cultivars) (Gooding and Tolin 1973). In these tobacco cultivars, the necrotic response to PVY infection and the resistance to *M. incognita* seem to be pleiotropic effects of the *Rk* gene (Rufty et al. 1983). Secondly, PVY isolates causing different symptoms have been described in relation to tobacco cultivars carrying different alleles of the recessive resistance gene *va*, the resistance gene used most commonly to control PVY infection in tobacco. The *va*-mediated resistance acts by slowing down the cell-to-cell movement of PVY during infection (Acosta-Leal and Xiong 2008). Sequencing of the tobacco genome and genetic analyses have now shown that *va* resistance to PVY (and other potyviruses) corresponds to a deletion of the *Nicotiana sylvestris*-form eIF4E-1 (eukaryotic initiation factor 4E-1) locus (Sierro et al. 2014; Julio et al. 2015). Inoculation of reference tobacco cultivars carrying the va^0, va^1 and va^2 resistance alleles with PVY isolates revealed the occurrence of the eight expected pathotypes, with two possible phenotypes for each reference cultivar, which were the occurrence or absence of systemic PVY infection (Lacroix et al. 2010).

2.2 PVY/Capsicum *spp. (Pepper) Interactions*

PVY infection is also frequent in pepper (*Capsicum* spp.) crops and can induce considerable economic losses (Fereres et al. 1996; Budnik et al. 1996). In genotypes of pepper devoid of resistance genes, PVY induces systemic mosaic or systemic necrotic symptoms, depending on the genotype and the PVY isolate (Dogimont et al. 1996). The most commonly used source of PVY resistance in pepper is the recessive *pvr2* gene, which comprises at least 30 alleles with varying specificities of action against diverse PVY isolates (Kang et al. 2005; Charron et al. 2008; Ibiza et al. 2010; Moury et al. 2014; J.-L. Gallois, personal communication). As with the *va* gene in tobacco, *pvr2* corresponds to eIF4E-1 (Ruffel et al. 2002). The different *pvr2* alleles correspond to different amino acid substitutions in the encoded eIF4E-1. The phenotype of resistance in pepper is somewhat different to that in tobacco, since the *pvr2* resistance acts very early in the PVY infection process and no virus is detectable in inoculated leaves and cotyledons, at least for the $pvr2^1$ and $pvr2^2$ alleles (Ruffel et al. 2002), which are the most widely used to create PVY-resistant pepper cultivars and have been exploited extensively for more than 50 years. $pvr2^2$-breaking PVY isolates are extremely rare (Moury and Verdin 2012), whereas $pvr2^1$-breaking PVY isolates are much more frequent but usually less prevalent than avirulent isolates (Luis-Arteaga and Gil-Ortega 1986). Consequently, cultivars carrying $pvr2^1$ are still economically viable in many growing regions. Reference pepper cultivars carrying $pvr2^1$ or $pvr2^2$ allowed three PVY pathotypes to be defined: (0), (0,1) and (0,1,2). Only PVY (0,1,2) isolates are able to infect plants of genotype $pvr2^2/pvr2^2$, while (0,1) and (0,1,2) isolates are able to infect plants of genotype $pvr2^1/pvr2^1$. Isolates of the three pathotypes can infect plants devoid of resistance alleles ($pvr2^+/pvr2^+$). A very large number of PVY pathotypes could be defined by taking into consideration additional *pvr2* alleles. For example, 11 different pathotypes were observed by considering alleles $pvr2^+$, $pvr2^1$, $pvr2^2$, $pvr2^3$, $pvr2^5$ and $pvr2^8$ (Ayme et al. 2007; Moury et al. 2014). A resistance similar to *pvr2* in pepper has also been identified in *Solanum habrochaites*, a relative of tomato (*S. lycopersicum*) (Parrella et al. 2002). Resistance is mediated by the *pot-1* gene, which is the orthologue of *pvr2* in pepper (Ruffel et al. 2005). However, the resistance has not been widely used in tomato crops, and PVY isolates breaking the *pot-1*-mediated resistance are rare (Legnani et al. 1995).

Importantly, most PVY genetic factors that interact with these resistance genes or that determine the particular symptoms used for PVY classification have been identified (Table 9.1). This has allowed a better understanding of the relationships between the various classification schemes for PVY based on the reaction of hosts. Three major PVY genome regions, encoding the HcPro, VPg and NIb, are responsible for pathogenicity (Table 9.1). Since these regions are scattered throughout the PVY genome, it appears that they can evolve quite independently and can be exchanged by recombination between different PVY isolates (Ogawa et al. 2008; Schubert et al. 2007). In this context, any combination of pathogenicity characteristics conferred by these three genome regions could be expected to occur. In contrast, when the same genome region is involved in determining several pathogenicity characteristics, their possible combinations will depend on the mutations involved and on the constraints acting on the evolution of these regions.

3 Relationships Between PVY Host Origin and Phylogeny

For potyviruses, the links between the host of origin of an isolate and virus phylogeny are quite complex. For some potyviruses, a link between host origin (and/or host adaptation) and phylogeny has been demonstrated, *e.g. Plum pox virus* (Glasa et al. 2013), *Turnip mosaic virus* (Ohshima et al. 2002) and *Bean yellow mosaic virus* (Kehoe et al. 2014). However, no such link was found with some other potyviruses, *e.g. Papaya ringspot virus* (Olarte-Castillo et al. 2011). In PVY, non-recombinant isolates in the N, O and C groups that were defined on the basis of phenotypic reactions in reference potato cultivars correspond to three main PVY phylogenetic groups, which could, therefore, be named by the same letters (Fig. 9.1). In addition, almost all isolates of the N group (*i.e.* no HR and no resistance reactions on potato cultivars carrying Ny_{tbr} or Nc_{tbr}) were shown to cause systemic veinal necrosis in tobacco cultivars that did not carry the *Rk* resistance gene (*i.e.* the vast majority of tobacco cultivars). The stability of the agreement between PVY phylogeny and interaction with potato resistance genes and tobacco could be due to the co-localisation of the molecular determinants of pathogenicity properties in the HcPro and to the quite high evolutionary constraint acting on this genome region (Tian and Valkonen 2015).

In worldwide surveys, almost all PVY isolates from potato have been found to belong to clades N, O or correspond to N × O recombinants (Blanco-Urgoiti et al. 1998b; Dedic et al. 2007; Schubert et al. 2007; Ogawa et al. 2008; Ogawa et al. 2012; Djilani-Khouadja et al. 2010; Kerlan et al. 1999; Lorenzen et al. 2006; Rolland et al. 2008; Lindner and Billenkamp 2005; Rolot 2007). In contrast, almost all PVY isolates from pepper belonged to clade C (Gebre Selassie et al. 1985; Blanco-Urgoiti et al. 1998a; Romero et al. 2001; Ayme et al. 2007; Ben Khalifa et al. 2012; Moury et al. 2014). There were, however, some exceptions to this situation. Two subclades, C1 and C2, were distinguished within clade C (Blanco-Urgoiti et al. 1998a). In this study, PVY isolates of clade C2 originated from potato and did

Table 9.1 Properties of plant species and genotypes used to classify PVY isolates into groups and corresponding PVY pathogenicity determinants

Species	Potato		Tobacco			Pepper	Tomato
Resistance gene or phenotype	Ny_{tbr}	Nc_{tbr}	*Rk*	Necrosis in non-*Rk* genotypes	*va*	*pvr2*	*pot-1*
Pathogenicity determinant	HcPro	HcPro	NIb	HcPro, CI, NIaPro	VPg	VPg	VPg
Mutations involved	Positions 227–327	Not determined	Not determined	Positions 205, 339, 400, 419 in HcPro	Positions 101–119	Positions 101–123	Position 119
References	Moury et al. (2011) and Tian and Valkonen (2013)	Moury et al. (2011)	Fellers et al. (2002)	Tribodet et al. (2005), Hu et al. (2009), and Faurez et al. (2012).	Janzac et al. (2014)	Moury et al. (2004) and Ayme et al. (2006, 2007)	Moury et al. (2004, 2014)

not infect pepper plants, whereas isolates of clade C1 were from pepper. Isolates from clade C have only rarely been detected in potato crops (Blanco-Urgoiti et al. 1998b; Browning et al. 2004; Rolland et al. 2008). Interestingly, the oldest potato PVY isolates characterised to date (collected in the Netherlands in 1938 in cv. 'Zeeuwse Blauwe' and in the United Kingdom in 1943 in cv. 'Edgecote Purple') belong to clade C1 (Dullemans et al. 2011; Kehoe and Jones 2011). These data suggest several shifts in the prevalence of the different groups of PVY isolates in potato crops, with a sharp decrease of the C1 and C2 groups and a more recent and/or slower decrease of O isolates, accompanied by an increase of N and N × O recombinant isolates (Rolland et al. 2008). The decrease in prevalence of PVY isolates of groups C and O in potato crops in the last decades could be due to the increased use of the resistance genes Nc_{tbr} and Ny_{tbr} (Cockerham 1970; Jones 1990).

Although most PVY isolates from pepper characterised to date belong to clade C1, there are also exceptions to this strong trend, such as the detection of two N-type isolates in Greece (Margaritopoulos et al. 2010) and suspicions of an O-type isolate in South Africa (Ibaba and Gubba 2011).

Two additional PVY clades, corresponding to a small number of divergent sequences with a common geographical origin, have been identified recently. A 'Chilean' clade includes PVY isolates from tobacco (Sudarsono et al. 1993) and from pepper (Moury 2010), and a 'Brazilian' clade includes isolates from tobacco (Janzac et al. 2015) (Fig. 9.1).

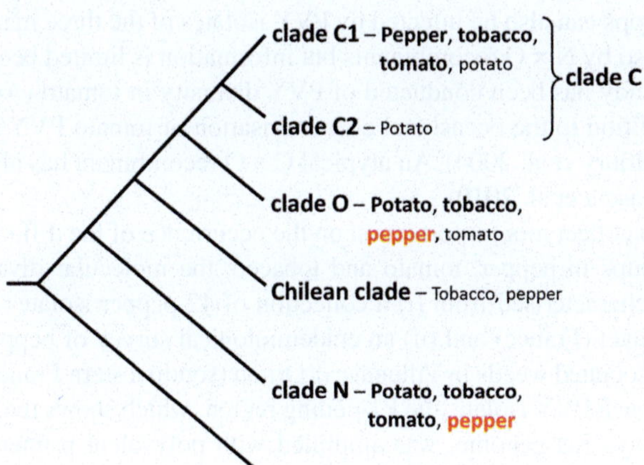

Fig. 9.1 Schematic phylogenetic tree representing the genetic diversity of the main PVY clades (excluding recombinant isolates) and the main crop species affected. Crops where PVY isolates of a given clade are frequent are indicated in *bold*. Crops where PVY isolates of a given clade were identified in the present study are indicated in *red*. Somewhat different tree topologies were obtained, depending on the phylogenetic method used, on the representative PVY isolates considered and on the outgroup(s) used to root the tree. Whereas the present topology was obtained in some studies (Visser et al. 2012; Janzac et al. 2015), the Chilean clade and clade N were switched, or their respective positions found uncertain, in others (Moury 2010; Cuevas et al. 2012)

Table 9.2 Primers used for PCR amplifications and sequencing

Name	Polarity	Primer sequence (5′ to 3′)[a]	Binding site[b]
P1-F[c]	+	tccccccTAAGAAAAACAACGCARAAAYACTYRYAAACGC[d]	25–57
P1-R[c]	–	AGGRTATCTCAWYYGYGCCC	1078–1097
P1-seq[e]	+	tccccccTAAGAAAAACAACGC[d]	25–39
VPg-F[c]	+	GATATGCAAARGRYTAYACYGC	5413–5434
VPg-R[c, e]	–	AARACAGGGAARTCYTTYGGC	6539–6559
CP-F[c]	+	TTCACTGARATGATGSTYGC	8502–8521
CP-R[c, e]	–	TAAAAGTAGTACAGGAAAAGCCA	9425–9447

[a]Y: C or T; R: A or G; S: C or G; W: A or T
[b]Referring to PVY isolate SON41p (accession number AJ439544)
[c]Primers used for RT-PCR
[d]Contains a C-rich 5′ overhang (lowercase letters) to improve PCR efficiency
[e]Primers used for sequencing

Serological or partial genome characterisation of isolates of PVY from tobacco crops revealed the prevalence of N, O and C clades (Li et al. 2006; Lacroix et al. 2010, 2011; Tian et al. 2011; Margaritopoulos et al. 2010; Janzac et al. 2014), in addition to the rarer occurrence of isolates of the 'Chilean' or 'Brazilian' clades.

Tomato crops can also be infected by PVY isolates of the three main clades N, O and C, and also by N × O recombinants but information is limited because only one large-scale study has been conducted of PVY diversity in tomato (Aramburu et al. 2006), in addition to the occasional characterisation of tomato PVY isolates (Behl et al. 1987; Moury et al. 2004). An atypical C × O recombinant has also been found in tomato (Mascia et al. 2010).

In order to collect more information on the occurrence of the different PVY phylogenetic groups in pepper, tomato and tobacco, the molecular diversity of PVY isolates was characterised from (i) a collection of 42 pepper isolates maintained at INRA Montfavet (France) and (ii) an epidemiological survey of pepper and tomato crops, and associated weeds in Villenave d'Ornon (southwestern France), from 2010 to 2012. For each PVY isolate, the P1-coding region, which shows the highest diversity in the potyvirus genome, was amplified with polyvalent primers (all primers used in this study are presented in Table 9.2), followed by direct sequencing of PCR products. Detection of recombination events was conducted using RDP software (version 2; Martin et al. 2005), which implements six different algorithms to detect recombinant sequences. The Montfavet collection was found to contain 32 isolates of clade C1, 3 of clade O, 4 of clade N and 3 isolates showing a recombination breakpoint at genome positions 495–496, with the 5′ end of the P1 cistron corresponding to clade O and the other end corresponding to clade N (Table 9.3). The recombination signature was strong ($P < 0.05$ with all six methods) and the breakpoint was

Table 9.3 Clustering of the sequence of P1-coding region of PVY isolates collected from pepper (*Capsicum* spp.) in the Montfavet collection among the main PVY clades

Origin	Clade	Number of isolates
France (continental)	C1	6
	N	2
	O	3
France (West Indies)	C1	6
France (La Réunion)	C1	2
Spain	C1	1
	N	1
Italy	C1	1
Poland	N	1
	O × N	3
Algeria	C1	3
Tunisia	C1	6
Turkey	C1	2
Israel	C1	1
Unknown	C1	4
Total	C1	32 (76.5%)
	N	4 (9.5%)
	O	3 (7%)
	O × N[a]	3 (7%)

[a]Recombinants with a 5′ part of P1 cistron clustering with clade O and a 3′ part of P1 cistron clustering with clade N. Recombination breakpoint was located at nucleotide positions 495–496, with reference to the genome of isolate N605 (accession number X97895)

located exactly at the same position as in the 'NTN' or 'W' recombinant isolates from *S. tuberosum* or *N. tabacum* in Poland or Germany (Schubert et al. 2007).

The PVY isolates collected from pepper and tomato crops grown outdoors, and from associated solanaceous weeds (*S. nigrum* and *S. villosum*) in Villenave d'Ornon (France), were very diverse, covering five major clades (Table 9.4). In total, of 131 isolates characterised, 53 (40.5%) belonged to clade O and 37 (28.2%) belonged to clade C1. Two groups of O × N recombinant isolates, with recombination breakpoints located at genome positions 630–631 (O × N #1) and 648–652 (O × N #2), were also identified (consistently detected by all six methods implemented in RDP). An exploration of sequence databases using Blastn identified PVY isolates with the same recombination breakpoint as group O × N #2, with potato isolate 261–4 from Germany (accession number AM113988) being genetically closest (98% identity) (Ogawa et al. 2008). None of the isolates shared the same breakpoint as group O × N #1. Finally, nine isolates clustering with clade N were recovered from weeds. Interestingly, the majority of pepper isolates (78%; 32/41) belonged to clade O, whereas the majority of tomato isolates (87.5%; 28/32) belonged to clade C1. The distribution of isolates from solanaceous weeds was more balanced among the various PVY phylogenetic groups. Two additional genome regions, the VPg and coat

Table 9.4 Clustering of the sequence of P1-coding region of PVY isolates collected in Villenave d'Ornon, France, in 2011–2012 among the main PVY clades

	PVY clade				
Plant of origin	O	Rec. O × N #1[a]	Rec. O × N #2[a]	N	C1
Pepper	32	2	6	0	1
Tomato	0	0	4	0	28
Solanum villosum	20	7	5	6	8
S. nigrum	1	5	3	3	0

[a]Recombinants with a 5′ part of P1 cistron clustering with clade O and a 3′ part of P1 cistron clustering with clade N. Recombination breakpoint was located at nucleotide positions 630–631 (O × N #1) or 648–652 (O × N #2), with reference to the genome of isolate N605 (accession number X97895)

Table 9.5 Clustering of three genome regions of PVY isolates from *Solanum villosum* collected in Villenave d'Ornon, France, in 2011–2012 among the main PVY clades

P1 cistron	VPg cistron	CP cistron	Number of isolates	Reference recombinant PVY isolates with similar genome organisation (after Ogawa et al. 2008)[b]
C1	C1	C1	3	
O	O	O	2	
N	N	N × O	4	423–3 etc....; accession number AY884982
Rec. O × N #1[a]	N	N × O	6	Gr99, 12-94 and 34/01; accession numbers AJ890343, AJ889866 and AJ890342
Rec. O × N #2[a]	N	O	1	156 and 156vari; accession numbers AJ889867 and AJ889868
Rec. O × N #2[a]	N	N × O	1	Gr99, 12-94 and 34/01; accession numbers AJ890343, AJ889866 and AJ890342

[a]cf. Table 9.4
[b]Only the recombination breakpoint in the P1 cistron is at a slightly different position (see text)

protein (CP) cistrons, were sequenced for 17 isolates selected to represent the five clades defined on the basis of the P1 cistron diversity. The isolates identified as belonging to clade O or C1 on the basis of the P1 cistron also clustered with clades O and C1, respectively, for the other two sequenced regions. The isolates that grouped in clade N (on the basis of the P1 cistron) clustered with clade N for the VPg cistron and were N × O recombinants for the CP cistron (breakpoint at positions 9169–9182 detected by all six RDP methods). An N × O recombination breakpoint was observed at the same position in several potato PVY^{NTN} isolates (Ogawa et al. 2008; Schubert et al. 2007). Isolates from the recombinant group O × N #1 clustered with clade N for the VPg cistron and were N × O recombinants for the CP cistron (same 9169–9182 breakpoint). Finally, recombinant isolates of group O × N #2 clustered with clade N for the VPg cistron and, for the CP cistron, were either a N × O recombinant (same 9169–9182 breakpoint; one isolate) or belonged to clade O (the other isolate). Altogether, based on the sequences of three genome regions, isolates collected in Villenave d'Ornon had genome patterns similar to those of typical potato isolates described previously, with only a slightly different recombination breakpoint in the P1 cistron in the O × N #1 recombinant group (Ogawa et al. 2008; Schubert et al. 2007; Table 9.5).

In summary, the analysis of these two relatively large collections of PVY isolates demonstrated that isolates clustering with clades N and/or O were not uncommon in pepper crops (23.5% and 97.6% in the Montfavet and Villenave d'Ornon collections, respectively).

4 PVY Host Specificity and Barriers to Host Change

Controlled aphid or mechanical inoculations have revealed some degree of host specificity among PVY isolates. The most salient trend was the tendency of PVY isolates from potato to be unable to infect pepper plants (Valkonen et al. 1996; McDonald and Kristjansson 1993; Blanco-Urgoiti et al. 1998a, 1998b; d'Aquino et al. 1995; Gebre Selassie et al. 1985; Romero et al. 2001; Fereres et al. 1993; Stobbs et al. 1994; Moury 2010). Similarly, most isolates from pepper were unable to infect potato plants (Gebre Selassie et al. 1985; Moury 2010; Janzac et al. 2015), but this conclusion should be treated with caution because only a few pepper isolates have been tested so far and, generally, against a very limited number of potato cultivars. In contrast, plants of tobacco and tomato cultivars could generally be infected by PVY isolates, irrespective of the host from which they originated (Stobbs et al. 1994; Blancard 1998; Quenouille et al. 2013; Janzac et al. 2014).

Because the genetic diversity of PVY is at least partially influenced by the host of origin (part 2; Cuevas et al. 2012; Moury 2010; Janzac et al. 2015), the barriers to infection of potato and pepper plants by pepper and potato isolates, respectively, are also linked to phylogenetic classification of PVY. Isolates of the N, O or C2 clades originating from potato did not infect pepper plants, whereas isolates of the C1 or Chilean clades originating from pepper did not infect potato plants (reviewed in Quenouille et al. 2013; Fig. 9.1). Interestingly, isolates of the Brazilian clade originating from tobacco did not infect potato or pepper plants (Janzac et al. 2015). Reconstruction of the evolution of host range traits in PVY phylogeny suggested that the capacity to infect potato and pepper are derived traits, and that PVY has undergone expansions and shifts of its host range during its evolutionary history (Moury 2010; Janzac et al. 2015). However, it is not known if (i) the host of origin of PVY, (ii) the overall genetic diversity of PVY or (iii) both are the true determinants of host specificities and the corresponding barriers to host change. Unravelling the genetic determinants of infection of pepper and potato plants within the PVY genome would be necessary to discriminate between these three hypotheses.

To get a more exhaustive view of the infection barrier in potato against pepper PVY isolates, plants of potato cvs 'King Edward' and 'Safrane', carrying or not carrying the Nc_{tbr} resistance gene respectively, were inoculated mechanically with four PVY isolates of clade C1 originating from, or adapted to, *C. annuum* (Table 9.6).

Table 9.6 Pathogenicity of PVY isolates from *Capsicum annuum* or *Solanum tuberosum* in potato cvs Safrane and King Edward

PVY isolate	Accession number	Origin	PVY clade	Cv. Safrane	Cv. King Edward (Nc_{tbr})
CAA141	KF670580, JQ954317	*C. annuum*	C1	Ø/weak Mo; ELISA+[b]	HR/Ø; ELISA-
CAA82	JQ954315	*C. annuum*	C1	Ø/weak Mo; ELISA+	HR/Ø; ELISA-
CAA157	KF670592	*C. annuum*	C1	Ø/weak Mo; ELISA+	HR/Ø; ELISA-
SON41p	AJ439544	*Solanum nigrum* and *C. annuum*[a]	C1	Ø/Ø; ELISA+	HR/Ø; ELISA-
C-Adgen	AJ890348	*S. tuberosum*	C2	Ø/Mo; ELISA+	HR/Ø; ELISA-
N605	X97895	*S. tuberosum*	N	Ø/Mo; ELISA+	Ø/weak Mo; ELISA+

Ø: no symptoms, *Mo*: mosaic, *HR*: hypersensitive reaction characterised by local necrotic lesions
[a]Isolated from *S. nigrum* and passaged several times in *C. annuum* cv. Florida VR2 (Gebre Selassie et al. 1985)
[b]Symptoms observed in inoculated leaves/symptoms observed in apical uninoculated leaves; DAS-ELISA positive or negative reaction in apical leaves (20 plants were assayed in two independent experimental replicas per plant genotype-PVY isolate combination)

As expected from their clustering with clade C1, the four isolates induced necrotic lesions in inoculated leaves of cv. King Edward, but the virus could not be detected by DAS-ELISA in apical, uninoculated leaves. The same reaction was observed with thc reference PVY^{C} isolate (Adgen) from potato belonging to clade C2. In contrast, potato isolate PVY^{N} N605, representing clade N, did not induce any reaction in inoculated leaves but was able to infect the plants systemically, inducing symptoms of mild mosaic. All isolates systemically infected plants of potato cv. Safrane, in most cases inducing mosaic symptoms. This result, which needs to be tested further using a larger range of clade C1 isolates and potato cultivars, challenges the current view of a general barrier to infection of potato plants by PVY isolates from pepper and suggests, on the contrary, that this barrier may apply only to some potato cultivars carrying specific resistance(s).

5 Host Plant Species as Springboards for PVY Adaptation to Other Species

The previous section reviewed observations suggesting the existence of PVY infection barriers between host species. Recent results suggest the opposite, that evolution of PVY in a given host species may occasionally favour its adaptation to another host species. Such situations were observed with pepper, tomato and tobacco genotypes carrying eIF4E-1-based resistances conferred by the *pvr2*, *pot-1* and *va* genes. It was

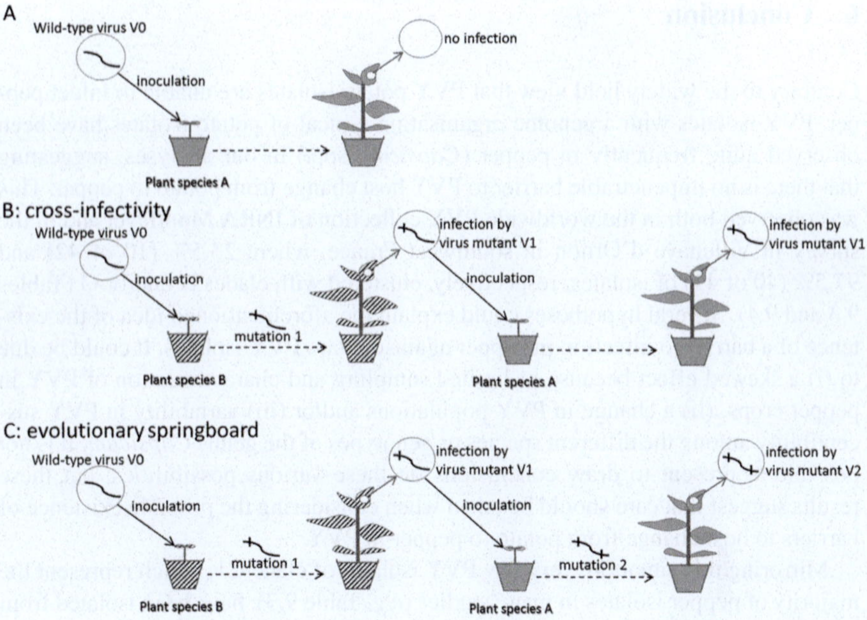

Fig. 9.2 Schematic representation of cross-infectivity and evolutionary springboard effects in a virus exposed to genotypes of two plant species. (**a**) Wild-type virus isolate V0 is unable to infect a plant genotype of species A. (**b**) Cross-infectivity: Previous exposure to a plant genotype of species B allowed the selection of mutation 1 in the virus genome (leading to mutant V1), which was sufficient for consecutive infection of the plant genotype of species A. (**c**) Evolutionary springboard: Previous exposure to a plant genotype of species B allowed the selection of mutation 1 in the virus genome (leading to virus V1), which, in turn, favoured the selection of mutation 2, which is necessary for infection of the plant genotype of species A. Mutation 1 is therefore necessary but not sufficient for infection of the plant genotype of species A

observed that a gain in the ability to overcome the resistance conferred by an allele of one of these three genes could confer the ability to overcome another of these resistance genes in another species (Moury et al. 2014). This is because these three resistance systems are based on the same host factor and target the same region in the VPg-coding region of PVY, with the consequence that nonsynonymous substitutions in this genome region may therefore have pleiotropic effects on PVY pathogenicity in several of the host species. The first kind of pleiotropic effect, named 'cross-infectivity', corresponds to the situation when a single VPg mutation in PVY leads to the simultaneous breakdown of two resistance genes in two different species (Fig. 9.2). The second kind of pleiotropic effect, named 'evolutionary springboard', occurs when acquisition of a mutation leading to resistance breakdown in a first species further favours the accumulation of additional mutations needed for breakdown in another species. By contrast, the direct confrontation of the initial PVY population with the second resistance gene does not lead to resistance breakdown and, therefore, to infection. Both effects are not uncommon and each accounted for 17.5% of the VPg mutations and resistance genes/alleles combinations examined (Moury et al. 2014).

6 Conclusion

Contrary to the widely held view that PVY potato isolates are unable to infect pepper, PVY isolates with a genome organisation typical of potato isolates have been observed quite frequently in pepper (*Capsicum* spp.) in our analyses, suggesting that there is no impenetrable barrier to PVY host change from potato to pepper. This was observed both in the worldwide PVY collection of INRA Montfavet and in the survey in Villenave d'Ornon in southwest France, where 23.5% (10 of 42) and 97.5% (40 of 41) of isolates, respectively, clustered with clades N and/or O (Tables 9.3 and 9.4). Several hypotheses could explain the aforementioned idea of the existence of a barrier to infection in pepper against potato PVY isolates. It could be due to (i) a skewed effect because of limited sampling and characterisation of PVY in pepper crops, (ii) a change in PVY populations and/or (iii) variability in PVY susceptibility among the different species or genotypes of the genus *Capsicum*. It is not possible at present to draw conclusions on these various possibilities, but these results suggest that care should be taken when considering the possible existence of barriers to host change from potato to pepper in PVY.

Mirroring this situation, very few PVY isolates of clade C1, which represent the majority of pepper isolates in many studies (*e.g.* Table 9.3), have been isolated from potato crops, except two isolates collected in the 1930s and 1940s. Clade C1 is frequently presented as the 'non-potato' clade in many phylogenetic studies (*e.g.* Blanco-Urgoiti et al. 1998a; Romero et al. 2001; Karasev et al. 2011; Visser et al. 2012). However, in this case too, no impenetrable, general barrier seems to exist in potato against pepper and/or clade C1 isolates of PVY, because the four randomly selected C1 isolates that we tested were able to infect systemically potato cv. Safrane (Table 9.6). The reason why such isolates were not found in potato crops in recent surveys could be due to the presence of the Nc_{tbr} resistance gene in many current potato cultivars (Cockerham 1970; Jones 1990), rather than to poor adaptation of clade C1 isolates to *S. tuberosum*. Another reason may be that potato and pepper production areas are often separated geographically, so C1 isolates may have limited opportunities to be present in potato-producing areas. In addition, recent analyses of eIF4E-mediated resistance in tobacco, pepper and tomato not only revealed the lack of PVY infection barriers among these three species but showed, on the contrary, the possible occurrence of an evolution springboard effect between them.

In summary, the elements and results discussed here challenge the current vision of the existence of broad and general barriers to host change applying to PVY clades. This in turn questions (i) the interest in using host range as a criterion for classifying PVY isolates and (ii) the traditional division of the different PVY clades or host groups by epidemiologists working on potato or those working on vegetable crops such as pepper and tomato. The analysis of PVY evolution, epidemiology and disease management strategies should, on the contrary, consider all plant species which can be a host for PVY, including the various solanaceous crops known as hosts and the many weed species acting as reservoirs. The advocacy for such a general view of PVY diversity and evolution integrating all potential hosts is particularly relevant if we consider solanaceous weeds which can host PVY isolates belonging to all phylogenetic groups (Table 9.4).

Acknowledgements We thank Christelle Lacroix for her comments on the manuscript and Dr. Jörg Schubert and Alain Buffière for providing PVY isolates.

References

Acosta-Leal R, Xiong Z (2008) Complementary functions of two recessive R-genes determine resistance durability of tobacco 'virgin a mutant' (VAM) to *Potato virus Y*. Virology 379:275–283

Aramburu J, Galipienso L, Matas M (2006) Characterization of potato virus Y isolates from tomato crops in northeast Spain. Eur J Plant Pathol 115:247–258

Ayme V, Souche S, Caranta C, Jacquemond M, Chadœuf J, Palloix A, Moury B (2006) Different mutations in the genome-linked protein VPg of potato virus Y confer virulence on the *pvr2*^{3} resistance in pepper. Mol Plant-Microbe Interact 19:557–563

Ayme V, Petit-Pierre J, Souche S, Palloix A, Moury B (2007) Molecular dissection of the *Potato virus Y* VPg virulence factor reveals complex adaptations to the *pvr2* resistance allelic series in pepper. J Gen Virol 88:1594–1601

Bhat AI, Varma A, Pappu HR, Rajamannar M, Jain RK, Praveen S (1999) Characterization of a potyvirus from eggplant (*Solanum melongena*) as a strain of *Potato virus Y* by N-terminal serology and sequence relationships. Plant Pathol 48:648–654

Behl MK, Chowfla SC, Khurana SMP (1987) Tomato shoe string disease incited by necrotic strain of *Potato virus Y* in Himachal Pradesh. A new record. Curr Sci 56:27–29

Ben Khalifa M, Simon V, Fakhfakh H, Moury B (2012) Tunisian *Potato virus Y* isolates with unnecessary pathogenicity towards pepper: support for the matching allele model in eIF4E resistance–potyvirus interactions. Plant Pathol 61:441–447

Blancard D (1998) Maladies du Tabac. Observer, Identifier, Lutter. INRA, Paris

Blanco-Urgoiti B, Sanchez F, de San Pérez Román C, Dopazo J, Ponz F (1998a) PVY-C isolates are a homogeneous pathotype but two different genetic strains. J Gen Virol 79:2037–2042

Blanco-Urgoiti B, Tribodet M, Leclere S, Ponz F, de San Pérez Román C, Legorburu FJ, Kerlan C (1998b) Characterization of potato potyvirus Y (PVY) isolates from seed potato batches. Situation of the NTN, Wilga and Z isolates. Eur J Plant Pathol 104:811–819

Browning I, Fox A, Holmes R, Mulholland V, Darling D (2004) Variability in *Potato virus Y* in seed potato crops in Scotland. Page 3 in: 12th EAPR Virol. Sect. Meet. Rennes

Budnik K, Laing MD, da Graca JV (1996) Reduction of yield losses in pepper crops caused by potato virus Y in KwaZulu-Natal, south Africa, using plastic mulch and yellow sticky traps. Phytoparasitica 24:119–124

Celebi-Toprak F, Slack SA, Jahn MM (2002) A new gene, *Ny*tbr, for hypersensitivity to *Potato virus Y* from *Solanum tuberosum* maps to chromosome IV. Theor Appl Genet 104:669–674

Charron C, Nicolaï M, Gallois J, Robaglia C, Moury B, Palloix A, Caranta C (2008) Natural variation and functional analyses provide evidence for co-evolution between plant eIF4E and potyviral VPg. Plant J 54:56–68

Cockerham G (1970) Genetical studies on resistance to potato viruses X and Y. Heredity 25:309–348

Cuevas JM, Delaunay A, Visser JC, Bellstedt DU, Jacquot E, Elena SF (2012) Phylogeography and molecular evolution of *Potato virus Y*. PLoS One 7:e37853

d'Aquino L, Dalmay T, Burgyán J, Ragozzino A, Scala F (1995) Host range and sequence analysis of an isolate of *Potato virus Y* inducing veinal necrosis in pepper. Plant Dis 79:1046–1050

Dedic P, Ptacek J, Cerovska N (2007) A shift of PVY strain spectrum on potatoes in CR in the course of past years. Page 59 in: 13th EAPR Virol. Sect. Meet. Coylumbridge

Djilani-Khouadja F, Glais L, Tribodet M, Kerlan C, Fakhfakh H (2010) Incidence of potato viruses and characterisation of *Potato virus Y* variability in late season planted potato crops in northern Tunisia. Eur J Plant Pathol 126:479–488

Dogimont C, Palloix A, Daubèze AM, Marchoux G, Gebre Selassie K, Pochard E (1996) Genetic analysis of broad spectrum resistance to potyviruses using doubled haploid lines of pepper (*Capsicum annuum* L.) Euphytica 88:231–239

Dullemans AM, Cuperus C, Verbeek, M. van der Vlugt, R.A.A. (2011) Complete nucleotide sequence of a potato isolate of strain group C of *Potato virus Y* from 1938. Arch Virol 156:473–477

Edwardson JR, Christie RG (1997) Potyviruses. In: In: *Florida Agricultural Experiment Station Monograph Series 18-II* – viruses infecting pepper and other Solanaceous crops. University of Florida, Gainesville

Faurez F, Baldwin T, Tribodet M, Jacquot E (2012) Identification of new *Potato virus Y* (PVY) molecular determinants for the induction of vein necrosis in tobacco. Mol Plant Pathol 13:948–959

Fellers JP, Tremblay D, Handest MF, Lommel SA (2002) The potato virus Y (MNR)-N-S NIb-replicase is the elicitor of a veinal necrosis-hypersensitive response in root knot nematode resistant tobacco. Mol Plant Pathol 3:145–152

Fereres A, Perez P, Gemeno C, Ponz F (1993) Transmission of Spanish pepper-PVY and potato-PVY isolates by aphid (Homoptera, Aphididae) vectors-epidemiological implications. Environ Entomol 22:1260–1265

Fereres A, Avilla C, Collar JL, Duque M, Fernandez-Quintanilla C (1996) Impact of various yield-reducing agents on open-field sweet peppers. Environ Entomol 25:983–986

Gebre Selassie K, Marchoux G, Delecolle B, Pochard E (1985) Variabilité naturelle des souches du virus Y de la pomme de terre dans les cultures de piment du sud-est de la France. Caractérisation et clas-sification en pathotypes. Agronomie 5:621–630

Gibbs A, Ohshima K (2010) Potyviruses and the digital revolution. Annu Rev Phytopathol 48:10.1–10.19

Gibbs AJ, Ohshima K, Phillips MJ, Gibbs MJ (2008) The prehistory of potyviruses: their initial radiation was during the dawn of agriculture. PLoS One 3:e2523

Glasa M, Prikhodko Y, Predajňa L, Nagyová A, Shneyder Y, Zhivaeva T, Šubr Z, Cambra M, Candresse T (2013) Characterization of sour cherry isolates of *Plum pox virus* from the Volga basin in Russia reveals a new cherry strain of the virus. Phytopathology 103:972–979

Gooding GV, Tolin SA (1973) Strains of *Potato virus Y* affecting flue-cured tobacco in the South-eastern United States. Plant Dis Rep 57:200–204

Hu X, Meacham T, Ewing L, Gray SM, Karasev AV (2009) A novel recombinant strain of *Potato virus Y* suggests a new viral genetic determinant of vein necrosis in tobacco. Virus Res 143:68–76

Ibaba JD, Gubba A (2011) Diversity of *Potato virus Y* isolates infecting solanaceous vegetables in the province of KwaZulu-Natal in the Republic of South Africa. Crop Prot 30:1404–1408

Ibiza VP, Canizares J, Nuez F (2010) EcoTILLING in capsicum species: searching for new virus resistances. BMC Genomics 11:631

Janzac B, Tribodet M, Lacroix C, Moury B, Verrier J-L, Jacquot E (2014) Evolutionary pathways to break down the resistance of allelic versions of the PVY resistance gene. Plant Dis 98:1521–1529

Janzac B, Willemsen A, Cuevas JM, Glais L, Tribodet M, Verrier J-L, Elena SF, Jacquot E (2015) Brazilian potato virus Y isolates identified as members of a new clade facilitate the reconstruction of evolutionary traits within this species. Plant Pathol 64:799–807.

Jones RAC (1990) Strain group specific and virus specific hypersensitive reactions to infection with potyviruses in potato cultivars. Ann Appl Biol 117:93–105

Julio E, Cotucheau J, Decorps C, Volpatti R, Sentenac C, Candresse T, Dorlhac de Borne F (2015) An eukaryotic translation initiation factor 4E (eIF4E) is responsible for the 'va' tobacco recessive resistance to potyviruses. Plant Mol Biol Report 33:609–623

Kang B-C, Yeam I, Frantz JD, Murphy JF, Jahn MM (2005) The *pvr1* locus in *Capsicum* encodes a translation initiation factor eIF4E that interacts with *Tobacco etch virus* VPg. Plant J 42:392–405

Karasev AV, Hu X, Brown CJ, Kerlan C, Nikolaeva OV, Crosslin JM, Gray SM (2011) Genetic diversity of the ordinary strain of *Potato virus Y* (PVY) and origin of recombinant PVY strains. Phytopathology 101:778–785

Kehoe MA, Jones RAC (2011) A proposal to help resolve the disagreement between naming of *Potato virus Y* strain groups defined by resistance phenotypes and those defined by sequencing. Arch Virol 156:2273–2278

Kehoe MA, Coutts BA, Buirchell BJ, Jones RAC (2014) Split personality of a potyvirus: to specialize or not to specialize? PLoS One 9:e105770

Kerlan C (2006) Potato virus Y. AAB/CMI Descriptions of Plant Viruses 414

Kerlan C, Tribodet M, Glais L, Guillet M (1999) Variability of *Potato virus Y* in potato crops in France. J Phytopathol 147:643–651

Lacroix C, Glais L, Kerlan C, Verrier J-L, Jacquot E (2010) Biological characterization of French *Potato virus Y* (PVY) isolates collected from PVY-susceptible or -resistant tobacco plants possessing the recessive resistance gene *va*. Plant Pathol 59:1133–1143

Lacroix C, Glais L, Verrier J-L, Jacquot E (2011) Effect of passage of a *Potato virus Y* isolate on a line of tobacco containing the recessive resistance gene va^2 on the development of isolates capable of overcoming alleles 0 and 2. Eur J Plant Pathol 130:259–269

Legnani R, Gebre Selassie K, Nono Womdim R, Gognalons P, Moretti A, Laterrot H, Marchoux G (1995) Evaluation and inheritance of the *Lycopersicon hirsutum* resistance against potato virus Y. Euphytica 86:219–226

Li N, Wang X, Zhou G, Dong J (2006) Molecular variability of the coat protein gene of *Potato virus Y* from tobacco in China. Acta Virol 50:107–113

Lindner K, Billenkamp N (2005) Changes in the spectrum of PVY strains and strain groups: could this be a cause for the increase of virus susceptibility of potato and tobacco varieties? Nachrichtenblatt des Deutschen Pflanzenschutzdienstes 57:245–253

Lorenzen JH, Meacham T, Berger PH, Shiel PJ, Crosslin JM, Hamm PB (2006) Whole genome characterization of *Potato virus Y* isolates collected in the western USA and their comparison to isolates from Europe and Canada. Arch Virol 151:1055–1074

Luis-Arteaga M, Gil-Ortega R (1986) Biological characterization of PVY as isolated from pepper in Spain. VI Meeting on Capsicum and eggplant, Zaragoza, October 21–24, 183–188

Margaritopoulos JT, Dovas CI, Gounaris J, Skouras PJ, Kanavaki OM, Katis NI, Tsitsipis JA (2010) Molecular analysis of the coat protein of *Potato virus Y* isolates in Greece suggests multiple introduction from different genetic pools. J Phytopathol 158:73–80

Martin DP, Williamson C, Posada D (2005) RDP2: recombination detection and analysis from sequence alignments. Bioinformatics 21:260–262

Mascia T, Finetti-Sialer MM, Cillo F, Gallitelli D (2010) Biological and molecular characterization of a recombinant isolate of *Potato virus Y* associated with a tomato necrotic disease occurring in Italy. J Pl Pathol 92:131–138

McDonald JG, Kristjansson GT (1993) Properties of strains of potato virus Y^N in North America. Plant Dis 77:87–89

Moury B (2010) A new lineage sheds light on the evolutionary history of *Potato virus Y*. Mol Plant Pathol 11:161–168

Moury B, Morel C, Johansen E, Guilbaud L, Souche S, Ayme V, Caranta C, Palloix A, Jacquemond M (2004) Mutations in *Potato virus Y* genome-linked protein determine virulence towards recessive resistances in *Capsicum annuum and Lycopersicon hirsutum*. Mol Plant-Microbe Interact 17:322–329

Moury B, Caromel B, Johansen E, Simon V, Chauvin L, Jacquot E, Kerlan C, Lefebvre V (2011) The helper component proteinase cistron of *Potato virus Y* induces hypersensitivity and resistance in potato genotypes carrying dominant resistance genes on chromosome IV. Mol Plant-Microbe Interact 24:787–797

Moury B, Verdin E (2012) Viruses of pepper crops in the Mediterranean basin: a remarkable stasis. Adv Virus Res 84:127–162

Moury B, Janzac B, Ruellan Y, Simon V, Ben Khalifa M, Fakhfakh H, Fabre F, Palloix A (2014) Interaction patterns between *Potato virus Y* and eIF4E-mediated recessive resistance in the *Solanaceae*. J Virol 88:9799–9807

Ogawa T, Tomitaka Y, Nakagawa A, Ohshima K (2008) Genetic structure of a population of *Potato virus Y* inducing potato tuber necrotic ringspot disease in Japan; comparison with north American and European populations. Virus Res 131:199–212

Ogawa T, Nakagawa A, Hataya T, Ohshima K (2012) The genetic structure of populations of *Potato virus Y* in Japan based on the analysis of 20 full genomic sequences. J Phytopathol 160:661–673

Olarte-Castillo XA, Fermin G, Tabima J, Rojas Y, Tennant PF, Fuchs M, Sierra R, Bernal AJ, Restrepo S (2011) Phylogeography and molecular epidemiology of *Papaya ringspot virus*. Virus Res 159:132–140

Ohshima K, Yamaguchi Y, Hirota R, Hamamoto T, Tomimura K, Tan Z, Sano T, Azuhata F, Walsh JA, Fletcher J, Chen J, Gera A, Gibbs A (2002) Molecular evolution of turnip mosaic virus: evidence of host adaptation, genetic recombination and geographical spread. J Gen Virol 83:1511–1521

Parrella G, Ruffel S, Moretti A, Morel C, Palloix A, Caranta C (2002) Recessive resistance genes against potyviruses are localised in colinear genomic regions of the tomato (*Lycopersicon* spp.) and pepper (*Capsicum* spp.) genomes. Theor Appl Genet 105:855–861

Quenouille J, Vassilakos N, Moury B (2013) *Potato virus Y*: a major crop pathogen that has provided major insights into the evolution of viral pathogenicity. Mol Plant Pathol 14:439–452

Rolland M, Lacroix C, Blanchard A, Baldwin T, Kerlan C, Jacquot E (2008) Potato virus Y (PVY): from its discovery to the latest outbreaks. Virologie 12:261–273

Rolot JL (2007) Balance between PVYN and PVYO strains in Belgium: first approaches. Page 70 in: 13 EAPR Virology Section Meeting, Coylumbridge

Romero A, Blanco-Urgoiti B, Soto M, Fereres A, Ponz F (2001) Characterization of typical pepper-isolates of PVY reveals multiple pathotypes within a single genetic strain. Virus Res 79:71–80

Ruffel S, Dussault MH, Palloix A, Moury B, Bendahmane A, Robaglia C, Caranta C (2002) A natural recessive resistance gene against potato virus Y in pepper corresponds to the eukaryotic initiation factor 4E (eIF4E). Plant J 32:1067–1075

Ruffel S, Gallois JL, Lesage ML, Caranta C (2005) The recessive potyvirus resistance gene *pot-1* is the tomato orthologue of the pepper *pvr2*-eIF4E gene. Mol Gen Genomics 274:346–353

Rufty RC, Wernsman EA, Powell NT (1983) A genetic analysis of the association between resistance to *Meloidogyne incognita* and a necrotic response to infection by a strain of *Potato virus Y* in tobacco. Phytopathology 73:1413–1418

Sadeghi MS, Behjatnia SAA, Masumi M, Izadpanah K (2008) Characterisation of a strain of *Potato virus Y* causing eggplant mosaic in southern Iran. Australasian Plant Pathol 37:79–86

Schubert J, Fomitcheva V, Sztangret-Wisniewska J (2007) Differentiation of *Potato virus Y* strains using improved sets of diagnostic PCR-primers. J Virol Methods 140:66–74

Sierro N, Battey JND, Ouadi S, Bakaher N, Bovet L, Willig A, Goepfert S, Peitsch MC, Ivanov NV (2014) The tobacco genome sequence and its comparison with those of tomato and potato. Nat Commun 5:3833

Singh RP, Valkonen JPT, Gray SM, Boonham N, Jones RAC, Kerlan C, Schubert J (2008) Discussion paper: the naming of *Potato virus Y* strains infecting potato. Arch Virol 153:1–13

Stobbs LW, Poysa V, Van Schagen JG (1994) Susceptibility of cultivars of tomato and pepper to a necrotic strain of *Potato virus Y*. Can J Plant Pathol 16:43–48

Sudarsono, Woloshuk SL, Xiong Z, Hellmann GM, Wernsman EA, Weissinger AK, Lommel SA (1993) Nucleotide sequence of the capsid protein cistrons from six *Potato virus Y* (PVY) isolates infecting tobacco. Arch Virol 132:161–170

Tian YP, Valkonen JPT (2013) Genetic determinants of *Potato virus Y* required to overcome or trigger hypersensitive resistance to PVY strain group O controlled by the gene Ny in potato. Mol Plant-Microbe Interact 26:297–305

Tian YP, Valkonen JPT (2015) Recombination of strain O segments to HCpro-encoding sequence of strain N of *Potato virus Y* modulates necrosis induced in tobacco and in potatoes carrying resistance genes *Ny* or *Nc*. Mol Plant Pathol 16:735–747

Tian YP, Liu JL, Zhang CL, Liu YY, Wang B, Li X-D, Guo ZK, Valkonen JPT (2011) Genetic diversity of *Potato virus Y* infecting tobacco crops in China. Phytopathology 101:377–387

Tribodet M, Glais L, Kerlan C, Jacquot E (2005) Characterization of *Potato virus Y* (PVY) molecular determinants involved in the vein necrosis symptom induced by PVY^N isolates in infected *Nicotiana tabacum* cv. Xanthi. J Gen Virol 86:2101–2105

Valkonen JPT, Kyle MM, Slack SA (1996) Comparison of resistance to potyviruses within *Solanaceae* : infection of potatoes with tobacco etch potyvirus and peppers with potato a and Y potyviruses. Ann Appl Biol 129:25–38

Visser JC, Bellstedt DU, Pirie MD (2012) The recent recombinant evolution of a major crop pathogen, *Potato virus Y*. PLoS One 7:e50631

The manufacturer's authorised representative in the EU is Springer Nature Customer Service Centre GmbH, Europaplatz 3, 69115 Heidelberg, Germany. If you have any concerns regarding our products, please contact ProductSafety@springernature.com

Printed and bound by CPI Group (UK) Ltd, Croydon, CR0 4YY
01/10/2026
02225871-0002